Atmospheric Science: Principles, Processes and Applications

Atmospheric Science: Principles, Processes and Applications

Edited by Smith Paul

SYRAWOOD
PUBLISHING HOUSE

New York

Published by Syrawood Publishing House,
750 Third Avenue, 9th Floor,
New York, NY 10017, USA
www.syrawoodpublishinghouse.com

Atmospheric Science: Principles, Processes and Applications
Edited by Smith Paul

© 2017 Syrawood Publishing House

International Standard Book Number: 978-1-68286-449-4 (Hardback)

Cataloging-in-publication Data

Atmospheric science : principles, processes and applications / edited by Smith Paul.
     p. cm.
Includes bibliographical references and index.
ISBN 978-1-68286-449-4
1. Atmosphere. 2. Atmospheric physics. 3. Atmospheric chemistry. 4. Meteorology. I. Paul, Smith.
QC861.3 .A86 2017
551.5--dc23

Printed in the United States of America.

TABLE OF CONTENTS

PREFACE

Atmospheric science deals with the study of the layers of the atmosphere, and studies the atmospheric composition of those regions. It branches out into various sub-fields such as atmospheric chemistry, climatology, atmospheric physics, etc. Topics included herein deal with atmospheric dynamics, climatology and meteorology. This book covers in detail some existent theories and innovative concepts revolving around atmospheric science. The various advancements in this discipline are glanced at along with their applications as well as ramifications. Different approaches, evaluations, methodologies and advanced studies in this field have also been included. This book on atmospheric science will serve as a guide for researchers and scholars in the fields of earth sciences, meteorology and atmospheric physics. Scientists and students actively engaged in this field will find this book full of crucial and unexplored concepts.

After months of intensive research and writing, this book is the end result of all who devoted their time and efforts in the initiation and progress of this book. It will surely be a source of reference in enhancing the required knowledge of the new developments in the area. During the course of developing this book, certain measures such as accuracy, authenticity and research focused analytical studies were given preference in order to produce a comprehensive book in the area of study.

This book would not have been possible without the efforts of the authors and the publisher. I extend my sincere thanks to them. Secondly, I express my gratitude to my family and well-wishers. And most importantly, I thank my students for constantly expressing their willingness and curiosity in enhancing their knowledge in the field, which encourages me to take up further research projects for the advancement of the area.

Editor

Restoring Coastal Plants to Improve Global Carbon Storage: Reaping What We Sow

Andrew D. Irving*, Sean D. Connell, Bayden D. Russell

Southern Seas Ecology Laboratories, School of Earth and Environmental Sciences, The University of Adelaide, Adelaide, South Australia, Australia

Abstract

Long-term carbon capture and storage (CCS) is currently considered a viable strategy for mitigating rising levels of atmospheric CO_2 and associated impacts of global climate change. Until recently, the significant below-ground CCS capacity of coastal vegetation such as seagrasses, salt marshes, and mangroves has largely gone unrecognized in models of global carbon transfer. However, this reservoir of natural, free, and sustainable carbon storage potential is increasingly jeopardized by alarming trends in coastal habitat loss, totalling 30–50% of global abundance over the last century alone. Human intervention to restore lost habitats is a potentially powerful solution to improve natural rates of global CCS, but data suggest this approach is unlikely to substantially improve long-term CCS unless current restoration efforts are increased to an industrial scale. Failure to do so raises the question of whether resources currently used for expensive and time-consuming restoration projects would be more wisely invested in arresting further habitat loss and encouraging natural recovery.

Editor: Andrew Hector, University of Zurich, Switzerland

Funding: Funding support was provided by the Australian Research Council. The funders had no role in study design, data collection and analysis, decision to publish, or preparation of the manuscript.

Competing Interests: The authors have declared that no competing interests exist.

* E-mail: andrew.irving@adelaide.edu.au

Introduction

As the varied consequences of a changing climate continue to challenge our technical capacity [1] and political will [2] to globally stabilize and manage greenhouse gas emissions, numerous potential solutions have gained traction among the scientific community [3] and general public [4]. Among these is the industrial-scale artificial capture and long-term storage of anthropogenic CO_2 before it is released to the atmosphere and contributes to the greenhouse effect. We already possess the technical expertise to engineer such feats by pumping liquefied CO_2 into porous geological formations more than 1 km underground [5]. However, uncertainty remains over the considerable expense and potentially damaging side-effects of such operations, including unforeseen geological de-stabilisation and chronic CO_2 leakage into marine and terrestrial environments, and ultimately into the atmosphere [6]. Additionally, routine use of such procedures will be unlikely until at least 2025 [7].

Compared to the attention given to methods of artificial carbon capture and storage (CCS), natural carbon sinks such as terrestrial and aquatic vegetation have often been overlooked or considered supplementary for management [8]. This disparity probably stems from the quantified inability of such biological reservoirs to compensate for the sheer volume of anthropogenic carbon currently produced ($\sim 440 \times 10^6$ vs $\sim 8500 \times 10^6$ t C yr^{-1}, respectively; [9,10]). Nevertheless, natural means of CCS are immediately available, cost-effective, publicly supported, and offer many complementary benefits such as the preservation of biodiversity and other natural resources. When coupled with current uncertainties regarding artificial CCS techniques, natural approaches appear to warrant serious consideration as an important contributor to managing the carbon problem.

As a carbon sink, the ocean can absorb up to one-third of anthropogenic CO_2 in the atmosphere [11]. It therefore seems fortunate that coastal vegetation such as seagrasses, mangroves, and salt marshes (Fig. 1A–C) capture and store carbon at non-trivial amounts of 60–210 t C km^{-2} yr^{-1} [9,12], and do so with far greater efficiency than their terrestrial counterparts (e.g. tropical forests only store 2.3–2.5 t C km^{-2} yr^{-1}; [9]). Under favorable conditions, the majority of captured carbon may be stored as below-ground biomass (e.g. peat) for decades to possibly thousands of years.

Like most biological resources, however, coastal vegetation has undergone extensive declines in global distribution and abundance [13], culminating in the loss of $\sim$ one-third of the world's seagrass meadows and mangrove forests, and more than one-half of salt marshes, during the past century [14,15,16]. Losses have been, and continue to be, largely driven by anthropogenic stressors, including pollution (e.g. eutrophication, turbidity), altered sedimentation regimes, and direct physical disturbance (e.g. reclaiming coasts). To date, losses are thought to have reduced global CCS rates by at least 25% [13], and continued losses may exacerbate the carbon problem by exposing below-ground biomass that can release hundreds to thousands of years worth of stored carbon as it erodes and degrades (Fig. 1D). Precise numbers on the potential magnitude of such 're-activation' of stored carbon are scarce, yet Cebrian [17] conservatively estimated that the loss of the world's mangrove forests to date has resulted in the release of 3.9×10^8 tonnes of previously stored carbon.

Figure 1. Major carbon-storing habitats on tropical and temperate coasts. Degradation and loss of (A) seagrass meadows, (B) mangrove forests, and (C) salt marshes may release hundreds to thousands of years worth of stored carbon through exposure and breakdown of below-ground biomass, shown in (D) for seagrasses. Photo credits: Andrew Irving.

Given the enormity of the carbon problem, anything less than a thorough consideration of all possible methods of mitigation could appear neglectful. Thus, it is worth asking whether trends of coastal habitat loss can be reversed to increase global CCS. If future losses of habitat can be prevented such that they do not occur at the expense of any gains, which is a significant challenge given the magnitude of global exploitation of coastal environments [18], facilitating natural recovery and expansion of habitats may be a critical first step. However, many species of coastal plants require decades to centuries to recover from disturbance because they depend primarily on clonal expansion rather than sexual reproduction for population growth [19]. Therefore, direct intervention through habitat restoration may represent a way to more rapidly improve rates of natural CCS. Habitat restoration is a potentially powerful approach, but the task of re-creating complex ecosystems presents many challenges that so far have typically produced viable self-sustaining populations well below parity of effort (e.g. 35–50% successful establishment of planted seagrass units [20]) and have generally confined restoration projects to small spatial scales (≤1 ha: [21]).

The purpose of this study was twofold. Firstly, we quantified the decline in global rates of CCS by mangroves, seagrasses, and salt marshes due to their historical decline in abundance. Based on this information, we (secondly) compared how rates of CCS might improve under future scenarios of habitat recovery and restoration. Of particular interest was to compare the long-term benefits of using different intensities of habitat restoration to provide some indication of the effort needed to produce a sizeable effect. While we recognize that restoration of habitats provides numerous benefits additional to CCS (e.g. nutrient cycling, coastal stabilization, preservation of biodiversity), we focus on CCS as a current topic of significant global concern and debate. Given the

considerable amounts of time and expense involved in most restoration programs (e.g. seagrass averages US\$48,700 ha^{-1}: [20], adjusted to 2010 dollar value), coupled with their often modest chances of success (35–50%: [20]), it is debatable whether the investment of finite resources would instead be better directed toward approaches that limit further habitat loss and promote natural recovery.

Materials and Methods

Data used to quantify changes to global rates of CCS by coastal vegetation were sourced from literature describing historical habitat abundance and/or calculated rates of CCS. We focused on seagrasses, mangroves, and salt marshes because they are well-known for their capacity to store significant amounts of carbon for long periods [8,13], and also because of their near-global presence on tropical and temperate coasts. We note that forests of kelp and other macroalgae also constitute a major coastal habitat, particularly at temperate latitudes, but they are excluded from this study because they are essentially ephemeral in their CCS capacity; i.e., carbon captured in their tissue is released as the plant decays or dies, with none of it stored below-ground since the plants possess no root structure.

For each habitat, global CCS was calculated by multiplying estimates of global habitat area, often averaged across several data sources, by quantified rates of CCS per unit area. Few estimates of CCS rates for each habitat are available in the literature, and so we used the highest and lowest rates we could find to provide some indication of variance in historical decline of CCS. While this is a relatively simple method that does not include compounding influences such as variation among species and latitudes, it nonetheless provides similar estimates to studies using more complex methods (e.g. [13]), and forms a standard baseline for relative predictions of increasing CCS under different scenarios of future habitat recovery. Where possible, our calculated global CCS was partitioned into decadal increments (i.e. the annual rate of global CCS by that habitat in each decade) because of instances of rapid habitat decline in particular decades (see Results). However, sparse data describing habitat abundance sometimes limited this approach, particularly in earlier years. Additionally, data describing changes in CCS by salt marshes were restricted to the continental USA because poor estimates of salt marsh abundance elsewhere precluded calculations of their global abundance. Even so, patterns observed in the USA are likely to be representative of many locales since salt marshes are one of the most common habitats "reclaimed" during coastal development [14]. Table 1 provides a summary of key values and data sources used in calculations.

Forecasting improvements to global CCS under different scenarios of habitat recovery and restoration was done by multiplying quantified rates of CCS per unit area by estimates of global habitat area resulting from either natural recovery alone, or natural recovery combined with restoration. Initially, it was hoped that improvements could be calculated for all three habitats, yet literature searches soon revealed that calculations could only be reliably done for seagrasses because good estimates of global rates of seagrass expansion (taken as indicative of recovery), as well as a key synthesis of rehabilitation efforts [20], were available. While results are therefore focused on seagrasses, similar patterns are likely for mangroves and salt marshes, though probably over different time scales (e.g. mangroves are slower growing and therefore may take longer to recover).

The rate of natural seagrass recovery was based on global measures of seagrass expansion presented in Table 1 of [16]. Note that seagrass is in global decline because the overall rate of expansion is overwhelmed by the rate of loss, yet some locations exhibit greater rates of expansion than loss, which can be used to calculate a nominal rate of recovery for forecasts described herein. Over a 70-year period, [16] quantified 879 km^2 of seagrass expansion (against losses exceeding 9,000 km^2) across study areas totaling 11,592 km^2 around the world. These measures equate to an average global seagrass expansion rate of 1.08% per decade. This value may be an underestimate since small-scale studies have shown rates of recovery up to 7.5% yr^{-1} [22]. Indeed, recovery is likely to be greater that 1.08% per decade under favourable water quality and physical conditions, but given the uncertainty regarding favourable future coastal conditions [23] such values are yet to be reliably determined. Furthermore, we use the value of 1.08% recovery per decade because this estimate is based on the truly global synthesis of [16], which provides necessary parity for the global scale calculations described herein.

The total global seagrass restoration effort has never been fully quantified, but a comprehensive synthesis by [20] tallied a total of 0.78 km^2 for the USA since the 1960s. Assuming an average restoration success rate of 42% (after [20]), one can expect the establishment of ~0.33 km^2 of seagrass from the 0.78 km^2 planted. Extrapolating efforts in the USA (supporting 7.1% of the world's seagrass: [24]) to the remainder of the world would give a successful global seagrass restoration effort of 4.59 km^2. In other words, if the restoration efforts of the USA were replicated throughout the world, 4.59 km^2 of seagrass would have been restored globally. This is certainly an overestimate because the USA has a long history of seagrass restoration relative to many other countries, but it nevertheless provides a quantifiable benchmark for future restoration efforts.

Table 1. Summary of key data sources and values used in calculations of global historical CCS rates and future changes under different habitat recovery and restoration scenarios.

(a) Historical global habitat abundance ×10^3 km^2	Time interval	Seagrass (16)	Mangrove (12, 15)	Salt Marsh (USA only) (14, 34)
	1879–1930	174.75		
	1930–1940	174.84		7721.609
	1940–1950	174.57		7296.080
	1950–1960	172.64		6474.491
	1960–1970	173.04		5257.010
	1970–1980	170.51	36957.64	4366.717
	1980–1990	126.02	30567.51	3629.156
	1990–2000	128.19	24177.38	
	2000–2006	125.54		

(b) Quantified CCS rates: t C km^{-2} yr^{-1}	Seagrass	Mangrove	Salt Marsh	
	83 (9)	139 (9)	210 (9)	
	83 (13)	139 (13)	151 (13)	
	133 (8)	60 (12)		

(c) Seagrass restoration	Area planted (km^2, USA only)	Average restoration success rate	Expected area restored (km^2)	Potential global area restored* (km^2)
	0.78 (20)	42% (20)	0.33	4.59

*Potential global area restored is based on extrapolating the amount of successful seagrass restoration in the USA, using the relative proportion of the world's seagrass contained within the USA (~7.1%). Given restoration efforts in the USA are likely greater than many other countries, this may over-estimate current global restoration effort.
Data sources are listed in parentheses.

For the current study, future changes to global CCS by seagrass meadows were calculated under a 'recovery only' scenario, where seagrasses were allowed to recover at a rate of 1.08% per decade from 2010 to 2100. Results were then compared to CCS rates when this level of recovery was combined with seagrass restoration at i) the current effort per decade (i.e. recovery + restoration of 4.59 km^2 per decade), ii) 10-times the current effort per decade (i.e. recovery+45.94 km^2 per decade), and iii) 100-times the current effort per decade (i.e. recovery+459.4 km^2 per decade). Finally, these forecasts were compared to a 'continued decline' scenario where loss of seagrass persists at the 1980–2000 average rate of 0.02% per decade (calculated from Table S2 in [16]).

Results

Historical habitat losses and decline in CCS

Seagrasses, mangroves, and salt marshes have all experienced substantial declines in abundance that has reduced the global CCS achieved by these coastal habitats. The greatest changes have occurred within mangrove forests, where world-wide habitat losses exceeding 90,000 km^2 since the 1970s have reduced their average global rate of CCS from $\sim 26.5 \times 10^6$ t C yr^{-1} to $\sim 17.3 \times 10^6$ t C yr^{-1} (Fig. 2A). Seagrass loss has occurred steadily since at least the early 1900s, but rates of loss peaked dramatically during the 1970s and 1980s, producing a rapid areal decline of over 44,000 km^2 during this period alone. Concomitantly, rates of global CCS by seagrass declined by $\sim 4 \times 10^6$ t C yr^{-1} to current average CCS of $\sim 16.7 \times 10^6$ t C yr^{-1}. Lastly, data available for salt marshes in the continental USA show a sustained rate of decline of $\sim 5\%$ cover per decade since the 1930s, equating to losses approaching 20,000 km^2 and a decline in average rates of CCS from $\sim 6.6 \times 10^6$ t C yr^{-1} to $\sim 3.1 \times 10^6$ t C yr^{-1}.

Improvements to global CCS

Using a global seagrass recovery rate of 1.08% per decade [16] and a conservative CCS rate of 83 t C km^{-2} yr^{-1} [9], natural seagrass recovery alone may produce global rates of CCS of $\sim 11.4 \times 10^6$ t C yr^{-1} by the year 2100, an increase of $\sim 10\%$ above current rates (Fig. 2B). Continuing along this trajectory, rates of CCS would reach 1920 levels (see 'historical loss' data: Fig. 2B) sometime around the year 2340. This timeframe may be shortened if rising CO$_2$ increases seagrass productivity [25] and reduces covers of calcareous epiphytic algae that smother seagrasses [26].

Combining natural recovery with a global replication of current seagrass restoration efforts (4.59 km^2) each decade until 2100 would improve rates of CCS by just 0.1% above benefits provided by natural recovery alone. Increasing seagrass restoration efforts 10-fold (45.94 km^2 per decade) would provide a 0.9% improvement over natural recovery, but a 100-fold increase to what would likely require industrial-scale operations (459.4 km^2 per decade) boosts CCS by a further 9.3%, resulting in rates of $\sim 11.8 \times 10^6$ t C captured yr^{-1} (Fig. 2B). Such large-scale efforts would generate a return to 1920 levels of CCS around the year 2260, ~ 80 years sooner than relying on natural recovery alone.

Discussion

The link between rising concentrations of atmospheric CO$_2$ and associated impacts of climate change has been argued to be one of the greatest challenges facing our understanding and management of the world's natural resources [27]. The sheer volume of anthropogenic carbon produced from fossil fuels and industry, $\sim 8,500 \times 10^6$ t C yr^{-1} [9,10], outweighs the CCS capacity of any

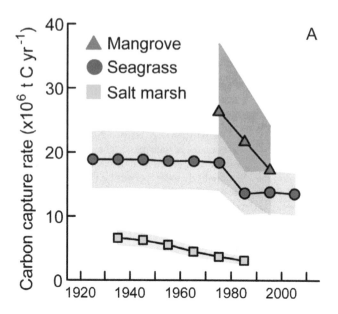

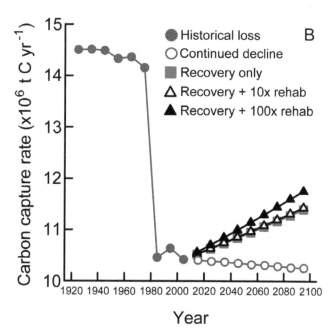

Figure 2. Historical and future carbon capture and storage rates (CCS) of coastal vegetation. (A) Extensive historical losses of seagrasses, mangroves, and salt marshes have reduced the CCS capacity of the coast. Points plotted represent the mean CCS for each habitat over time, and are bounded by lines of maximum and minimum rates of CCS published in the literature. Note that minimum rates for mangroves overlaps with the range of values for seagrass (depicted with purple shading). (B) Historical rates of CCS by seagrass are compared to rates under future scenarios of natural habitat recovery, as well as recovery combined with different intensities of restoration. Increasing restoration efforts to 100-times current levels will produce benefits to CCS that are similar to natural recovery alone. Rates of CCS following current trends in continued global seagrass decline are also plotted for reference. Data for calculating CCS rates were primarily sourced from [8,9,12,13,15,16,34] (also see Table 1).

natural habitat by at least an order of magnitude and immediately suggests that artificial methods of CCS are the only realistic CCS management option. However, artificial CCS methods, like most large-scale artificial management strategies [28], are currently burdened with uncertainty regarding their cost-effectiveness, long-term viability, and environmental dormancy. Natural CCS by aquatic and terrestrial vegetation offers a sustainable, low-risk, and potentially significant contribution toward managing the carbon problem, provided alarming historical trends in habitat degradation and loss can be slowed, arrested, or ideally, reversed.

Coastal vegetation such as mangroves, seagrasses, and salt marshes, have undergone extensive declines in abundance, distribution, and CCS over the past century. Based on the data available, we calculated a cumulative failure to capture at least 434×10^6 t of carbon due to habitat loss since the 1920s. This value is certainly an underestimate, however, since data describing historical losses are typically limited (Fig. 2A), the baseline values used for calculations likely represent already impacted habitats (especially for mangroves where the earliest reliable data comes from the 1970s), and also because data for salt marshes, which have the greatest CCS potential of the three habitats examined [8], are restricted to the continental USA.

Can we rely on natural habitat recovery alone to regain lost CCS capacity among coastal habitats? Will coastal habitat restoration help improve global CCS? Using seagrasses as a model system, global recovery at a modest rate of 1.08% per decade over the next century may increase global CCS by seagrasses by ~10% above current levels. However, this estimate depends on there being no further decline in net seagrass abundance, as well as the risky proposition that no major and as yet unforeseen future events will impact seagrass abundance during the decades to centuries needed for recovery [19]. Habitat restoration is a challenging but potentially powerful tool for further improving rates of CCS, yet the data suggest that efforts would need to be dramatically increased above current levels in order to contribute any significant effect beyond that gained through natural recovery alone. Industrial-scale restoration operations, in the order of 100s of kms restored per decade, may provide a substantial boost of ~9% greater CCS above natural recovery alone (Fig. 2B). In isolation, such improvements to seagrass CCS would equate to the capture of ~0.14% of predicted annual carbon emissions in 2100 (based on emission scenario A1B) [29]. While this proportion could certainly be improved by considering recovery and restoration of additional aquatic and terrestrial habitats, it still would not compensate for anthropogenic emissions [9,10]. The cost of restoring such large amounts of seagrass could average ~US$224 million yr^{-1} (based on restoring 459.4 km^2 per decade), which may become cheaper if better techniques reduce costs from the current estimate of ~US $48,700 ha^{-1}, and if financial incentives can be provided through carbon trading schemes.

If such restoration is possible, maintaining abundant and optimally functioning coastal habitats may provide benefits against a changing climate that go beyond improved CCS. The effects of dense stands of terrestrial vegetation on local climatic conditions, such as reducing temperatures and desiccation stress, are well-known [30,31]. Recent evidence suggests that such effects may represent disproportionately large buffers against forecast impacts of climate change both on land [32] and in the sea (e.g. kelp forests: Falkenberg, Russell and Connell unpubl. data). Thus,

present-day investment in the maintenance and expansion of vegetation appears likely to not only improve global rates of CCS, but may also provide additional future rewards by lessening impacts of climate change.

Slowing or even reversing trends in net global habitat loss to improve the natural CCS capacity of the Earth is far easier said than done since CCS, and the benefits it may provide, is only one aspect of a complex issue. Often, the original reasons for habitat loss centre on direct tangible economic and social improvements. For example, one of the greatest threats to mangrove forests is the clear-felling of extensive areas to create space for commercial pond aquaculture of fish and crustaceans [15]. Such practices are most common in developing countries, particularly in SE Asia, and although mangrove removal radically alters the local ecology [12], the resulting land-use provides significant and much-needed socio-economic benefits. While coastal environments around the world have a long history of exploitation [18], it would ideally be managed to minimize long-term environmental impacts while still providing sustainable socio-economic rewards. Currently, such outcomes appear in the minority, yet there is encouraging evidence that it is achievable (e.g. sustainable rotating harvest of ~1000 ha yr^{-1} of mangroves for wood since 1906: [12]).

Recognition of the value in reversing alarming trends of habitat loss is certainly not new, yet the importance of achieving such goals becomes clearer as we continue to learn about the numerous benefits that optimally functioning habitats can provide. CCS appears to be an increasingly valuable function of natural habitats, and habitat restoration may offer a solution to increase natural CCS at faster rates than through natural recovery alone. For coastal habitats such as seagrasses, it appears that increases in long-term CCS could be negligible unless restoration efforts can be increased to industrial-scale operations and/or restoration success rates can be improved through greater investment in research and methodology [33]. If such outcomes are beyond our technical expertise and political resolve, then questions must be asked about whether resources currently used for expensive and time-consuming restoration projects may instead be more wisely invested in arresting further habitat loss and encouraging natural recovery by mitigating pollutants and other impacts. While restoration even on a small scale certainly provides many benefits beyond CCS (e.g. habitat for other plants and animals, nutrient cycling, etc.), it appears that one of the most effective opportunities for mitigating climate effects is to reduce non-climate human impacts that are under local control, and thereby encourage natural habitat recovery.

Acknowledgments

The authors thank A. Hector and three anonymous reviewers for their constructive comments on an earlier version of this manuscript.

Author Contributions

Conceived and designed the experiments: ADI SDC BDR. Performed the experiments: ADI SDC BDR. Analyzed the data: ADI SDC BDR. Contributed reagents/materials/analysis tools: ADI SDC BDR. Wrote the paper: ADI SDC BDR.

References

1. Hoffert MI, Caldeira K, Benford G, Criswell DR, Green C, et al. (2002) Advanced technology paths to global climate stability: Energy for a greenhouse planet. Science 298: 981–987.
2. Rogelj J, Nabel J, Chen C, Hare W, Markmann K, et al. (2010) Copenhagen accord pledges are paltry. Nature 464: 1126–1128.
3. Wigley TML (2006) A combined mitigation/geoengineering aproach to climate stabilization. Science 314: 452–454.
4. Walsh B (2009) Can geoengineering help slow global warming? TIME.com.
5. Chalmers H, Gibbins J (2010) Carbon capture and storage: the ten year challenge. Proceedings of the Institute of Mechanical Engineering 224: 505–518.

6. Bachu S (2008) CO_2 storage in geological media: Role, means, status and barriers to deployment. Progress in Energy and Combustion Science 34: 254–273.

7. Lal R (2008) Carbon sequestration. Philosophical Transactions of the Royal Society of London B 363: 815–830.

8. Laffoley DdA, Grimsditch G, eds. (2009) The Management of Natural Coastal Carbon Sinks. Gland, Switzerland: IUCN. 53 p.

9. Pidgeon E (2009) Carbon sequestration by coastal marine habitats: Important missing sinks. In: Laffoley DdA, Grimsditch G, eds. The Management of Natural Coastal Carbon Sinks. Gland, Switzerland: IUCN. pp 47–51.

10. Raupach MR, Marland G, Ciais P, Le Quere C, Canadell JG, et al. (2007) Global and regional drivers of accelerating CO_2 emissions. Proceedings of the National Academy of Sciences of the United States of America 104: 10288–10293.

11. Sabine CL, Feely RA, Gruber N, Key RM, Lee K, et al. (2004) The ocean sink for anthropogenic CO_2. Science 305: 367–371.

12. Alongi DM (2002) Present state and future of the world's mangrove forests. Environmental Conservation 29: 331–349.

13. Duarte CM, Middelburg JJ, Caraco N (2005) Major role of marine vegetation on the oceanic carbon cycle. Biogeosciences 2: 1–8.

14. Kennish MJ (2001) Coastal salt marsh systems in the U.S.: A review of anthropogenic impacts. Journal of Coastal Research 17: 731–748.

15. Valiela I, Bowen JL, York JK (2001) Mangrove forests: One of the world's threatened major tropical environments. Bioscience 51: 807–815.

16. Waycott M, Duarte CM, Carruthers TJB, Orth RJ, Dennison WC, et al. (2009) Accelerating loss of seagrasses across the globe threatens coastal ecosystems. Proceedings of the National Academy of Sciences of the United States of America 106: 12377–12381.

17. Cebrián J (2002) Variability and control of carbon consumption, export, and accumulation in marine communties. Limnology and Oceanography 47: 11–22.

18. Vitousek PM, Mooney HA, Lubchenco J, Melillo JM (1997) Human domination of Earth's ecosystems. Science 277: 494–499.

19. Kirkman H, Kuo J (1990) Pattern and process in southern Western Australian seagrasses. Aquatic Botany 37: 367–382.

20. Fonseca MS, Kenworthy WJ, Thayer GW (1998) Guidelines for the conservation and restoration of seagrasses in the United States and adjacent waters. Silver Spring, Maryland: NOAA Coastal Ocean Program. 222 p.

21. Orth RJ, Carruthers TJB, Dennison WC, Duarte CM, Fourqrean JW, et al. (2006) A global crisis for seagrass ecosystems. Bioscience 56: 987–996.

22. Orth RJ, Harwell MC, Inglis GJ (2006) Ecology of seagrass seeds and seagrass dispersal processes. In: Larkum AWD, Orth RJ, Duarte CM, eds. Seagrasses: Biology, Ecology and Conservation. Dordrecht, The Netherlands: Springer. pp 111–133.

23. Harley CDG, Hughes AR, Hultgren KM, Miner BG, Sorte CJB, et al. (2006) The impacts of climate change in coastal marine systems. Ecology Letters 9: 228–241.

24. Green EP, Short FT, eds. (2003) World Atlas of Seagrasses. Berkeley: University of California Press. 310 p.

25. Palacios SL, Zimmerman RC (2007) Resonse of eelgrass *Zostera marina* to CO_2 enrichment: possible impacts of climate change and potential for remediation of coastal habitats. Marine Ecology Progress Series 344: 1–13.

26. Russell BD, Thompson J-AI, Falkenberg LJ, Connell SD (2009) Synergistic effects of climate change and local stressors: CO_2 and nutrient-driven change in subtidal rocky habitats. Global Change Biology 15: 2153–2162.

27. Kennedy D, Norman C (2005) What don't we know? Science 309: 75.

28. Russell BD, Connell SD (2010) Honing the geoengineering strategy. Science 327: 144–145.

29. Meehl GA, Stocker TF, Collins WD, Friedlingstein P, Gaye AT, et al. (2007) Global climate projections. In: Solomon S, Qin D, Manning M, Chen Z, Marquis M, et al., eds. Climate Change 2007: The Physical Science Basis Contribution of Working Group I to the Fourth Assessment Report of the Intergovernmental Panel on Climate Change. Cambridge: Cambridge University Press. pp 747–845.

30. Callaway RM (1995) Positive interactions among plants. The Botanical Review 61: 306–349.

31. Goldenheim WM, Irving AD, Bertness MD (2008) Switching from negative to positive density-dependence among populations of a cobble beach plant. Oecologia 158: 473–483.

32. McAlpine CA, Syktus J, Ryan JG, Deo RC, McKeon GM, et al. (2009) A continent under stress: interactions, feedbacks and risks associated with impact of modified land cover on Australia's climate. Global Change Biology 15: 2206–2223.

33. Irving AD, Tanner JE, Seddon S, Miller D, Collings GJ, et al. (2010) Testing alternate ecological approaches to seagrass rehabilitation: links to life-history traits. Journal of Applied Ecology 47: 1119–1127.

34. Watzin MC, Gosselink JG (1992) The fragile fringe: Coastal wetlands of the continental United States. Washington DC: US Fish and Wildlife Service, and National Oceanic and Atmospheric Administration. 19 p.

2

Anthropogenic Chromium Emissions in China from 1990 to 2009

Hongguang Cheng[1]*, Tan Zhou[1,2], Qian Li[1], Lu Lu[1], Chunye Lin[1]

1 School of Environment, Beijing Normal University, Beijing, China, **2** Spatial Science Laboratory, Texas A&M University, College Station, Texas, United States of America

Abstract

An inventory of chromium emission into the atmosphere and water from anthropogenic activities in China was compiled for 1990 through to 2009. We estimate that the total emission of chromium to the atmosphere is about 1.92×10^5 t. Coal and oil combustion were the two leading sources of chromium emission to the atmosphere in China, while the contribution of them showed opposite annual growth trend. In total, nearly 1.34×10^4 t of chromium was discharged to water, mainly from six industrial categories in 20 years. Among them, the metal fabrication industry and the leather tanning sector were the dominant sources of chromium emissions, accounting for approximately 68.0% and 20.0% of the total emissions and representing increases of15.6% and 10.3% annually, respectively. The spatial trends of Cr emissions show significant variation based on emissions from 2005 to 2009. The emission to the atmosphere was heaviest in Hebei, Shandong, Guangdong, Zhejiang and Shanxi, whose annual emissions reached more than 1000t for the high level of coal and oil consumption. In terms of emission to water, the largest contributors were Guangdong, Jiangsu, Shandong and Zhejiang, where most of the leather production and metal manufacturing occur and these four regions accounted for nearly 47.4% of the total emission to water.

Editor: Vipul Bansal, RMIT University, Australia

Funding: This research work was funded by the National Natural Science Foundation of China (key project 40930740 and general project 41171384) and Special Environmental Research Funds for Public Welfare (No. 201009046 and No. 201109064). The funders had no role in study design, data collection and analysis, decision to publish, or preparation of the manuscript.

Competing Interests: The authors have declared that no competing interests exist.

* E-mail: chg@bnu.edu.cn

Introduction

Chromium is a naturally occurring metal that is present in small amounts throughout the environment. Human activities have increased the levels of chromium pollution in the air [1,2]and water [3,4,5] and may cause adverse effects on human health and the environment [6,7], especially in the case of hexavalent chromium Cr(VI) and its compounds.

In China, hexavalent chromium Cr(VI) and its compounds are released into the air as by-products of fossil fuel combustion, waste incineration [8,9] and various industrial processes (e.g., aerospace products and parts manufacturing, pulp and paper mills, ferrochromium or chromium metal production) [10,11] as diffused pollution. Meanwhile, chromium is also discharged into the water in the form of wastewater from industries such as leather tanning, metal fabrication and chromium plating [12]. Additional areas of application of chromium include wood preservatives, production of chrome pigments (e.g., lead chromate) which are used in paints, printing inks and anti-corrosive materials.

The extensive use of chromium in various industrial processes also results in the introduction of the metal into soil or landfills. The chromium in the soil may come from atmospheric deposition, sediment accumulation and the potential leachability of chromium slag. According to previous study [13], there is the possibility that chromium contained in the chromium slag could be leached into the soil. However, the leachability of chromium is very low, and it may take 15 to 20 years to lead to a 10% decrease in the total concentration of chromium in a morhorizon through leaching

[14][15,16] [15,16]. In addition, chromium salt production bases are relatively concentrated, and there are only approximately 20 factories in China. The pattern of this kind emission can be seen as point pollution, which is easier to control than emission to the air and water. Some previous studies [1,17] have also compiled Chinese studies related to chromium pollution in the soil. Therefore, we only offer a brief summary of the sources of chromium discharge to the atmosphere and water in this paper.

There have been extensive studies regarding the emission inventories of several volatile trace elements, such as Hg [18,19,20,21], As and Se [22] and Pb [23]. While most of these studies concentrate on the emission from coal combustion or gasoline combustion, a comprehensive investigation of Chromium discharge in China is lacking.

To improve our knowledge and understanding of the status of chromium pollution, a historical perspective can be indispensable when addressing the issues related to long-term heavy metal pollution. Hence, the purpose of this paper is to present a new inventory of emissions of chromium to the atmosphere and water from various sources from 1990 to 2009 and to analyze the temporal and spatial variations of chromium emissions in China. It is anticipated that this study will provide useful information related to the environmental health risks of chromium emissions for policy development in China.

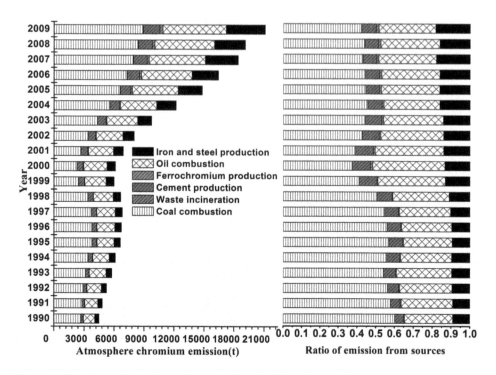

Figure 1. The atmospheric chromium emissions in China from 1990 to 2009.

Methods and Materials

According to previous studies [24,25], the main sources of chromium emission to the atmosphere are coal combustion, oil combustion, iron and steel production, cement production and waste incineration. In China, the significant sources of chromium discharge to the aquatic system mainly come from industrial wastewater effluent, which includes the leather industry, the chromite mining industry, the chemical manufacturing industry, the non-ferrous smelting industry, the fabricated metal industry and other industries [26,27].

The methods for estimating emissions include material balances, emission factors and source measurements. Chromium emissions are estimated by using fuel consumption data and detailed chromium emission factors. The basic concept of the chromium emission calculation is described by the following equation:

$$E_{i,j} = Q_{i,j} \times F_{i,j}$$

where E is the emission of chromium of one source, Q is the consumption data or industrial production, F is the specific emission factor, i is the sector type, j is the year.

The detailed methods of estimation of chromium for different source can be seen in the first section of Supporting Information where available chromium emission factors are also presented for each source: File S1. Tables–S2 illustrate the average content of chromium in coal as consumed by province and the release rates and control devices of coal combustion; File S1. Tables S10–S11 illustrate the emission factors of different types of oil combustion, ferrochromium production and related industries.

In this study, there are two major categories of all emission sources: atmosphere emission and water effluent emission. The annual consumption data for coal, petroleum products (gasoline, kerosene, diesel, cement production iron and steel production and

waste incineration by province are from the China National Bureau of Statistics from 1990–2009. The annual activity level of industries such as the ferrochromium production (File S1. Table S3), the chromites mining (File S1. Table S4) and the leather industry (File S1. Table S5) come from references and online data. In the sectors that emitted chromium to water, such as the fabricated metal industry, non-ferrous smelting, chemical manufacturing and other industries (File S1. Table S6), the sources of these data come from the China Statistical Yearbook. Their specific emission coefficients can be obtained from the Manual for Coefficients of Pollutant Generation and Discharge in Industrial Pollution Sources in China. These data do not include the special administrative regions of Hong Kong or Macao, and they also exclude Taiwan because the data are kept separately for these regions. The detailed information about sources, emission factors and activity data can be seen in Support information.

Results and Discussion

Atmospheric Emission

The total and annual atmospheric chromium emissions from various sources in China from 1990 to 2009 are summarized in Table 1 and Figure 1. The results show that the accumulated total emissions of chromium in China from 1990 to 2009 are estimated to be 1.92×10^5 t. Of the major sources of the total Cr emissions to air shown in Figure 1, coal combustion was the largest emission source, contributing approximately 46.6% of the total. Oil combustion was the second largest source, releasing nearly 5.85×10^4 t of chromium over twenty years, which represented approximately 30.5% of the total chromium emission, followed by the iron and steel industry, the cement industry and ferrochromium production.

Temporal trend of chromium emissions by sector. The total emission of chromium from 1990 to 2009 gradually increased, except in 1998. The reduced economic growth caused by the Asian financial crisis [28] in 1998 may be the primary

Table 1. The atmospheric chromium emissions from various sources in China from 1990 to2009.

Source	Emissions (t)	Ratio
Coal combustion	8.93×10^4	46.6%
Power plants	1.56×10^4	8.1%
Industrial sector	7.37×10^4	38.4%
Residential sector	4.00×10^1	0.02%
Other sector	2.60×10^1	0.01%
Oil combustion	5.85×10^4	30.5%
Diesel	4.05×10^4	21.1%
Kerosene	4.50×10^3	2.4%
Gasoline	1.34×10^4	7.0%
Ferrochromium production	1.81×10^3	0.9%
Iron and steel industry	2.75×10^4	14.3%
Cement production	1.46×10^4	7.6%
Waste incineration	8.50×10^1	0.04%
Total	1.92×10^5	100.0%

reason for this phenomenon because the chromium-related production activities decreased significantly in this year. This finding also coincides with the inventories of other trace metals, such as Hg [29], Pb [23], Cd and Cr from coal combustion [8]. Generally, the national emission showed a peak value in 1996. Since 2001, the rapid growth of the economy and the increased energy consumption contributed to significant emission increases.

Coal is the primary energy source in China, and coal combustion has increased dramatically during the past two decades [30]. As is shown in the Figure 1, chromium emissions from coal combustion increased from 2.70×10^3 t in 1990 to 8.88×10^3 t in 2009 at an annual average rate of 7.3%. Chromium emissions from coal combustion declined from 1998 to 2000, and the probable explanation is the economic decline that occurred during this period, which is also consistent with the study of Tian [8]. The detailed information about coal combustion and oil combustion by sub-sector are illustrated in File S1. Figures S1–S2.

The annual emission from oil combustion is continuously increasing, which is consistent with the increased demand from civilian-owned motor vehicles [31]. For oil combustion, most of emission comes from diesel and gasoline combustion which account for 69.2% and 23.0% of oil combustion. These two

Table 2. The chromium discharge into water from various sources in China from 1990 to 2009.

Sources	Emission(t)	Ratio
Nonferrous Metals Smelting and Pressing industry	8.00	0.1%
Fabricated metal industry	9.10×10^3	68.0%
Leather industry	2.68×10^3	20.0%
Chemical manufacturing industry	9.90×10^1	0.7%
Nonferrous Metals Mining and Dressing industry	1.02×10^3	7.6%
Other Industries	4.82×10^2	3.6%
Total	1.34×10^4	100.0%

categories comprise the majority of China's Cr emissions from oil combustion, and their shares kept relatively constant during the past two decades. The emission of kerosene is relative for which the demand and consumption of kerosene is lower than other types of oil. The detailed information about oil combustion can be seen in the Supplementary Information.

Ferrochromium production emitted approximately 1.81×10^3t of chromium during the period 1990–2009. It is noteworthy that the emissions presented some fluctuations in the first 10 years and continuously increased in the following years. The overall general trend is characterized by growth at 18.1% each year. There were no available data about ferrochromium production from 1990 to 1999, and the estimate seems smaller compare with the measured data in the second ten years. Additionally, the production of other chromium-related alloys, such as nickel chrome, will also emit chromium into the air. In the present study, there were no accurate national data about these alloys. All these factors could lead to an underestimation of chromium emission from ferro-chromium production.

It is important to note that the emissions of chromium from the iron and steel industry increased until 2008, which may be due to factors related to the 2008 Olympic Games in China and the global financial crisis of 2008. The ratio of chromium emission from other sources, such as cement production and waste incineration, also increased during this period to a different extent. According to some studies [22,32], with the development of dust emission requirements and advancement of technology, electrostatic precipitators(ESTs) were gradually substituted by fabric filters (FFs), which are used in cement kilns in China. However, FFs are still not widely used, along with the continuously rapid growth of economic activity and cement consumption. These factors may both contribute to the yearly increase in the emission from cement production in China.

As shown in Figure 1, the emission from iron and steel production is increasing with the high rate of growth at an annual average rate of 12.9%, especially for the recent years. The production and consumption of iron and steel has substantially increased since the 1990s. It has become the world's largest production base at the beginning of the 21st century, accounting for nearly half of the world's iron and steel production. Thus, the substantial production of iron may the main reason for the high emission and growth.

As waste incineration is gradually developing, the emissions are also increasing. Although the emission from this sector has been relatively small in recent years, it may have a significant impact on the total emission in the future because municipal solid waste (MSW) has recently been recognized as a renewable source of energy, and waste incineration is playing an increasingly important role in MSW management in China [33]. The detailed information about six sectors' emission to air is summarized in File S1. Table S7.

Spatial variation of chromium emission to air. The emission of chromium to the air demonstrates significant differences at the province level as is shown in Figure 2. The contributions to the five-year emissions from the top four emission-producing provinces were as follows: 7.95×10^3t from Hebei, 7.71×10^3 t from Shandong, 5.28×10^3 t from Jiangsu and 5.28×10^3t from Guangdong. The four largest emission regions produced nearly one-third of the total emissions. These regions are located in the east and are traditional industrial bases and economically intensive areas in China. In contrast, Tibet (6t), Qinghai (262 t), Hainan (317 t), Ningxia (644 t) and Gansu (911 t) are the regions with the lowest emissions. Further, notable unevenness can be seen among the provincial inventories.

Emissions from the eastern and central provinces are much higher than those in the western regions, except for Yunnan province, in which coal combustion emissions are quite high, corresponding to the rapid increase in the demand for coal-based industries, such as power plants, metallurgy and chemical industries in recent years [34]. In short, the low emission areas are mainly concentrated in areas that are less populated and that have lower economic strength (GDP), i.e., the Northwestern area. The pie chart in Figure 2 represents the emission structure in each region. In most regions, coal combustion or oil combustion contributed more than half of the total emissions. Specifically, chromium emissions from coal consumption are the major cause of pollution in the western area as well as in central China, representing over 50% of the total emissions of the region. Emission from oil combustion occurs in the eastern and southern parts of China, which are more developed and richer regions.

Generally, the majority of the provinces show a peak value of chromium emissions in 2007, followed by a subsequent decline in 2008. It is noteworthy that a negative growth rate appeared in 19 provinces in 2008, including Beijing, Hunan, Jilin, Guizhou and other regions. These findings may imply that the 2008 Olympic Games in China and the global financial crisis in 2008 are the factors lead to the emission decline.

However, the general trend among the provinces shows a high rate of growth. The fastest growth occurred in Qinghai province in 2007 (Tibet is excluded due to a lack of data), with an annual rate of increase of 38.2%, which coincides with the results for lead emission reported by Li [23]. For four municipalities, emissions from Shanghai were nearly twice those of Beijing, Tianjin and Chongqing which is also consistent with Hao's study of NO_X in China [28]. The population, GDP, and gross industrial output value of Shanghai are the highest of these four regions, which may explain this result.

Discharge into Water

Chromium is a common pollutant introduced into natural waters mainly due to the discharge of wastewater from chromite mining, leather tanning and stainless steel production and other related chromium industries, such as wood preservation, pigment production and electronics manufacturing [1,14,35,36]. Based on national environmental statistical data and the specific chromium emission factors of different sectors, the national gross chromium emission to the water from 1990–2009 was approximately 1.34×10^4t and detail information about the discharge of chromium to the aquatic environment were summarized in File S1. Table S9.

Inventories of chromium emissions by sector. There is a general trend for the annual emissions (Figure 3) demonstrating slow growth before the 21th century and a fast increase in recent years. This observation can be partly explained by the rapid growth of industry activities mixed with the management of wastewater discharge by the government [37]. Importantly, there was no general trend for growth in 1997–1999 for any of the sectors, and negative growth was observed in the fabricated metal industry, the nonferrous mining industry and the chemical manufacturing industry despite continuous robust GDP growth of 8.8% in 1997 and 7.8% in 1998. Contrary to earlier expectations, the years of 1997 and 1998 witnessed a decline in China's output and the reduction of fabricated metal production and other industries, mostly driven by the financial crisis of Asia.

The trend was continuously decreasing over 20 years based on the statistics data from the China Environment Statistical Yearbook. A strict set of standards of wastewater discharge was assumed to be the major reason for this decrease. However, compared with the result of First National Census of Pollution Sources [26]and previous research [26,38], approximately 2057t heavy metals, including Cd, Cr, As, Pb and Hg, were discharged into the water in 2007. Among them, chromium was the largest contributor with 1643t, which is comparable to the 1502t of chromium emissions of 2007 in the present study.

Of the major source contributions to the total Cr emissions to the water shown in Table 2, the fabricated metal industry contributed over 68.0% of the total emissions, and the leather industry contributed 20.0%. These two sectors comprise over

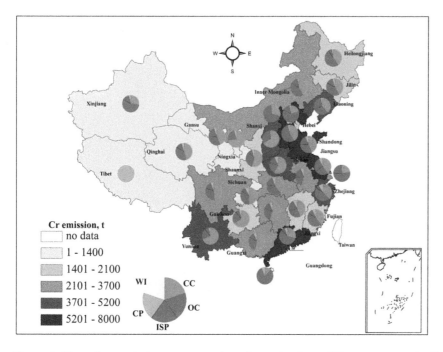

Figure 2. Chromium emissions to atmosphere by region from different sources in China in 2005 to 2009.

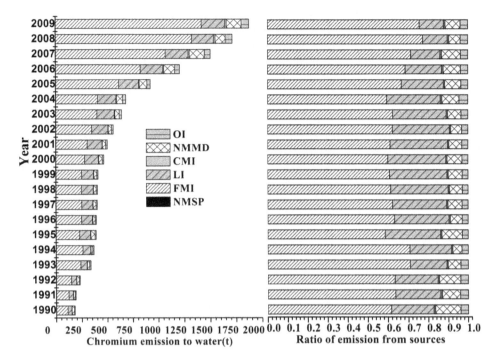

Figure 3. The chromium discharge into the water in China from 1990 to 2009.

88.0% of total chromium emissions to the water in China. The share contributed by the fabricated metal industry remained relatively constant in the past decades, while it was the largest chromium emission contributor during this period. Chromium emissions to water from this sector increased from 111t in 1990 to 1.42×10^3t in 2009 with an annual growth rate of 15.6% in 20 years. From 2005 to 2009, the annual growth rate reached more than 29.1%, and this may be attributed to the drastic development of the automobile industry, the construction of highways and railways and the related industries in these years during which China's GDP doubled [39].

The contribution of the leather industry experienced a gradual increase followed by a continuous decline in recent years, accounting for 12.0% of the total chromium emissions in 2009. The emissions increased from 38t in 1990 to 226t in 2009, and the average annual growth rate was 10.3%. Specifically, the emission from heavy leather production was approximately 1.6×10^3t, which is higher than the emission from light leather production with 1.08×10^3t. The low utilization of chromium and relatively backward tanning process [40,41,42]may be the primary reasons for the increased emission of chromium from the leather tanning industry. It is reported that only 60%–70% of the chromium is used in the tanning process, and 10% of the residual liquids are directly discharged into the river [40]. Interestingly, there was a decline in 1997–1998 for nonferrous mining. The nonferrous mining and dressing industry pressing industry was the next largest contributor, and the emissions increased rapidly in recent years, from 76t in 2005 to 143t in 2009, with an approximately 23.5% annual growth rate.

Particular attention should also be paid to other industries, such as manufacturing of textile, electronic equipment and furniture etc., which released 482t chromium during the period 1990 through 2009and account for 7.6% of the total emissions. With the recently attention and intensity of the government's control of heavy metal pollution in China, the emission of chromium from different sources has been gradually reduced. For the non-ferrous

smelting and chemical manufacturing industries, the annual emissions were relatively stable in the beginning and gradually increased in recent years, accounting for less than 1.0% altogether. Although the emission from these sources is small, the concentration of emissions in the absence of effective control equipment and standard process represents a significant health hazard for the surrounding population [23]. Additionally, the rapid economic growth mixed with the over standard discharge of wastewater in local areas [43] will present a challenge to the prospect of reduction.

Therefore, to achieve reductions in chromium emissions into the water, China should mainly aim to eliminate emissions from the fabricated metal industry and leather tanning by employing advanced process and treatment technologies. Additionally, the government should enforce standards of wastewater emission to reduce local emissions.

Spatial variations in the emission of chromium to water. The national annual emission of chromium was combined with the ratio for each specific province to yield estimations of the emissions at the provincial level (Table 3). From 2005 to 2009, the emissions of the provinces showed a peak value in 2007, followed by a decline in 2008. The trend is consistent with the national and provincial emission to air from 2005 to 2009. On the whole, the emissions of chromium are highly concentrated in the provinces of the eastern, central and southern regions, such as Guangdong, Zhejiang, Jiangsu, Hubei and Hunan. The chromium emissions of the different provinces in present study are very similar to the results of Wu (2012), who also singled out eight provinces, including Guangdong, Zhejiang, Jiangsu and Hunan, as preferential control areas for heavy metal pollution from wastewater.

The heavy emissions from these areas are driven by high levels of industrial activities and large population. Guangdong is the largest producer of fabricated metal, Hunan is rich with mining activities, and most of the leather tanning industries are located in Zhejiang, according to survey data [26]. On the contrary, most of

Table 3. Summary of Total Chromium Emission Estimates (t) to water by Province from 2005 to 2009.

Region	2005	2006	2007	2008	2009	Total
Northern Region	**120**	**154**	**187**	**219**	**233**	**912**
Beijing	25	31	36	35	38	165
Tianjin	25	33	37	42	45	182
Hebei	42	51	63	78	82	317
Shanxi	17	23	29	34	32	134
Inner Mongolia	10	16	22	30	37	113
Northeastern Region	**69**	**93**	**115**	**138**	**155**	**570**
Liaoning	38	54	68	84	96	340
Jilin	15	18	24	28	34	120
Heilongjiang	16	21	23	26	25	111
Eastern Region	**435**	**565**	**698**	**782**	**852**	**3332**
Shanghai	61	71	83	85	82	381
Jiangsu	123	158	198	229	250	958
Zhejiang	88	111	134	138	140	611
Anhui	18	23	29	38	46	153
Fujian	31	38	46	51	57	225
Jiangxi	11	16	23	29	33	113
Shandong	103	148	185	213	243	892
Central and Southern	**221**	**291**	**368**	**418**	**455**	**1752**
Henan	38	53	76	88	95	350
Hubei	22	28	36	45	53	185
Hunan	18	23	31	39	46	158
Guangdong	131	170	205	221	233	961
Guangxi	9	13	17	21	24	83
Hainan	2	2	4	4	4	15
Southwestern Region	**49**	**64**	**83**	**97**	**115**	**407**
Chongqing	11	12	16	19	23	82
Sichuan	22	30	41	50	62	205
Guizhou	6	8	9	11	12	46
Yunnan	10	13	16	17	18	74
Tibet	0	0	0	0	0	1
Northwestern Region	**31**	**42**	**52**	**60**	**64**	**251**
Shaanxi	13	17	21	25	29	105
Gansu	7	9	12	12	13	54
Qinghai	2	2	3	4	4	15
Ningxia	3	3	4	5	5	19
Xinjiang	7	10	12	14	14	57

provinces in the western part of China had low chromium emissions into the water, except Shaanxi and Sichuan. This may be mainly due to the increasing labor cost in the developed eastern coastal areas. Motivated by the policies of the western local governments, some highly polluted manufacturing enterprises are moving to these provinces.

Interestingly, the chromium emissions in Shaanxi in2008 differed from the normal levels, which reached suddenly85t. It is difficult to identify the exact reason for this result because we have no detailed data for this province. The relative backward water pollution controlling facilities that lead to ineffectively treated industrial wastewater and urban sewage being directly discharged into the rivers may have contributed to this [44]. Consequently, attention should also be concentrated on the emissions of western regions due to the potentially harmful impacts of emission on the local environment and human health.

Release of Chromium to Ecosystems

Estimations of the chromium emissions to the atmosphere and water from1990 to 2009 are presented in Table 4. To better predict the trend, the two decades are divided into four periods, which could provide a more typical reflection of the emission status.

Emissions into the air increased during these four periods, and the most rapid increased occurred from 2005 to 2009, and rate of growth over these five years (there were no available data about waste incineration before 2003) reached over 100.0%, which coincides with the economy and energy use during this period. China's GDP increased by a double-digit annual average during the period of 2005 to 2009, reaching as high as 10.2%. Specifically, the most drastic change came from ferrochromium production, and the trend became increasingly obvious at the beginning of the 21st century because the majority of ferrochromium is used in the stainless steel industry, reaching 85% [45]. The increasing demand for stainless steel may be the major reason for this dramatic change. In contrast, coal combustion presented the lowest growth rate, while it remains the largest contributor to emissions.

For emissions into water, all sectors experienced gradual growth in the first three periods and drastic increases in the last period. This increase from 2005 to 2009 is also consistent with the industrial production and rapid development in recent years. The drastic change is associated with the fabricated metal industry, which nearly tripled its emission in the third period (2000–2004). For the non-ferrous smelting industry and the chemical manufacturing industry, the growth rates were much smaller. Based on the China Statically Yearbook from 2000–2009, the development of these sectors has been less obvious. Additionally, the attention to environmental protection and the implementation of discharge standards has contributed to this smaller growth [46,47].

It is also noteworthy that emissions into the air were much larger than emissions into the water. However, it is not correct to say that the emission of chromium pollution into the atmosphere is more severe than the emission into water because the emission of chromium into the air is dissipated over a large area, while chromium discharge to water flows along rivers, and which has a line distribution, making it much easier for chromium contamination to enter the food chain and have a more direct effect on human life and health.

Speciation of Cr Compounds

The speciation of Cr compounds for each emission source sector is pivotal to accurately assess the environmental and human health impacts since toxicity of Cr(III) (non-carcinogenic) and Cr(VI) (carcinogenic) to organism varies considerably. There are a large amount of chromium with +2, +3, +6 or in combination are introduced into environment. It is well established that chromium is mobilized in association with airborne particles derived from high-temperature combustion sources, like coal, oil combustion and waste incineration. For chromium in industrial wastes, cement-producing, it predominantly occurs as the hexavalent form in chromate (CrO_4^{2-}) and dichromate $(Cr_2O_7^{2-})$ ions. Emissions of chromium from the processing of raw material like smelting or mining, chromium plating would predominantly be in the trivalent oxidation state. However, the limited amounts of data

Table 4. Summary of chromium emission to atmosphere and water in different periods (t).

Category	Sources	1990–1994	1995–1999	2000–2004	2005–2009	sum
To air	Coal combustion	1.50×10^4	1.71×10^4	1.83×10^4	3.90×10^4	8.93×10^4
	Waste incineration	0.00	0.00	9.01	7.65×10^1	8.55×10^1
	Cement production	1.56×10^3	2.59×10^3	3.78×10^3	6.70×10^3	1.46×10^4
	Ferrochromium production	1.40×10^2	1.40×10^2	4.18×10^2	1.11×10^3	1.81×10^3
	Oil combustion	7.24×10^3	9.45×10^3	1.42×10^4	2.76×10^4	5.85×10^4
	Iron and steel production	2.50×10^3	3.50×10^3	6.25×10^3	1.52×10^4	2.75×10^4
	Total per year	5.28×10^3	6.55×10^3	8.59×10^3	1.80×10^4	3.84×10^4
To water	Nonferrous smelting and Pressing	0.52	0.68	1.20	5.25	7.65
	Fabricated metal industry	8.79×10^2	1.22×10^3	1.74×10^3	5.25×10^3	9.10×10^3
	Leather industry	2.68×10^2	5.54×10^2	7.90×10^2	1.07×10^3	2.68×10^3
	Chemical manufacturing industry	9.00	1.30×10^1	2.02×10^1	5.70×10^1	9.92×10^1
	Nonferrous Mining and Dressing	1.04×10^2	1.48×10^2	2.03×10^2	5.65×10^2	1.02×10^3
	Other industries	4.83×10^1	5.99×10^1	1.01×10^2	2.73×10^2	4.82×10^2
	Total per year	2.62×10^2	4.00×10^2	5.72×10^2	1.44×10^3	2.68×10^3

related to chromium speciation of each source make the detailed emissions estimation unavailable.

Uncertainty Analysis

Because the emission of chromium comes from so many sources and most of the activity data indirectly come from references or statistical reports, an uncertainty analysis is necessary to better understand the estimation of the emissions.

In general, there are three aspects of the uncertainty of the emissions that need to be considered: commercial energy use, emission factors and emission control [48]. In the present study, these three factors apply to atmosphere emission, while only the first two contribute to wastewater emissions because emission control data are incorporated into the discharge coefficient in the Manual for Coefficients of Pollutant Generation and Discharge in Industrial Pollution Sources.

For atmosphere emission, the commercial energy use data come from China's official statistics yearbook, and the figures are somewhat underestimated according to some studies [18,19]. There are two reasons for this underestimation: one is that the data may have been adjusted for political reason and using data from a multitude of sources could reduce the uncertainty of the output. Another reason is that many of the activities that release large amounts of chromium occur in remote parts of the country (and may actually be illegal). Residential coal use in rural areas may be under-reported in Chinese statistics.

With regard to the emission factors, most of them were cited from EPA and EEA because there are no data available in China. The uncertainty may come from differences in boiler combustion efficiency and process technologies used in China. However, information about the chromium contents of raw materials by province can improve the accuracy of the estimation. The control devices used by combustion industries are also diverse across different provinces, and most industries installed advanced devices after entering 21th century. Therefore, to adopt the average values of the measurements from foreign and domestic studies to balance this change could make it more reasonable. There is less detailed information about the activities of industries that discharge chromium into the water, and the statistical data on chromium

emissions into the water were indirectly obtained though the industrial output in China Statistical Yearbook. While the emission coefficients are choose from China's research which may enhance the accuracy of estimation. Compared with measured data of First National Census of Pollution Sources, it is suggested that the estimations are located in the reasonable ranges. In addition, the two major sources, namely the leather industry and chromite mining, were recalculated by combing the activity data with the discharge coefficient to reduce the possible underestimation of the emission.

It is inevitable that estimations of emissions will contain uncertainties. Additional detailed investigations and field tests for all chromium-related industries will be helpful for a complete understanding of the emissions of chromium to the environment. Furthermore, there is little information about the emission of chromium, which could provide a basis for policies aiming for emission control in China.

Conclusion

We present, for the first time, a detailed estimation of chromium emissions to the environment from different sources by province during the period from 1990 to 2009. The national total atmospheric emissions of chromium were estimated at approximately 1.92×10^5t, at annual growth rates of 8.8% since 1990. Coal combustion was identified as the largest contributor, accounting for 45.6% of the total atmospheric emission. However, the contribution of coal decreased from 60.2% in 1990 to 42.2% in 2009. Conversely, the contribution of oil combustion to chromium emissions has grown slightly during these years, which may be because coal has been replaced with cleaner fuels, such as oil, natural gas and LPG, in recent years. Other sources of emission have also increased at different rates, coinciding with the rapid economic development in China. As for emission into water, the national total emission has undergone fluctuations that may have been caused by a combination of industrial growth and the implementation of wastewater discharge standards. The fabricated metal industry and the leather tanning industry were singled out as the major emission sources, accounting for 68.0% and 20.0% of the total emission to water. Other industries, such as non-ferrous smelting, mining and fabricated metals, were also not negligible sources.

The spatial characteristics of emissions to air and water are based on estimation from 2005 through 2009. Guangdong, Zhejiang, Jiangsu, Shandong and Hebei are the provinces with the largest emissions in China. The emissions are concentrated in the costal and central areas of China, and the regions with severe pollution are slowly moving toward western areas, such as Gansu, Yunnan and Shaanxi. This phenomenon may be due to the policies of the local governments and environmental pressure of the coastal region. The magnitude of these emissions may be debated, but their growing importance is unquestionable. To improve our knowledge and understanding of the current scenario, it is necessary to establish a complete field testing system and pay more attention to the forms of chromium in the environment.

Supporting Information

Figure S1   The temporal trend of emission of chromium from coal combustion. *For four sectors of coal combustion have drastic difference of the industry sector and power plant sector based on the left vertical axis, and others and residential sector based on right vertical axis. Among all of the coal consuming sectors, the chromium emissions from the power sector are increasing the fastest, with an average 7.91% annual increase, reaching 1407 t in 2009. However, the main contributor to coal combustion is the industry sector, which remained nearly constant throughout study period, representing over 80% of coal combustion. Another highlighted feature is that the emissions of chromium from the industry have increased substantially since 2005. The rapid expansion of energy-intensive manufacturing industries, such as steel and cement production, and coal consumption by industrial sector after the recovery of the economic decline may explain this rapid growth. The growth rate of power plants is lower compared with that of the industrial

sector, and negative growth was observed in 2004 and 2008. This may due to the improvement of PM and SO_2 control devices in coal-fired power plants [29]. Emission from the residential sector and other sectors declined at the beginning of the study period and then increased after 2003. This fluctuation may be a co-effect of the substitution of coal with cleaner fuels, such as natural gas, and the gradual increase of coal consumption in recent years.

Figure S2   The temporal trend of emission of chromium from oil combustion.

File S1   Table S1. Average contents of Cr in raw coals consumed in China by provinces. Table S2. Release rates and control devices of Cr for coal combustion. Table S3. The estimated emission of ferrochromium production into atmosphere, 1990–2009. Table S4. The estimated discharge into water from nonferrous mining industry. Table S5. The estimated discharge into water from leather tanning industry. Table S6. Industrial output values of sectors and discount rate of China from 1990 to 2009. Table S7. The estimated atmospheric chromium emissions in China, 1990–2009. Table S8. The estimated atmospheric chromium emissions by region in China, 2005–2009. Table S9. The estimated chromium emissions to water in China, 1990–2009. Table S10. Emission factors of different types of oil combustion. Table S11. Emission factors of ferrochromium production and related industries.

Author Contributions

Conceived and designed the experiments: HC TZ. Performed the experiments: TZ QL. Analyzed the data: TZ HC QL. Contributed reagents/materials/analysis tools: TZ. Wrote the paper: TZ. Data analysis: LL. Data collection: CL.

References

1. Cheng SP (2003) Heavy metal pollution in China: Origin, pattern and control. Environmental Science and Pollution Research 10: 192–198.
2. Kampa M, Castanas E (2008) Human health effects of air pollution. Environmental Pollution 151: 362–367.
3. Zhang JD, Li XL (1987) Chromium pollution of soil and water in Jinzhou. Chinese Journal of Preventive Medicine 21: 262.
4. Tziritis E, Kelepertzis E, Korres G, Perivolaris D, Repani S (2012) Hexavalent Chromium Contamination in Groundwaters of Thiva Basin, Central Greece. Bulletin of Environmental Contamination and Toxicology 89: 1073.
5. Richard MS, Beaumont J, Thomas AM, Reynolds S, Krowech G, et al. (2006) Review of the evidence regarding the carcinogenicity of hexavalent chromium in drinking water. Journal of Environmental Science and Health Part C 24: 155–182.
6. Rowbotham AL, Levy LS, Shuker LK (2000) Chromium in the environment: an evaluation of exposure of the UK general population and possible adverse health effects. Journal of Toxicology and Environmental Health Part B: Critical Reviews 3: 145–178.
7. Kuo CY, Wong RH, Lin JY, Lai JC, Lee H (2006) Accumulation of chromium and nickel metals in lung tumors from lung cancer patients in Taiwan. Journal of Toxicology and Environmental Health, Part A 69: 1337–1344.
8. Tian H, Cheng K, Wang Y, Zhao D, Lu L, et al. (2012) Temporal and spatial variation characteristics of atmospheric emissions of Cd, Cr, and Pb from coal in China. Atmospheric Environment 50: 157–163.
9. Cong Z, Kang S, Zhang Y, Li X (2010) Atmospheric wet deposition of trace elements to central Tibetan Plateau. Applied Geochemistry 25: 1415–1421.
10. Su D, Wang T, Liu L, Bai L (2010) Research on the spatio-temporal variation of pollutant discharged from industrial wastewater in the Liaohe River Basin. Ecology Environment Science, 19: 2953–2959.
11. Wang HB, Wang X, Ue ND, Du XM (2006) Characteristics of Typical Pollutants in Wastewater from Different Industrial Sectors in the Northeast Provinces of China." Journal of Agro-Environment Science 25: 1685–1690.
12. Fishbein L (1981) Sources, transport and alterations of metal compounds: an overview. I. Arsenic, beryllium, cadmium, chromium, and nickel. Environmental health perspectives, 40, 43.

13. Du L, Wang JS (2004) Environmental impact of chromic slag and analysis of chromic slag's output. Journal of Safety and Environment, 2, 010.
14. Bergbäck B, Anderberg S, Lohm U (1989) A reconstruction of emission, flow and accumulation of chromium in sweden 1920–1980. Water, Air, & Soil Pollution 48: 391–407.
15. Sheng C, Chai L, Wang Y, Li X (2006) Research Status and Prospect of Detoxification of Chromium-containing Slag By Hydro-based Process. Industrial Safety and Environmental Protection, 2: 000.
16. Tian HZ, Cheng K, Wang Y, Zhao D, Chai FH, et al. (2011) Quantitative Assessment of Variability and Uncertainty of Hazardous Trace Element (Cd, Cr, and Pb) Contents in Chinese Coals by Using Bootstrap Simulation. Journal of the Air & Waste Management Association 61: 755–763.
17. Huamain C, Chunrong Z, Cong T, Yongguan Z (1999) Heavy metal pollution in soils in China: status and countermeasures. Ambio: 130–134.
18. Feng X (2005) Mercury pollution in China–an overview. In Dynamics of Mercury Pollution on Regional and Global Scales: (pp. 657–678). Springer US.
19. Streets DG, Hao J, Wu Y, Jiang J, Chan M, et al. (2005) Anthropogenic mercury emissions in China. Atmospheric Environment 39: 7789–7806.
20. Pacyna EG, Pacyna JM, Steenhuisen F, Wilson S (2006) Global anthropogenic mercury emission inventory for 2000. Atmospheric Environment 40: 4048–4063.
21. Wu Y, Wang S, Streets DG, Hao J, Chan M, et al. (2006) Trends in anthropogenic mercury emissions in China from 1995 to 2003. Environmental science & technology 40: 5312–5318.
22. Tian HZ, Wang Y, Xue ZG, Cheng K, Qu YP, et al. (2010) Trend and characteristics of atmospheric emissions of Hg, As, and Se from coal combustion in China, 1980–2007. Atmospheric Chemistry and Physics 10: 11905–11919.
23. Li Q, Cheng H, Zhou T, Lin C, Guo S (2012) The estimated atmospheric lead emissions in China, 1990–2009. Atmospheric Environment 60: 1–8.
24. Nriagu JO, Pacyna JM (1988) Quantitative assessment of worldwide contamination of air, water and soils by trace metals. Nature 333: 134–139.
25. Shular J (1989) Locating and Estimating Air Emissions from Sources of Chromium: Supplement: US Environmental Protection Agency, Office of Air Quality Planning and Standards.

26. Agriculture MO (2010) Bulletin on the First National Census on Pollution Sources.
27. Wu W, Jiang H (2012) Equal Standard Pollution Load of Heavy Metals from Industrial Wastewater in China. Environmental Science & Technology 35: 180–185.
28. Hao J, Tian H, Lu Y (2002) Emission inventories of NOx from commercial energy consumption in China, 1995–1998. Environmental science & technology 36: 552–560.
29. Wu Y, Wang S, Streets DG, Hao J, Chan M, et al. (2006) Trends in Anthropogenic Mercury Emissions in China from 1995 to 2003. Environmental Science & Technology 40: 5312–5318.
30. Crompton P, Wu Y (2005) Energy consumption in China: past trends and future directions. Energy Economics 27: 195–208.
31. China Automotive Technology and Research Center (2009) China Automotive Industry Yearbook. China Automotive Industry Yearbook House, 613.
32. You CF, Xu XC (2010) Coal combustion and its pollution control in China. Energy, 35(11), 4467–4472.
33. Cheng H, Hu Y (2010) Municipal solid waste (MSW) as a renewable source of energy: Current and future practices in China. Bio-resource Technology 101: 3816–3824.
34. Peng J, Wang Y, Ye M, Chang Q (2005) Research on the change of regional industrial structure and its eco-environmental effect: A case study in Lijiang city, Yunnan province. Acta Geographica Sinica-Chinese Edition, 60, 798.
35. Anderberg S, Bergbäck B, Lohm U (1989) Flow and distribution of chromium in the Swedish environment: A new approach to studying environmental pollution. Ambio: 216–220.
36. Owlad M, Aroua MK, Daud WAW, Baroutian S (2009) Removal of hexavalent chromium-contaminated water and wastewater: a review. Water, Air & Soil Pollution 200: 59–77.
37. Liang S, Sun H (2007) Industrial Wastewater in China: Pollution and Affecting Factors Analysis. Environmental Science & Technology 5: 17.
38. Wu W, Jiang H (2012) Equal Standard Pollution Load of Heavy Metals from Industrial Wastewater in China. Environmental Science & Technology 35: 180–185.
39. Wang B, Song W, Chen W, Wang J (2012) The main steel wire products and its equipment development situation analysis in China. Metal Products 38: 10–19.
40. Liu W, Zhang X, Liu M (2001) Chromiun Pollution and Control in the Leather-making Industry. Leather Science and Engineering 11: 1–6.
41. Huang S, Yang H (2011) Research Progress on Treatment of Chrome Wastewater from Tannery. West Leather 33: 47–50.
42. Huo X, Liu C (2009) Review of Present Situation of Chrome Tanning Wastewater Treatment. West Leather 31.
43. Yu F, Guo X, Zhang Q (2004) Wastewater pollution situation and countermeasures for Chinese Mineral Industry. Resources Science 26: 46–53.
44. Feng X (2010) Water pollution assessment and research of quality evolution rule in Guanzhong section of Weihe river.
45. Wang W (2003) Summarized introduction of Chromite mining and Chromi-mum-iron alloy production. Mining Engineering 1: 10–13.
46. Li J, Zhao Z (2011) Performance Evaluation and Trend Expectation of Chinese Environmental Governance under Green Development. Environmental Science and Management 36: 162–168.
47. Yang F (2007) Survey on the State of Environmental Protection Consciousness of the Public. Journal of Hohai University (Philosophy and social sciences) 2: 31–35.
48. Jozef MP, Elisabeth GP (2001) An assessment of global and regional emissions of trace metals to the atmosphere from anthropogenic sources worldwide. Environmental Reviews 9: 269.

3

Synthesis and Detection of Oxygen-18 Labeled Phosphate

Eric S. Melby, Douglas J. Soldat*, Phillip Barak

Department of Soil Science, University of Wisconsin-Madison, Madison, Wisconsin, United States of America

Abstract

Phosphorus (P) has only one stable isotope and therefore tracking P dynamics in ecosystems and inferring sources of P loading to water bodies have been difficult. Researchers have recently employed the natural abundance of the ratio of $^{18}O/^{16}O$ of phosphate to elucidate P dynamics. In addition, phosphate highly enriched in oxygen-18 also has potential to be an effective tool for tracking specific sources of P in the environment, but has so far been used sparingly, possibly due to unavailability of oxygen-18 labeled phosphate (OLP) and uncertainty in synthesis and detection. One objective of this research was to develop a simple procedure to synthesize highly enriched OLP. Synthesized OLP is made up of a collection of species that contain between zero and four oxygen-18 atoms and, as a result, the second objective of this research was to develop a method to detect and quantify each OLP species. OLP was synthesized by reacting either PCl_5 or $POCl_3$ with water enriched with 97 atom % oxygen-18 in ambient atmosphere under a fume hood. Unlike previous reports, we observed no loss of oxygen-18 enrichment during synthesis. Electrospray ionization mass spectrometerty (ESI-MS) was used to detect and quantify each species present in OLP. OLP synthesized from $POCl_3$ contained 1.2% $P^{18}O^{16}O_3$, 18.2% $P^{18}O_2{}^{16}O_2$, 67.7% $P^{18}O_3{}^{16}O$, and 12.9% $P^{18}O_4$, and OLP synthesized from PCl_5 contained 0.7% $P^{16}O_4$, 9.3% $P^{18}O_3{}^{16}O$, and 90.0% $P^{18}O_4$. We found that OLP can be synthesized using a simple procedure in ambient atmosphere without the loss of oxygen-18 enrichment and ESI-MS is an effective tool to detect and quantify OLP that sheds light on the dynamics of synthesis in ways that standard detection methods cannot.

Editor: Meni Wanunu, University of Pennsylvania, United States of America

Funding: This research was funded by the University of Wisconsin-Madison Graduate School. The funders had no role in study design, data collection and analysis, decision to publish, or preparation of the manuscript.

Competing Interests: The authors have declared that no competing interests exist.

* E-mail: djsoldat@wisc.edu

Introduction

The lack of multiple stable isotopes of P has led researchers to study the ratio of ^{16}O and ^{18}O in phosphate, the dominant form of P in the natural environment. The naturally occurring ratio of $^{18}O/^{16}O$ bonded to phosphorus in phosphate, expressed as $\delta^{18}O_p$, has been used to infer the source of P in water bodies [1–6]. Gruau et al. [7] found substantial variability of $\delta^{18}O_p$ within P sources and considerable overlap of $\delta^{18}O_p$ among different P sources and therefore questioned the value of this method as a research tool. Based on the variability and overlap of $\delta^{18}O_p$ in natural sources, some researchers have used a method that enriches phosphate in oxygen-18, making it very distinct from any other P sources. The synthesis of oxygen-18 labeled phosphate (OLP) is generally carried out by hydrolyzing either PCl_5 or $POCl_3$ with water enriched with oxygen-18. This reaction is typically carried out in an atmosphere devoid of moisture to prevent the highly reactive PCl_5 or $POCl_3$ from reacting with atmospheric water [8–11]. Middleboe and Saaby Johansen [12] reported that up to one-half of the ^{18}O enrichment was sacrificed when $POCl_3$ was hydrolyzed with $H_2{}^{18}O$ in ambient atmosphere.

Analysis of the $^{18}O/^{16}O$ ratio of OLP has been previously carried out through the use of optical emission spectroscopy [9,13,14], in which the OLP present in solution, plant tissue (after being extracted with 0.5 M HCl), or soil (after being extracted with anion exchange resin) is precipitated as struvite ($MgNH_4PO_4 \cdot 6H_2O$) [9,14]. This precipitate is pyrolyzed to carbon monoxide, which is analyzed for its $^{18}O/^{16}O$ ratio by optical emission spectroscopy [14]. Synthesized OLP is made up of a collection of species that contain between zero to four oxygen-18 atoms. Optical emission spectroscopy is only capable of determining an overall $^{18}O/^{16}O$ ratio and is not able to distinguish between the various OLP species. The use of electrospray ionization mass spectrometry (ESI-MS) has also been presented as a viable means to determine the $^{18}O/^{16}O$ ratio of OLP and as a detection tool that can differentiate between OLP species. OLP dissolved in water was analyzed by ESI-MS and the results were compared to those obtained with ^{31}P NMR analysis. The authors reported good agreement between the ESI-MS and ^{31}P NMR results [11]. However, ESI-MS is sensitive at mg L^{-1} P levels, which is lower than ^{31}P NMR limits of detection.

The first objective of this research was to develop a simpler, efficient technique for the synthesis of OLP. The second objective was to use ESI-MS to quantify the individual OLP species present in synthesized OLP compounds. A final objective included determining the stability of OLP stored in sterile conditions over a period of several months.

Methods

The synthesis of OLP was carried out after consultation with the University of Wisconsin-Madison Environment, Health and Safety

Department (Madison, Wisconsin) as $POCl_3$ and PCl_5 are both highly toxic chemicals. A chemical resistant suit, chemical resistant gloves, and a positive pressure, self-contained breathing apparatus (Scott Health and Safety, Monroe, North Carolina) were worn during the handling of these chemicals and during their reaction with water to prevent any contact or inhalation of fumes.

OLP Synthesis from $POCl_3$

The synthesis of OLP from $POCl_3$ was carried out in a 25×200 mm glass test tube contained within a 250 mL Erlenmeyer flask filled with 150 mL of water to moderate the temperature. The flask and test tube were placed on a combination stir/hot plate within a fume hood. OLP was synthesized through the following spontaneous and highly exothermic reaction:

$$POCl_3 + 3H_2O \rightarrow H_3PO_4 + 3HCl$$

2.00 g of water with 97 atom % ^{18}O (Sigma-Aldrich, St. Louis, Missouri) were added to the test tube, followed by 4.87 g of $POCl_3$ (Pfaltz and Bauer Inc., Waterbury, Connecticut) added drop wise while stirring with a magnetic stir bar. Labeled water was added in 10% excess to ensure that the $POCl_3$ reacted completely with water containing ^{18}O. After the vigorous reaction subsided, the test tube was heated for 2 h at 100°C with stirring to remove HCl produced during the reaction. Following heating, the test tube was allowed to cool to room temperature and 5 M KOH was added to raise the pH to 4.7. The KH_2PO_4 product was diluted with ultrapure water, and the P concentration of this solution was determined using an IRIS Advantage inductively coupled plasma optical emission spectrometer (ICP-OES) (Thermo Jarrell Ash, Franklin, Massachusetts). The reaction was carried out with pure $POCl_3$, and with $POCl_3$ stored for a period of 6 and 17 months after it was initially opened to determine if there was any effect of long term $POCl_3$ storage on the OLP species composition.

OLP Synthesis from PCl_5

The synthesis of OLP from PCl_5 was carried out using the same procedure as described above, but this time reacting 2.50 g of PCl_5 (Sigma-Aldrich, St. Louis, Missouri) with 1.00 g of water containing 97 atom % ^{18}O (Sigma-Aldrich, St. Louis, Missouri)

through the following spontaneous and highly exothermic reaction:

$$PCl_5 + 4H_2O \rightarrow H_3PO_4 + 5HCl$$

The reaction was carried out with freshly opened PCl_5, and PCl_5 that was stored for a period of 17 months after it was initially opened.

Quantification of Species Using ESI-MS and Long-Term Stability of OLP in Sterile Water

Electrospray ionization mass spectrometry (ESI-MS) was used to establish a ratio of each of the species present in the synthesized OLP. All samples were analyzed using an Agilent 1100 LC-MSD SL single-quadrupole model 1946D mass spectrometer (Agilent Technologies, Santa Clara, California). Table 1 lists the liquid chromatography (LC) and mass spectrometer (MS) conditions used for this analysis. The samples were diluted to approximately 60 mg L^{-1} P prior to analysis and analyzed in duplicate injections. Oxygen labeled phosphate eluted from the LC beginning at $\sim$4.2 min and ending at 4.5 min after injection. During OLP analysis by ESI-MS, the m/z range monitored was 90–120, as the peaks of interest were m/z 97 ($H_2P^{16}O_4^-$), 99 ($H_2P^{16}O_3^{18}O^-$), 101 ($H_2P^{16}O_2^{18}O_2^-$), 103 ($H_2P^{16}O^{18}O_3^-$), and 105 ($H_2P^{18}O_4^-$). All other peaks within this range were negligible (Fig. 1). The area of each spectra for individual phosphate species were compared to the cumulative area of all the phosphate species allowing for the calculation of the relative contribution of each phosphate species shown in Figure 1. This relative information was used with the total P concentration of the sample to calculate the concentration of phosphate species in solution.

To monitor the stability of OLP solutions in a sterile environment, aqueous OLP solutions were diluted with autoclaved ultrapure water, placed in autoclaved polypropylene containers, and stored in darkness at 4°C.

Results

Data from the ESI-MS analysis of OLP synthesized from $POCl_3$ and PCl_5 are shown in Tables 2 and 3, respectively, along with the theoretical synthesis values. Theoretical values were calculated

Table 1. Liquid chromatography and mass spectrometry conditions used for electrospray ionization mass spectrometer analysis of oxygen-18 labeled phosphate.

Parameter	Condition
Liquid chromatography	
Column	Waters IC-Pak Anion HR, 4.6×75 mm, 6 μm particle size (Waters Corp., Milford, Massachusetts)
Mobile phase	Isocratic: 25% acetonitrile, 75% 50 mM ammonium bicarbonate (pH 10)
Flow rate	0.5 mL min^{-1}
Column temperature	30°C
Mass spectrometry	
Injection volume	15 μL
Source voltage	2250 V
Drying gas	Nitrogen, 350°C, 40 L min^{-1}, 40 psi
Ionization	Electrospray
MS detection	Single ion recording (SIR) mode

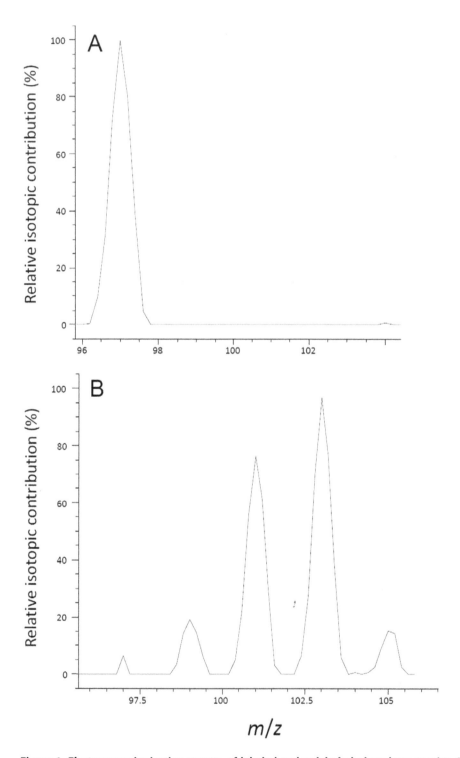

Figure 1. Electrospray ionization spectra of labeled and unlabeled phosphate species. Relative abundance of (a) unlabeled phosphate (m/z = 97) and (b) ^{18}O-labeled phosphate species (m/z = 99, 101, 103, and 105) as determined by ESI-MS.

assuming completely random arrangement of the oxygen atoms from labeled water and assuming no P-O bonds of $POCl_3$ were broken. We observed almost no loss of oxygen-18 enrichment when synthesizing OLP from pure $POCl_3$ (Table 2). However, OLP synthesized from $POCl_3$ stored six months had an oxygen-18 content of 62.2% and $POCl_3$ stored 17 months had an oxygen-18 content of 55.3%. This enrichment loss can probably be attributed

to water vapor entering the $POCl_3$ container either when the bottle was opened or throughout the storage period. There was no loss of oxygen-18 enrichment when synthesizing OLP from PCl_5 that had been stored for 17 months after it was initially opened (Table 3).

Based on the results from the reaction of PCl_5 with 97 atom % ^{18}O water carried out in ambient atmosphere (Table 3) in

Table 2. Amount of each oxygen labeled phosphate (OLP) species present in OLP synthesized with pure $POCl_3$ during two separate synthesis events with either fresh $POCl_3$ or with $POCl_3$ previously opened and stored containers as compared to the theoretical values.

OLP species	Amount of species present in OLP				
	Theoretical†	June 2009		May 2010	
		Fresh	Stored‡	Fresh	Stored§
---- m/z ----	----------------------------------- % -----------------------------------				
97	0.0	0.0	1.0	0.0	21.7
99	0.3	1.2	8.7	3.7	4.0
101	8.5	18.2	36.3	23.6	19.36
103	91.1	67.7	48.5	52.6	51.6
105	0.2	13.0	5.58	20.2	13.4
Total ^{18}O %	72.8	73.1	62.2	72.3	55.3

†Calculated assuming no P-O bonds in $POCl_3$ were broken during the reaction.
‡Opened and stored for 6 months.
§Opened and stored for 17 months.

Table 3. Amount of each oxygen labeled phosphate (OLP) species present in OLP synthesized from PCl_5 during two separate synthesis events (December 2008 and May 2010) as compared to theoretical values.

OLP species	Amount of species present in OLP		
	Theoretical	December 2008	May 2010†
---- m/z ----	---------------------------- % ----------------------------		
97	0.0	0.7	0.0
99	0.0	0.0	0.0
101	0.5	0.0	0.0
103	11.0	9.3	12.2
105	88.5	90.0	87.8
Total ^{18}O %	97.0	97.0	97.0

†The May 2010 synthesis event used PCl_5 which was opened and stored for 17 months, while the PCl_5 used in the December 2008 synthesis was opened immediately prior to synthesis.

December 2008, completely labeled $P^{18}O_4$ made up 90.0% of the species present and the overall oxygen-18 content of the OLP was 97.0%.

There was very little deviation between the theoretical values and actual values of the OLP species present in OLP synthesized from PCl_5 (Table 3). However, large differences between the theoretical values and actual values of the amount of each OLP species present in OLP synthesized from $POCl_3$ were observed (Table 2). One unexpected result of the synthesis of OLP from $POCl_3$ was the formation of a significant amount of $P^{18}O_4$ (m/z 105), which made up 13.0% of the OLP synthesized from fresh $POCl_3$. This is noteworthy because the $POCl_3$ starting material contains one oxygen atom that was expected to remain bonded to the P atom, preventing the formation of $P^{18}O_4$. The natural abundance of oxygen-18 at 0.2% does not explain this level of $P^{18}O_4$ production. After observing these results, we hypothesized that the amount of energy released during the reaction between $POCl_3$ and water was enough to break the P-O bond in $POCl_3$, thus resulting in a random assortment of the oxygen-16 and oxygen-18 atoms bonded to the P atoms. We predicted the expected synthesis outcome based on all the P-O bonds in $POCl_3$ breaking and the total amount of oxygen-18 (73.0%) and oxygen-16 (27.0%) contained in the synthesized OLP, (Table 4). Based on the differences between the expected and actual results, especially for the OLP species at m/z 103 ($P^{18}O_3{}^{16}O$) and 105 ($P^{18}O_4$), it appears that some of the O atoms present in the $POCl_3$ starting material remained bonded to the P atom during the reaction with water, while the bond between other $POCl_3$ O atoms and P was broken.

The extent of broken P-O bonds in $POCl_3$ during OLP synthesis was calculated using an implicit equation that reduced the overall 97% oxygen-18 enrichment by an amount equal to the extent of broken bonds supplying additional oxygen-16. This calculation allowed those $POCl_3$ molecules with broken P-O bonds to react with four O atoms from the O isotope pool and those with no broken P-O bonds to react with three O atoms. Solutions to the equation were iterated comparing the measured isotopic ratios with those predicted until the sum of squares of the

differences was minimized. The results show close agreement between the actual amount of each OLP species synthesized and the predicted values (Table 5). The results of the calculation suggest that 17.9% of the P-O bonds in $POCl_3$ were broken during the December 2008 synthesis, while 38.4% of the bonds were broken during the May 2010 synthesis.

The OLP solutions were synthesized with fresh $POCl_3$ and PCl_5 in December 2008 and they were analyzed by ESI-MS 16 months later in April 2010. After the 16 month period, the total oxygen-18 content of the OLP solutions synthesized from $POCl_3$ and PCl_5 were 73.0% and 97.0%, respectively. This indicates that there was no O exchange between phosphate and water, which is in agreement with Blake et al. [15] and McLaughlin and Paytan [16].

Discussion

These synthesis reactions were carried out in ambient air under a fume hood using a fairly simple procedure as compared to previous synthesis attempts. Middleboe and Saaby Johansen [12] reported a loss of about one-half of the expected oxygen-18 enrichment when synthesizing OLP in ambient atmosphere from $POCl_3$, which they suggested should be expected for any OLP synthesis carried out in ambient atmosphere. Based on the reaction of $POCl_3$ with water that contains 97 atom % ^{18}O, it would be expected that the maximum amount of oxygen atoms being oxygen-18 would be 72.8% (assuming that 0.2% of the $POCl_3$ oxygen atoms are oxygen-18 based on the natural abundance of oxygen-18). Additionally, based on the reaction between PCl_5 and 97 atom % ^{18}O water, one would expect the maximum amount of oxygen-18 atoms to be 97%. Following Middleboe and Saaby Johansen [12], we would have expected only 36.5% of the phosphate oxygen atoms in the OLP synthesized from $POCl_3$ and 48.5% of the phosphate oxygen atoms in the OLP synthesized from PCl_5 would be oxygen-18. However, our enrichment levels were substantially greater for both $POCl_3$ and PCl_5 (Table 2). In fact, we observed almost no loss of oxygen-18 enrichment by performing the reactions in ambient air. It is possible that the loss of oxygen-18 enrichment reported by Middleboe and Saaby Johansen [12] when synthesizing OLP in ambient atmosphere was the result of decreased $POCl_3$ purity as the result of reaction with atmospheric water prior to synthesis, as

Table 4. Calculation of the expected amount of each oxygen labeled phosphate (OLP) species formed with completely random assortment of the oxygen atoms and comparison to the actual results from OLP synthesized with fresh $POCl_3$ in December 2008.

OLP species	^{16}O	^{18}O	Combinations†	Probability‡	Expected composition§	Observed composition
m/z	# of atoms				——— % ———	
97	4	0	1	0.00531	0.53	0.00
99	3	1	4	0.01437	5.75	1.18
101	2	2	6	0.03885	23.31	18.18
103	1	3	4	0.10504	42.01	67.69
105	0	4	1	0.28398	28.39	12.95

†Number of different ways this combination of oxygen-16 and oxygen-18 atoms can be arranged around the phosphorus atom.
‡Probability was calculated by taking the amount of each oxygen atom (62.2 and 37.8% for oxygen-18 and oxygen-16, respectively) to the power of the number of that atom contained in the OLP species (e.g., the probability for m/z 99 is $0.378^3 * 0.622^1 = 0.033594$).
§Calculated by multiplying the number of combinations by the probability.

we noted after running the reactions with $POCl_3$ that had been opened and stored for several months.

Ray [10] carried out a reaction of water and PCl_5 in dioxane and took extreme care to exclude atmospheric water from the reaction; he reported a yield of $85\pm5\%$ $P^{18}O_4$. The results obtained using the current procedure resulted in a yield of $P^{18}O_4$ that was equal to or greater than that reported using the more difficult procedure of Ray [10] (Table 3). Similarly, Risley and Van Etten [8] also carried out this reaction, this time excluding atmospheric water using drying tubes, and obtained an overall oxygen-18 enrichment of 90%. This enrichment is 9% less than what would be expected from the reaction as 99 atom % ^{18}O water was used. Again, the simpler procedure detailed in this study resulted in an improved oxygen-18 enrichment of 97.0%, which represents no loss of enrichment as 97 atom % ^{18}O water was used in this study.

Natural abundance methods rely on a $\delta^{18}O_p$ range of about 10‰ to 25‰ between different phosphate sources, which represents a narrow range. Additionally, the $\delta^{18}O_p$ ranges of many phosphate sources overlap one another, making it very difficult to infer sources of phosphate. The OLP synthesized from fresh $POCl_3$ and PCl_5, on the other hand, had a $\delta^{18}O_p$ value around 1,350,000‰ and 16,120,000‰, respectively. This huge separation from other potential phosphate sources should make it possible to differentiate between OLP and other phosphate sources.

Our use of ESI-MS proved to be an effective means to quantify the products of OLP synthesis. ESI-MS analysis provides information about the individual OLP species composition, which sheds light on reaction dynamics that would not be known by simply analyzing an overall O^{18}/O^{16} ratio. For example, we determined that although the total ^{18}O composition of our synthesized compounds matched the predicted amount, the ^{18}O composition of the individual species differed, indicating that the assumptions of either complete, or no P-O bond-breakage in $POCl_3$ were not met. Bond breakage ranged from 18 to 38%, possibly due to different reaction temperatures during the very violent and exothermic reaction caused by drop wise additions of labeled water.

There have been several previous reports on the stability or decay of OLP in an aqueous solution. Results have ranged from the exchange of oxygen between water and phosphate reaching equilibrium in 3 h [17] to reports of a much slower half-life period of 100 months [12]. The latter result is in closer agreement with more recent publications that state oxygen exchange between phosphate and water under standard environmental conditions does not occur in the absence of biological mediation [15,16]. Our results support the findings of the later studies.

The results from the reactions carried out in ambient atmospheric conditions between $POCl_3$ or PCl_5 and 97 atom % ^{18}O water show that over 99% of the OLP molecules contained at

Table 5. Comparison between synthesis results and theoretical values of P-O bonds in $POCl_3$ broken during the synthesis of oxygen-18 labeled phosphate (OLP) using $POCl_3$.

OLP species	December 2008		May 2010	
	Theoretical†	Actual	Theoretical‡	Actual
—— m/z ——	——————————————— % ———————————————			
97	0.1	0.0	0.2	0.0
99	1.7	1.2	3.5	3.7
101	18.1	18.2	22.4	23.6
103	67.5	67.7	52.8	52.6
105	12.7	13.0	21.1	20.2

†Calculated assuming 17.9% of the $POCl_3$ P-O bonds were broken.
‡Calculated assuming 38.4% of the $POCl_3$ P-O bonds were broken.

least one oxygen-18 atom, and would therefore be potentially useful for creating and tracing compounds in the laboratory or possibly the environment. Overall, the results indicate that OLP highly enriched in oxygen-18 can be synthesized under ambient atmosphere using a simple procedure and this product does not lose its label over time in the absence of microbial mediation. ESI-MS can be used to easily and effectively gain information about the individual OLP species composition.

References

1. Markel D, Kolodny Y, Luz B, Nishri A (1994) Phosphorus cycling and phosphorus sources in Lake Kinneret: Tracing by oxygen isotopes in phosphate. Israel J Earth Sci 43: 165–178.
2. Colman AS, Blake RE, Karl DM, Fogel ML, Turekian KK (2005) Marine phosphate oxygen isotopes and organic matter remineralization in the oceans. Proc Natl Acad Sci USA 102: 13023–13028.
3. McLaughlin K, Cade-Menun BJ, Paytan A (2006a) The oxygen isotopic composition of phosphate in Elkhorn Slough, California: a tracer for phosphate sources. Estuar Coast Shelf S 70: 499–506.
4. McLaughlin K, Kendall C, Silva SR, Young M, Paytan A (2006b) Phosphate oxygen isotope ratios as a tracer for sources and cycling of phosphate in North San Francisco Bay, California. J Geophys Res 111: G03003.
5. Young M, McLaughlin K, Kendall C, Stringfellow W, Rollog M, et al. (2009) Characterizing the oxygen isotopic composition of phosphate sources to aquatic ecosystems. Environ Sci Technol 43: 5190–5196.
6. Elsbury KE, Paytan A, Ostrom NE, Kendall C, Young MB, et al. (2009) Using oxygen isotopes of phosphate to trace phosphorus sources and cycling in Lake Erie. Environ Sci Technol 43: 3108–3114.
7. Gruau G, Legeas M, Riou C, Gallacier E, ois Martineau F, et al. (2005) The oxygen isotope composition of dissolved anthropogenic phosphates: a new tool for eutrophication research? Water Res 39: 232–238.
8. Risley JM, Van Etten RL (1978) A convenient synthesis of crystalline potassium phosphate-$^{18}O_4$ (monobasic) of high isotopic purity. J Labelled Compd 15: 533–538.
9. Larsen S, Middleboe V, Saaby Johansen H (1989) The fate of ^{18}O labelled phosphate in soil/plant systems. Plant Soil 117: 143–145.
10. Ray WJ (1992) A simple procedure for producing oxygen-labelled inorganic phosphate from isotopically labelled water that optimizes yield based on isotope. J Labelled Compd 31: 637–639.
11. Alvarez R, Evans LA, Milham P, Wilson M (2000) Analysis of oxygen-18 in ortho-phosphate by electrospray ionization mass spectrometry. Int J Mass Spectrom 230: 177–186.
12. Middleboe V, Saaby Johansen H (1992) Facile oxygen-18 labelling of phosphate and its delabelling under various conditions. Appl Radiat Isot 43: 1167–1168.
13. Saaby Johansen H (1988) Determination of oxygen-18 abundances by analysis of the first negative emission spectrum of CO^+. Appl Radiat Isot 39: 1059–1063.
14. Saaby Johansen H, Larsen S, Middleboe V (1989) The stability of oxygen-18 labelled phosphate in aqueous solution. Appl Radiat Isot 40: 641.
15. Blake RE, O'Neil JR, Surkov AV (2005) Biogeochemical cycling of phosphorus: Insights from oxygen isotope effects of photoenzymes. Am J Sci 305: 596–620.
16. McLaughlin K, Paytan A (2007) The oceanic phosphorus cycle. Chem Rev 107: 563–576.
17. Blumenthal E, Herbert JBM (1937) Interchange reactions of water. 1. Interchange of oxygen between water and potassium phosphate in solution. T Faraday Soc 33: 849.

Acknowledgments

The authors thank Cameron Scarlett and the University of Wisconsin-Madison School of Pharmacy Mass Spectrometer Facility for assistance with sample analysis.

Author Contributions

Conceived and designed the experiments: DJS PB. Performed the experiments: ESM DJS PB. Analyzed the data: ES DJS PB. Contributed reagents/materials/analysis tools: DJS PB. Wrote the paper: ES DJS PB.

4

Proper Interpretation of Dissolved Nitrous Oxide Isotopes, Production Pathways, and Emissions Requires a Modelling Approach

Simon J. Thuss[¤], **Jason J. Venkiteswaran***, **Sherry L. Schiff**

Department of Earth and Environmental Sciences, University of Waterloo, Waterloo, Ontario, Canada

Abstract

Stable isotopes (δ^{15}N and δ^{18}O) of the greenhouse gas N_2O provide information about the sources and processes leading to N_2O production and emission from aquatic ecosystems to the atmosphere. In turn, this describes the fate of nitrogen in the aquatic environment since N_2O is an obligate intermediate of denitrification and can be a by-product of nitrification. However, due to exchange with the atmosphere, the δ values at typical concentrations in aquatic ecosystems differ significantly from both the source of N_2O and the N_2O emitted to the atmosphere. A dynamic model, SIDNO, was developed to explore the relationship between the isotopic ratios of N_2O, N_2O source, and the emitted N_2O. If the N_2O production rate or isotopic ratios vary, then the N_2O concentration and isotopic ratios may vary or be constant, not necessarily concomitantly, depending on the synchronicity of production rate and source isotopic ratios. Thus *prima facie* interpretation of patterns in dissolved N_2O concentrations and isotopic ratios is difficult. The dynamic model may be used to correctly interpret diel field data and allows for the estimation of the gas exchange coefficient, N_2O production rate, and the production-weighted δ values of the N_2O source in aquatic ecosystems. Combining field data with these modelling efforts allows this critical piece of nitrogen cycling and N_2O flux to the atmosphere to be assessed.

Editor: David William Pond, Scottish Association for Marine Science, United Kingdom

Funding: Funding was provided by a Natural Sciences and Engineering Research Council (NSERC) and BIOCAP grant 336807-06 (SLS), an NSERC scholarship (SJT), and Environment Canada's Science Horizons Youth Internship Program. The funders had no role in study design, data collection and analysis, decision to publish, or preparation of the manuscript.

* E-mail: jjvenkit@uwaterloo.ca

¤ Current address: Golder Associates, London, ON, Canada

Introduction

Nitrous oxide (N_2O) is a powerful greenhouse gas, 298 times more potent than CO_2 over a 100-year time line [1]. Atmospheric N_2O concentrations have been increasing at a rate of 0.25%/year over the last 150 years [2]. Consequently, the global N_2O budget has been the subject of intensive research efforts over the past few decades. N_2O is produced through multiple microbial pathways: hydroxylamine oxidation during nitrification and as an obligate intermediate during denitrification and nitrifier–denitrification. Because these pathways of N_2O production have different stable isotopic enrichment factors, isotopic analysis of N_2O can potentially distinguish N_2O produced through different pathways or from different sources [3]. Identifying N_2O sources will provide insights on the fate of N at the ecosystem-scale (e.g., [4–6]). The isotopic ratios of N_2O produced in soil environments (e.g., [7–11]), and in aquatic environments (e.g., [12–18]) have been measured to some extent. Although N_2O production in rivers and estuaries is a significant portion of the global N_2O budget (approximately 1.5 TgN/year, [19]), few studies report isotopic data for rivers [5,20,21].

In ice-free aquatic ecosystems, the δ^{15}N and δ^{18}O of dissolved N_2O is affected by gas exchange with the atmosphere. As a result, the isotopic ratios of dissolved N_2O are not equal to those of the N_2O produced within the aquatic ecosystem and continue to change as atmospheric exchange (both ingassing and outgassing) occurs. In addition, isotopic fractionation during influx and efflux causes the isotopic ratios of N_2O flux emitted to the atmosphere to be different than that of the dissolved N_2O [22]. Thus, the simple method of calculating the instantaneous isotopic ratios of the N_2O flux by taking measured dissolved isotopic ratios, adding an equilibrium isotope fractionation, and applying them to measured flux rates is inappropriate. Adjustments of measured isotopic ratios are necessary to understand the isotopic ratios of both produced and emitted N_2O.

In this paper, we present a dynamic model of the stable isotopic composition of both the dissolved and emitted N_2O in aqueous systems. We apply this model to two different measured diel patterns of the isotopic ratios of N_2O in an aquatic ecosystem. We use the model to elucidate the relationship between the isotopic ratios of source, dissolved, and emitted N_2O, to allow for improved interpretation of dissolved N_2O isotope data. Ultimately, a process-based understanding on N cycling with aquatic ecosystems may be developed based on interpretation of N cycling processes.

Materials and Methods

Stable Isotopes of N₂O

N_2O is an asymmetric molecule: the most abundant isotopologues of N_2O are $[^{14}N^{14}N^{16}O]$, $[^{15}N^{14}N^{16}N]$, $[^{14}N^{15}N^{16}O]$ and $[^{14}N^{14}N^{18}O]$. The isotopic ratios, ^{15}N: ^{14}N and ^{18}O: ^{16}O, are:

$$^{15}R = \frac{[^{15}N^{14}N^{16}O] + [^{14}N^{15}N^{16}O]}{2[^{14}N^{14}N^{16}O]} = \frac{[^{15}N_2O]}{[^{14}N_2O]} \quad (1)$$

$$^{18}R = \frac{[^{14}N^{14}N^{18}O]}{[^{14}N^{14}N^{16}O]} = \frac{[N_2{}^{18}O]}{[N_2{}^{16}O]} \quad (2)$$

where $[^{14}N^{14}N^{16}O]$, $[^{15}N^{14}N^{16}O]$, $[^{14}N^{15}N^{16}O]$ and $[^{14}N^{14}N^{18}O]$ represent the concentrations of the various N_2O isotopologues. Note that 15R is the bulk ^{15}N: ^{14}N ratio and represents an average ratio of the two ^{15}N isotopomers and isotopic ratios are reported as $\delta^{15}N$ relative to air and $\delta^{18}O$ relative to VSMOW. Although the isotopic ratio of the ^{15}N isotopomers can be measured (e.g., [23–25]), the gas exchange fractionation factors are not affected by the intramolecular distribution of ^{15}N [22]. Many laboratories cannot measure the intramolecular distribution of ^{15}N and analysis of the bulk ^{15}N: ^{14}N ratio of N_2O is more common [26]. Here, we confine our analysis to bulk ^{15}N: ^{14}N ratios and use $^{15}N_2O$ to represent the average abundance of the two ^{15}N isotopomers. The same approach could easily be extended to consider each isotopologue separately.

Dynamic Isotope Model for Dissolved N₂O

A simple three box model (SIDNO, Stable Isotopes of Dissolved Nitrous Oxide) was created using Stella modelling software (version 9.1.4, http://www.iseesystems.com) in order to study the relationships between the isotopic ratios of source, dissolved and emitted N_2O (model file is available at https://github.com/jjvenky/SIDNO and by contacting the corresponding author). This model is an adaptation of the isotopic gas exchange portion of the PoRGy model [27], which successfully modelled diel isotopic ratios of O_2 resulting from photosynthesis, respiration, and gas exchange in aquatic ecosystems. One key difference is photosynthetically produced O_2 in PoRGy has a $\delta^{18}O$ value fixed by the H_2O molecules, whereas SIDNO has N_2O production $\delta^{15}N$ and $\delta^{18}O$ values that can vary independently of each other and of N_2O production rate in order to simulate variability in nitrification and denitrification.

One box in SIDNO is used for the total mass of dissolved N_2O and two additional boxes for the dissolved masses of the two heavy isotopologues ($^{15}N_2O$ and $N_2{}^{18}O$). The boxes are open to the atmosphere for gas exchange, are depth agnostic, and each box can gain N_2O via a production term; there is no N_2O consumption term since the δ values of N_2O are largely controlled by the production pathways [28,29] though certain waters can exhibit significant N_2O reduction to N_2 [30,31]. The masses and magnitude of the flows of $^{15}N_2O$ and $N_2{}^{18}O$ relative to bulk N_2O are used to calculate the isotopic composition of source, dissolved, and emitted N_2O. Although isotopic ratios are used in the model, we discuss δ values that are common for reporting isotopic ratios. N_2O production rate and its δ values are user-defined and can be adjusted for diel patterns in N_2O production that may be caused by variable O_2 levels [32–37].

Stable Isotope Dynamics of Gas Exchange

The δ values of the net gas exchange flux are controlled by the kinetic fractionation factors for evasion ($\alpha_{ev} = R_{evaded}/R_{dissolved}$, 0.9993 for $\delta^{15}N$ and 0.9981 for $\delta^{18}O$) and invasion ($\alpha_{in} = R_{invaded}/R_{gas}$, 1.0000 for $\delta^{15}N$ and 0.9992 for $\delta^{18}O$) [22]. These two α values are related to the equilibrium fractionation factor: $\alpha_{eq} = R_{gas}/R_{dissolved} = \alpha_{ev}/\alpha_{in}$ (0.99925 for $\delta^{15}N$ and 0.99894 for $\delta^{18}O$) and are independent of temperature over the range of 0°C to 44.5°C [22].

The δ values of tropospheric N_2O are 6.72 ‰ ± 0.12‰ for $\delta^{15}N$ and 44.62‰ ± 0.21‰ for $\delta^{18}O$ [38]. Therefore, at equilibrium, dissolved N_2O has dissolved δ values slightly greater than these at 7.48‰ and 45.73‰, respectively.

In the model, net N_2O flux between the atmosphere and dissolved phase was calculated using the thin boundary layer approach as:

$$Flux = k\left(p_{N_2O} \times k_H - [N_2O]_{dissolved}\right) \quad (3)$$

where the N_2O flux is calculated in mol/m²/h, k is the user-modifiable gas exchange coefficient (m/h), is the partial pressure of tropospheric N_2O (assumed to be 320 ppbv from data provided by the ALE GAGE AGAGE investigators, [39,40]), k_H is the Henry constant for N_2O (mol/atm-m³), and $[N_2O]_{dissolved}$ is the dissolved concentration of N_2O (mol/m³). k_H is a function of water temperature [41]:

$$k_H = 0.025 e^{-2600\left(\frac{1}{T} - \frac{1}{298.15}\right)} \quad (4)$$

where T is temperature in kelvins.

Gas exchange is a two-way process. The net N_2O flux rate (the difference between the invasion and evasion rates) depends on the dissolved N_2O concentration. When a solution is at equilibrium with the atmosphere, the invasion and evasion rates will be equal, and the net flux will be zero.

As with the bulk N_2O flux, the flux of the heavy isotopologues ($^{15}N_2O$ and $N_2{}^{18}O$) can be calculated by including the kinetic fractionation factors for N_2O (adapted from [27]):

$$Flux\,^{15}N_2O = k\left(\alpha_{in}^{15} \times p_{15N_2O} \times k_H - \alpha_{ev}^{15}[^{15}N_2O]_{dissolved}\right) \quad (5)$$

$$Flux\,N_2{}^{18}O = k\left(\alpha_{in}^{18} \times p_{N_2{}^{18}O} \times k_H - \alpha_{ev}^{18}[N_2{}^{18}O]_{dissolved}\right) \quad (6)$$

where p_{15N_2O} and $p_{N_2{}^{18}O}$ are the partial pressures of $^{15}N_2O$ and $N_2{}^{18}O$.

Results

Test of Model Performance

To test the ability of SIDNO to reproduce observed isotopic data, input parameters (N_2O production rate, N_2O δ values, and k) were set to replicate a series of experiments designed to derive fractionation factors for N_2O gas exchange [22]. In these experiments, degassed water was exposed to N_2O gas of known isotopic ratios in a sealed container to varying degrees of saturation.

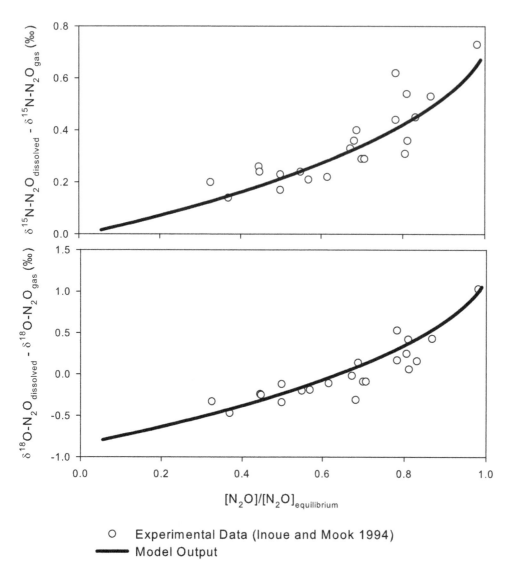

Figure 1. Comparing the model output to the experimental data of [22]. The coefficient of determination for experimental data and SIDNO model outputs were comparable to those of [22], $R^2 = 0.77$ for $\delta^{15}N$ and $R^2 = 0.82$ for $\delta^{18}O$. Precision of measurements for the experimental data was $\pm$ 0.05‰ for $\delta^{15}N$ and $\pm$ 0.1‰ for $\delta^{18}O$.

Modelled dissolved N_2O concentration and δ values increased in response to gas exchange (Figure 1). The model fit to the experimental data is comparable to the original best-fit derivations ($R^2 = 0.77$ for $\delta^{15}N$ and $R^2 = 0.82$ for $\delta^{18}O$ for both the original fit [22] and the SIDNO fit) (Figure 1). The initial isotopic composition of dissolved N_2O was identical to the gas phase $\delta^{15}N$ value, but the $\delta^{18}O$ of dissolved N_2O was slightly less than the gas phase $\delta^{18}O$ value. Ultimately, at 100% saturation the δ values of the dissolved N_2O were greater than those of the gas phase as a result of α_{eq}. The model successfully simulated the kinetic and equilibrium fractionations during gas exchange under the experimental conditions.

Next, SIDNO was used to provide insight into the effect of degassing on the δ values of dissolved and emitted N_2O. Here the results of two model runs with the same initial N_2O concentration but different initial δ values of dissolved N_2O were compared (Figure 2). As N_2O saturation declined both the dissolved $\delta^{15}N$ values and instantaneous $\delta^{15}N$ values of the emitted N_2O remained relatively constant, dissolved $\delta^{18}O$ values and instanta-

neous $\delta^{18}O$ values of the emitted N_2O varied by about 10, when the solution was very supersaturated (>300% saturation). The δ values rose quickly as the system approached 100% saturation. Because the light isotopologue diffuses out of solution faster than the heavy isotopologue, the instantaneous δ values of the emitted N_2O were always less than the concomitant δ values of dissolved N_2O. The isotopologues of N_2O reached equilibrium independently of each other and therefore the total mass emitted for each isotopologue and rate of change depended on the initial concentration and δ values. The retention of N_2O in the dissolved phase caused the δ values of the mass emitted to differ from those of total mass production. However, when initial dissolved N_2O concentrations were high (>1000% saturation) the δ values of the total N_2O emitted were similar to the δ values of dissolved N_2O because the mass of N_2O lost is very much larger than the N_2O that remained dissolved. The value of k did not affect the gas exchange trajectories only the speed at which the system reached equilibrium.

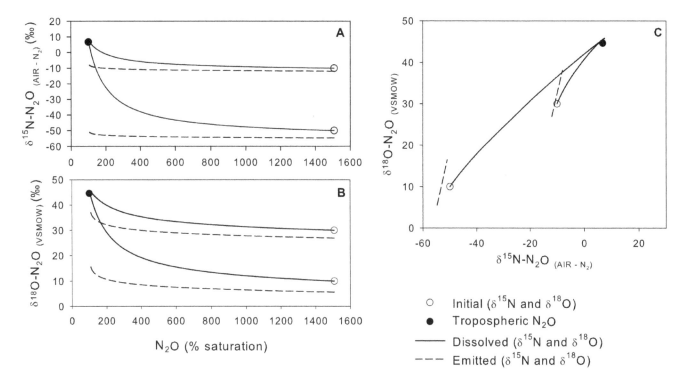

Figure 2. $\delta^{15}N$ and $\delta^{18}O$ trajectories for dissolved and emitted N_2O in two supersaturated solutions with zero N_2O production. Initial dissolved isotopic values for the two dissolved N_2O solutions were $\delta^{15}N = -50‰$, $\delta^{18}O = 10‰$, and $\delta^{15}N = -10‰$, $\delta^{18}O = 30‰$. Both runs used an initial dissolved N_2O concentration of 1500% saturation. Note that in the $\delta^{18}O$ versus $\delta^{15}N$ plot, the dissolved N_2O curves do not pass through the tropospheric N_2O value due to the small equilibrium isotope effect.

N_2O isotope data are often plotted as $\delta^{18}O$ versus $\delta^{15}N$ to elucidate relationships between the various sources and tropospheric N_2O [38]. The trajectories on these plots (Figure 2C) were dictated by the δ values of the source relative to the constant atmospheric value and the α values. Note that some plots in the literature differ due to different reference materials for the $\delta^{18}O$ scale (VSMOW and atmospheric O_2).

Modelling Scenarios with Steady State Production of N_2O

The SIDNO model can be used to probe the stable isotope dynamics of N_2O in a variety of situations that may be encountered in aquatic environments to elucidate the relationship between the N_2O source (a function of N cycling processes), dissolved (the easily measured component), and emitted (of consequence for greenhouse gas production and global N and N_2O cycle).

In the steady-state production of N_2O (constant rate and δ values), by definition, the δ values of N_2O production must be the same as those of the emitted N_2O. As a result, the δ values of the dissolved N_2O cannot equal that of the source (or emitted) N_2O at steady state because the dissolved N_2O must be offset from the emitted N_2O by at least the α_{ev} values. As the steady-state production rate was increased, the steady-state N_2O concentration increased and the dissolved δ values approached but did not equal the source (Figure 3). Even at moderate supersaturations (<1000%) the effect of atmospheric N_2O equilibration on the δ values of dissolved N_2O cannot be ignored.

At steady state, the δ values of the emitted N_2O must be equal to the source; the large difference between source/emitted and dissolved N_2O underscores the importance of adjusting the measured δ values of dissolved N_2O in order to determine aquatic contributions of N_2O to the atmosphere or N_2O sources. This is

critical when using dissolved measurements of N_2O to constrain the global isotopic N_2O budget, but not been done in most studies, e.g., [16,42–44] but see [45].

Modelling Scenarios with Variable Production of N_2O

The relationship between the δ values of source, dissolved, and emitted N_2O are much more complicated when N_2O production is variable rather than when it is constant. N_2O production may vary with respect to production rate and/or δ values; in many aquatic environments, N_2O production is not likely to be constant. The N_2O production processes, nitrification and denitrification, are sensitive to redox conditions, which can be highly variable, due to diel changes in dissolved O_2 concentration, flow regime, etc. For example, [34] observed diel changes in the denitrification rate in the Iroquois River and Sugar Creek (Midwestern USA) and found that the denitrification was consistently greater during the day than night. The relative importance of nitrification and denitrification can change in response to the diel oxygen cycle: e.g., [46] observed a change from daytime nitrification to nighttime denitrification in a subtropical eutrophic stream. Coupling of N_2O and O_2 diel cycles has been observed in agricultural and waste-water treatment plant (WWTP) impacted rivers [36]. Since fractionation factors and substrates are different for nitrification and denitrification, ecosystem-scale fractionation factors may be rate and process dependent, and the δ values of N_2O production in a given ecosystem may not be constant over a diel cycle.

To simulate the diel variability, various scenarios were modelled by adjusting either production rate and/or the associated δ values. The variabilities in these input parameters were driven by a sine function with a 24 h period similar to a dissolve O_2 curve. In all scenarios, the chosen range of production rates was based on published N_2O flux rates (Table 1) and varied from 1 to

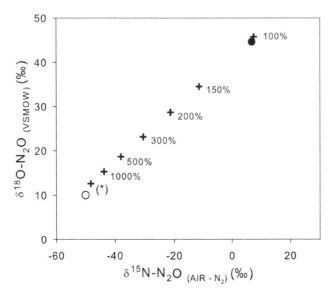

Figure 3. The relationship between $\delta^{15}N$, $\delta^{18}O$ and N_2O concentration in a system at steady state with constant N_2O production and open to gas exchange with the atmosphere. The point marked with a * represents the minimum difference between the isotopic composition of dissolved and source N_2O. The point at 100% saturation is the equilibrium value, the $\delta^{15}N$ and $\delta^{18}O$ of this point is controlled by the isotopic composition of tropospheric N_2O and the equilibrium enrichment factors.

5 mol/m^2/h[1] (Table 2), which was between the diel variation in N_2O flux observed by [33] and [46]. Temperature was held constant at 20°C. The value of k was varied between 0.1 and 0.3 m/d (Table 2), within the range observed in other river studies (Table 1). The combination of production rates and k values were chosen to produce N_2O between 150% and 500% saturation (Table 2) coinciding with the range of published data (Table 1). The range of δ values used for the N_2O source (Table 2) was within published values from various field studies [47]. For

scenarios where the δ values of source N_2O was variable, the sine function for the δ values was synchronized so that maxima and minima $\delta^{15}N$ values coincided with those of $\delta^{18}O$. This was done for simplicity, and because, in general, nitrification yields N_2O with lower $\delta^{15}N$ and $\delta^{18}O$ values than denitrification (e.g., [10,48]). Nevertheless, scenarios with greater amounts of N_2O reduction to N_2 can be modelled by increasing the source $\delta^{15}N$ and $\delta^{18}O$ values to those appropriate for any given ecosystem. Model scenarios were run until the output parameters (i.e., N_2O saturation and the δ values of dissolved, source, and emitted N_2O) reached dynamic steady state: model output was not constant over 24 h but the diel patterns on successive days were repeated.

Model Scenario #1: Variable Production Rate, Constant Isotopic Composition of Source

In scenario #1 (Table 2), the δ values of source N_2O were held constant and the production rate was variable. An example of such a system may be N_2O production via denitrification in river sediments with abundant NO_3^-. Denitrification rates in rivers have been observed to fluctuate in response to the diel O_2 cycle [34]. If the fractionation factors for denitrification are not rate dependent, the resulting N_2O production rate would be variable but the source δ values of N_2O could be constant.

Here, the maximum concentration lagged approximately 2.75 h behind the maximum N_2O production rate, a function of the magnitude of the gas exchange coefficient, cf. [49]. The δ values for the instantaneously emitted N_2O were relatively constant and very similar to the N_2O source (within 0.4‰ for $\delta^{15}N$ and 1.1‰ for $\delta^{18}O$, Figure 4, Table 3). However, the δ values of dissolved N_2O were more variable, spanning 16‰ for $\delta^{15}N$ and 10‰ for $\delta^{18}O$. Thus, a change in the δ values of dissolved N_2O can be driven simply by a change in production rate and not necessarily a change in the δ values of the source. Since the system was at dynamic steady state, the average δ values of the emitted N_2O were identical to the average δ values of the source. This must be true in all steady-state cases to conserve the mass of the N_2O isotopologues.

In some aquatic systems, the N_2O production rate may remain constant with time but the δ values of the source may change with time. In rivers or lakes without a strong diel O_2 cycle, sediment denitrification may produce N_2O at an approximately constant rate. Denitrification rate may also be independent of water column NO_3^- concentration if limited by factors other than diffusion in the sediments. The δ values of the source N_2O may thus change if the

Table 1. Summary of relevant published data on N_2O production in aquatic environments.

Location	Range of N_2O Saturation (%)	Range of N_2O Flux (μmol/m^2/h[1])	Range of k values (m/h)	Reference
Ohio River, OH, US	95 to 745	5 to 90	—	[12]
5 agricultural streams, ON, CA (over 2–3 years)	14 to 1700	−1 to 91.7	0.002 to 0.59	[5]
10 agricultural streams, ON, CA (over 17 diel cycles)	30 to 2570	−0.33 to 52.1	0.004 to 0.30	[45]
Bang Nara River, TH	170 to 2000	—	—	[20]
LII River, NZ	201 to 404	1.35 to 17.9	0.13 to 0.82	[58]
LII River, NZ	402 to 644	0.46 to 0.89	14.76	[33]
Seine River, FR	—	2.2 to 5.2	0.04 to 0.06	[59]
Canal Two, Yaqui Valley, MX	100 to 6000	0 to 34.9	0.3 to 0.6	[46]
agricultural stream, UK	100 to 630	0 to 37.5	—	[60]
Grand River, ON, CA	38 to 8573	−1.4 to 173.6	0.06 to 0.35	[36,37]

Table 2. Summary of input parameters for the SIDNO model scenarios for non-steady state production of N_2O.

Scenario #	Results Figure	k	N_2O Source Production Rate	N_2O Source $\delta^{15}N$ and $\delta^{18}O$ Values
		(m/k)	($\mu mol/m^2/h^1$)	(‰)
Variable Production Rate, Constant Isotopic Composition of Source				
1	4	0.3	1 to 5	−50, 10
Constant Production Rate, Variable Isotopic Composition of Source				
2	5	0.3	3	−50, 10 to −30, 10
3	6	0.1	3	−50, 10 to −30, 10
Variable Production Rate, Variable Isotopic Composition of Source				
4*	7	0.3	1 to 5	−50, 10 to −30, 10
5**	8	0.3	1 to 5	−50, 10 to −30, 10
6*	9	0.1	1 to 5	−50, 10 to −30, 10

*Maximum production rate coincides with the lowest source δ values.
**Maximum production rate coincides with the highest source δ values.

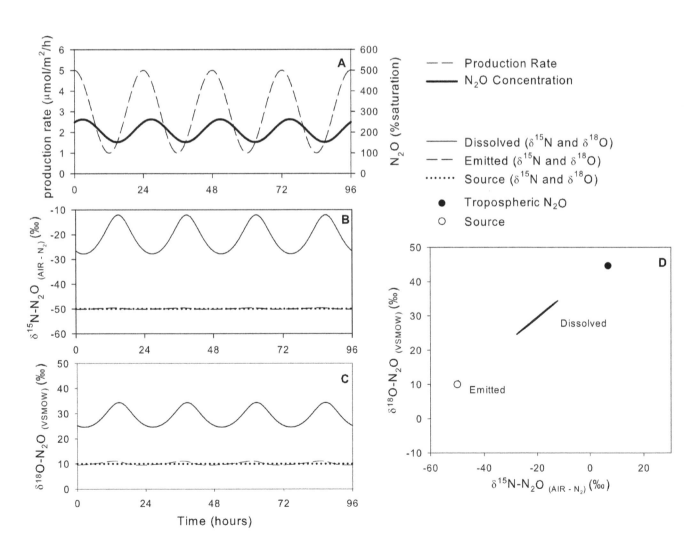

Figure 4. Model scenario #1 – isotopic composition of dissolved and emitted N_2O with a variable production rate and constant isotopic composition of the source. Note, in panel D, the data points for emitted N_2O are masked by the data point for source N_2O.

Table 3. Summary of SIDNO output for model scenarios simulating non-steady state production of N_2O.

Scenario #	Dissolved N_2O				Emitted N_2O		
	Saturation	$\delta^{15}N$, $\delta^{18}O$	$\Delta\delta^{15}N$	$\Delta\delta^{18}O$	$\delta^{15}N$, $\delta^{18}O$	$\Delta\delta^{15}N$	$\Delta\delta^{18}O$
	(%)	(‰)	(‰)	(‰)	(‰)	(‰)	(‰)
Variable Production Rate, Constant Isotopic Composition of Source							
1	153 to 263	−27.8, 24.6 to −12.1, 34.4	37.9	24.4	−49.6, 9.5 to −50.2, 11.1	0.2	0.5
Constant Production Rate, Variable Isotopic Composition of Source							
2	208	−19.6, 29.4 to −3.7, to 37.4	30.4	19.4	−45.3, 12.3 to −14.7, 27.7	4.7	2.3
3	423	−26.1, 24.8 to −15.1, 30.3	23.9	14.8	−37.2, 16.4 to −22.8, 23.6	12.8	6.4
Variable Production Rate, Variable Isotopic Composition of Source							
4*	153 to 263	−25.8, 25.6 to −1.2, 39.8	24.2	15.6	−46.9, 11.3 to −18.0, 26.5	8	3.5
5**	153 to 263	−11.2, 33.8 to −5.0, 36.1	38.8	23.8	−41.6, 14.6 to −13.4, 28.1	8.4	4.6
6*	345 to 501	−31.6, 21.6 to −18.4, 29.3	18.4	11.6	−42.1, 13.7 to −29.4, 20.6	19.4	9.4

Temporally variable parameters are given as a range. $\Delta\delta^{15}N$ and $\Delta\delta^{18}O$ ($\Delta = \delta - \delta$) are the maximum difference between the range of source N_2O and the range for the model output parameter.
*Maximum production rate coincides with the lowest source δ values.
**Maximum production rate coincides with the highest source δ values.

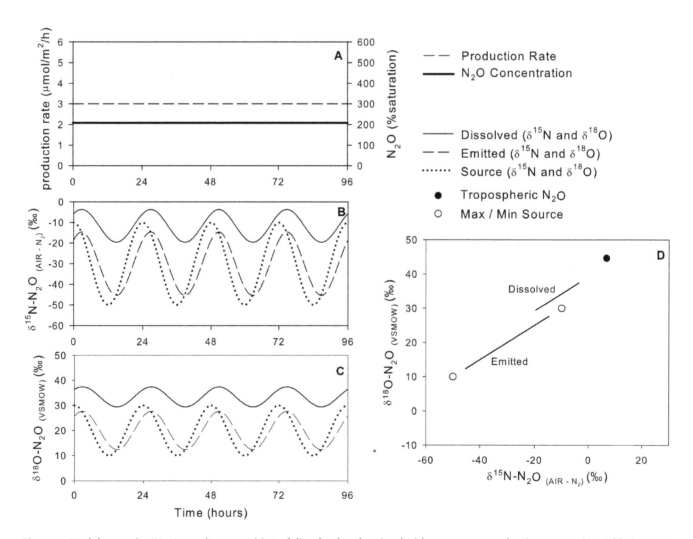

Figure 5. Model scenario #2 – Isotopic composition of dissolved and emitted with a constant production rate and variable isotopic composition of the source.

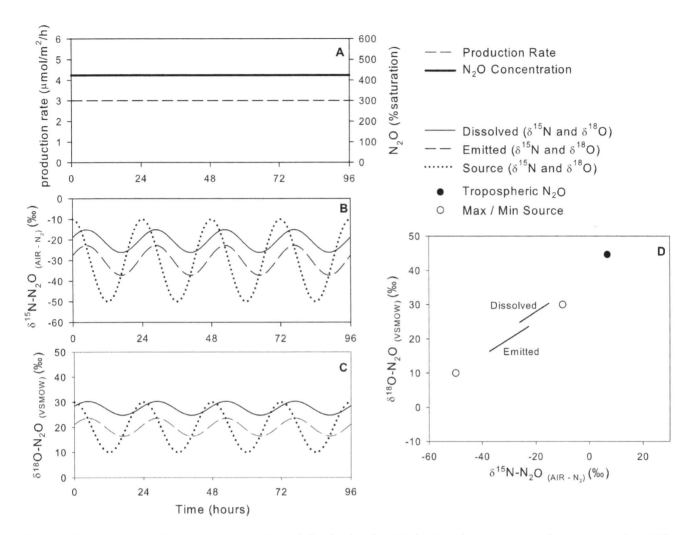

Figure 6. Model scenario #3 – Isotopic composition of dissolved and emitted N$_2$O with a constant production rate and variable isotopic composition of the source. k is reduced from 0.3 m/h to 0.1 m/h.

δ values of the NO$_3^-$ substrate changed with time. For example, many studies have shown that the δ values of residual NO$_3^-$ increase during denitrification [50]. Similarly, NO$_3^-$ from WWTPs may have different δ values than agricultural runoff and diel changes in WWTP release may result in changing δ values of NO$_3^-$. Changes in N cycling may also vary on a diel basis but result in fortuitously similar N$_2$O production rates due to, for example, changes in the N$_2$O:N$_2$ ratio of denitrification or changes in the relative importance of nitrification and denitrification. Thus changes in δ values of the N$_2$O source do not necessarily indicate changes in N$_2$O production rates.

Model Scenario #2: Constant Production Rate, Variable Isotopic Composition of Source

In scenario #2, when the N$_2$O production rate was held constant and the δ values of the source varied with time (from -50‰ to -10‰ for δ^{15}N and from 10‰ to 30‰ for δ^{18}O), the δ values of the dissolved N$_2$O was also much farther from that of the source than the dissolved N$_2$O due to the effects of atmospheric exchange and the emitted N$_2$O varies linearly between the two source values. In contrast, the dissolved N$_2$O is parallel but offset from the line connecting the two sources (Figure 5, Table 3). The maximum difference between emitted and source N$_2$O was 4.7‰

for δ^{15}N and 2.3‰ for δ^{18}O. The dissolved and emitted δ values also lagged 2.75 h behind the source as a result of gas exchange (as above). Since the system was at dynamic steady state, the average δ values of the emitted N$_2$O were identical to the average δ values of the source.

Model Scenario #3: Constant Production Rate, Variable Isotopic Composition of Source

To examine the effects of varying k on the scenario of constant N$_2$O production with variable isotopic signature of the source, k was reduced from 0.3 m/h (scenario #2) to 0.1 m/h (scenario #3; Figure 6, Table 3). The δ values for the emitted N$_2$O were centred between the sources N$_2$O values, but dissolved N$_2$O δ values were farther from tropospheric N$_2$O than the high-k scenario #2 (Figure 6 D).

The effect of reducing k was an increase in N$_2$O concentration with the same production rate and a shift in the δ values of dissolved N$_2$O toward the source values. Reducing k also dampened the response between the instantaneous δ values of the emitted N$_2$O and the δ values of the source. As above, the lag time between the δ values of the source and emitted N$_2$O increased as k decreased. The total range of the source and

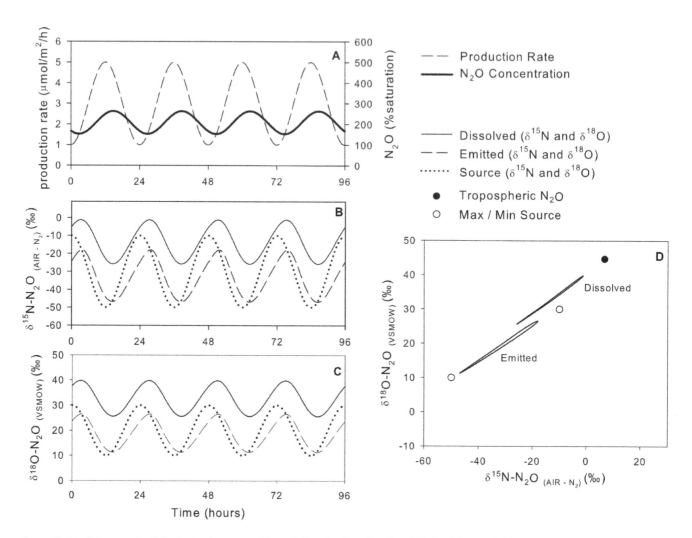

Figure 7. Model scenario #4 – Isotopic composition of dissolved and emitted N₂O with a variable production rate and variable isotopic composition of the source. Maximum production rate is in sync with the lowest $\delta^{15}N$ and $\delta^{18}O$ values of the source.

emitted δ values decreased. The difference between the source and emitted δ values was 12.8‰ for $\delta^{15}N$ and 6.4‰ for $\delta^{18}O$.

To simulate a system alternating between two N₂O production processes, such as differing relative contributions of nitrification and denitrification, with different rates of N₂O production and δ values, the model was run with both production rate and its δ values variable with time (scenarios #4, #5, and #6). The production rate and δ values were adjusted so that the maximum rate coincided with the lowest source δ values in scenarios #4 and #6 and so that maximum rate coincided with the highest source δ values in scenario #5.

Model Scenario #4: Variable Production Rate, Variable Isotopic Composition of Source

For scenario #4, the resulting N₂O concentrations were identical to those in model scenario #1, with the maximum concentration lagging approximately 2.75 h behind the maximum production rate (Figure 7, Table 3). The relationship between the δ values of the dissolved and emitted N₂O was more complex than in other scenarios. The lag time between the maximum source δ values and those of dissolved and emitted N₂O (when the production rate was minimum) was 3.75 h; however, the lag time between the minimum source δ values and those of the dissolved

and emitted N₂O (when the production rate was maximum) was only 2.25 h. The difference between the emitted and source N₂O was 3.1‰ to 8.0‰ for $\delta^{15}N$ and 1.3‰ to 3.4‰ for $\delta^{18}O$. The δ values of emitted N₂O were closer to those of the source during periods of high production rates (and thus higher concentrations) than periods of low production rates. However, the flux-weighted average δ values of emitted N₂O were equal to the average production-weighted source δ values because the system was at dynamic steady state.

Model Scenario #5: Variable Production Rate, Variable Isotopic Composition of Source

The isotopic counterpoint to scenario #4 is adjusting the timing of maximum N₂O production to coincide with the highest δ values of production (scenario #5). All other parameters were the same as scenario #4 (Table 3). The resulting pattern for the δ values of dissolved N₂O was very different than scenario #4 (Figure 8, Table 3). While the dissolved N₂O concentrations were identical to the model scenario #4, the δ values of dissolved N₂O were nearly constant with time. The relationship between the δ values of emitted and source N₂O was similar to scenario #4, although the instantaneous difference in δ values were slightly greater. The δ values of the dissolved N₂O were greatly dampened by the fact

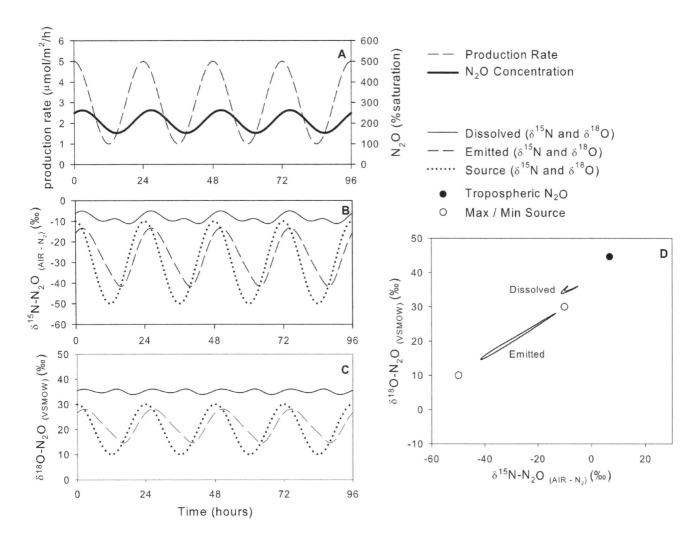

Figure 8. Model scenario #5 – Isotopic composition of dissolved and emitted N₂O with a variable production rate and variable isotopic composition of the source. Maximum production rate is in sync with the greatest $\delta^{15}N$ and $\delta^{18}O$ values of the source.

that maximum production rate coincided with source δ values that were closest to tropospheric N_2O. In scenario #4, the high rates of N_2O production at δ values very different than tropospheric N_2O increased the amplitude of the δ values of dissolved N_2O.

Model Scenario #6: Variable Production Rate, Variable Isotopic Composition of Source

To determine the effects of a lower k on model scenario #4, k was reduced from 0.3 m/h from 0.1 m/h for scenario #6. As shown above, lower k increased the dissolved N_2O concentrations and dampened the diel range of δ values of both dissolved and emitted N_2O (Table 3, Figure 9). Lower k also increased the lag time between the δ values of emitted and source N_2O and increased the difference between the δ values of emitted and source N_2O (Figure 9). As in all scenarios, the flux-weighted average δ values of emitted N_2O were equal to the average production-weighted source δ values.

Grand River

The ability of SIDNO to reproduce measured patterns of N_2O concentration and δ values in a human-impacted river was also assessed. The Grand River is a seventh-order, 300 km long river that drains 6800 km² in southern Ontario, Canada, into Lake

Erie, see [36,37,51]. There are 30 WWTPs in the catchment and their cumulative impact can be observed via the increase in artificial sweeteners in the river [52].

Samples were collected approximately hourly for 28 h at two sites in the central, urbanized portion of the river: sites 9 and 11 in [51,52]. The upstream site, Bridgeport, is where the river enters the urban section of the river at the city of Waterloo and is immediately above that city's WWTP. Blair is 26.6 km downstream of Bridgeport and below the cities of Waterloo and Kitchener. It is also 5.5 km downstream of the Kitchener WWTP. Average river depth at both sites was 30 cm. Values of k were determined by best-fit modelling of diel O_2 and $\delta^{18}O$-O_2 values at the sites [36]. N_2O concentration analyses were performed on a Varian CP-3800 gas chromatograph with an electron capture detector and isotopic ratio analyses were performed on a GV TraceGas pre-concentrator coupled to a GV Isoprime isotope ratio mass spectrometer, see [5] for analytical details.

Data from upstream and downstream of large urban wastewater treatment plants on the Grand River show diel patterns in N_2O saturation and δ values (Figures 10 and 11). At the Bridgeport site, the diel patterns of N_2O saturation and $\delta^{15}N$ values were opposite of each other, that is, when N_2O saturation was highest around sunrise the $\delta^{15}N$ values were lowest and when

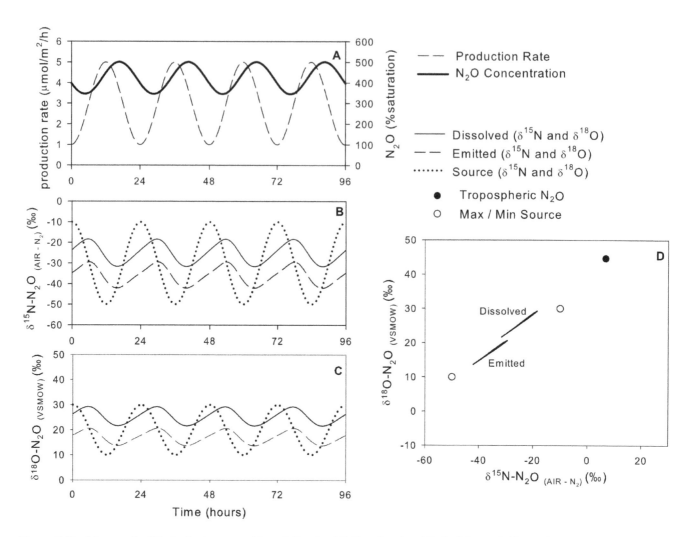

Figure 9. Model scenario #6 – Isotopic composition of dissolved N₂O and emitted N₂O with a variable production rate and variable isotopic composition of the source. Maximum production rate is in sync with the lowest $\delta^{15}N$ and $\delta^{18}O$ values of the source. k is reduced from 0.3 m/h to 0.1 m/h.

when N₂O saturation was lowest around before sunset the $\delta^{15}N$ values were greatest. R^2 values between field and model data for N₂O saturation, $\delta^{15}N$, and $\delta^{18}O$ values are 0.83, 0.68, and 0.30. Model results reproduce the range and sinusoidal patterns of the field data though the $\delta^{18}O$ fit was poor in the second half of the field data. The diel pattern in $\delta^{18}O$ values was similar to that of $\delta^{15}N$ but was shifted earlier by about 4 h. These patterns were similar to those of scenario #4 (variable N₂O production and variable δ values of the source N₂O coinciding when maximum production rates coincided with lowest source δ values) and the result of consistent diel five-fold variability in N₂O production and variability in the $\delta^{18}O$ of the N₂O produced in the river.

At the downstream Blair site, both $\delta^{15}N$ and $\delta^{18}O$ values were much lower and exhibited a greater range than at Bridgeport. R^2 values between field and model data for N₂O saturation, $\delta^{15}N$, and $\delta^{18}O$ values are 0.78, 0.53, and 0.03. Model results reproduce the range and peak-and-trough pattern of the N₂O saturation and $\delta^{15}N$ data. Model results reproduce the range of $\delta^{18}O$ values but the pattern is not well reproduced. While all data at Bridgeport exhibited smooth, sinusoidal diel changes, the data at Blair show rapid changes. The diel patterns of N₂O saturation and $\delta^{15}N$ values were opposite of each other, that is, when N₂O saturation

was highest around midnight, the $\delta^{15}N$ values were lowest and when when N₂O saturation was lowest during mid-day, the $\delta^{15}N$ values were greatest. The diel pattern in $\delta^{18}O$ values was more complex at Blair than at Bridgeport suggesting that daytime and nighttime were associated with different $\delta^{18}O$ values of N₂O production. These patterns were similar to those of scenario #5 (variable N₂O production and variable δ values of the source N₂O coinciding when maximum production rates coincided with highest source δ values) and the result of a five-fold variability in day-to-night N₂O production and variability in $\delta^{15}N$ and $\delta^{18}O$ of the N₂O produced in the river.

For both Bridgeport and Blair data, the cause of poorer fits for $\delta^{18}O$ than $\delta^{15}N$ deserve further research. Adding concomitant measurements of $\delta^{15}N$ and $\delta^{18}O$ values of NO_3^- may provide clues about N cycling and help explain some of the observed variability in N₂O [53]. Predicting $\delta^{18}O$-N₂O values from nitrification [11] and denitrification [54] is difficult because of the complex relationship between $\delta^{18}O$-H₂O values and $\delta^{18}O$-N₂O values. Additionally, diel variability in N₂O reduction to N₂ [45,46], may also manifest itself in $\delta^{18}O$-N₂O values because of the strong O isotope fractionation factor during denitrification [55].

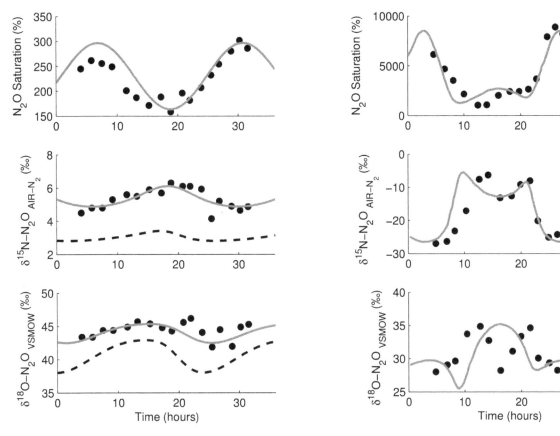

Figure 10. Diel variability in N₂O concentration and δ values at Bridgeport in the Grand River, Canada. The time axis begins at 00:00 on 2007-06-26. Maximum production rate is in sync with the greatest $\delta^{18}O$ values of the source, while $\delta^{15}N$ of the source was constant. R^2 values between field and model data for N₂O saturation, $\delta^{15}N$, and $\delta^{18}O$ values are 0.83, 0.68, and 0.30. This is similar to model scenario #4 (Figure 7).

Figure 11. Diel variability in N₂O concentration and δ values at Blair in the Grand River, Canada. The time axis begins at 00:00 on 2007-06-26. Maximum production rate is in sync with the lowest $\delta^{15}N$ and $\delta^{18}O$ values of the source. R^2 values between field and model data for N₂O saturation, $\delta^{15}N$, and $\delta^{18}O$ values are 0.78, 0.53, and 0.03. This is similar to model scenario #5 (Figure 8).

Discussion

Calculating the δ values of emitted or source N₂O is critical for regional and global N₂O isotopic budgets and also provides information about the source of N₂O and thus N cycling processes. However, SIDNO can simulate the relationships between the δ values of dissolved, source, and N₂O emitted from aquatic ecosystems to the atmosphere. In systems with N₂O production at dynamic steady state, the δ values of dissolved N₂O will not always be directly indicative of the δ values of the source N₂O. The difference between dissolved and source δ values increases as N₂O saturation decreases (as demonstrated in Figures 2 and 3). Even above 1000% saturation (from high production rates and/or low k), the δ values of dissolved N₂O will only approach δ values of the source but offset by 0.7‰ for $\delta^{15}N$ and 1.9‰ for $\delta^{18}O$, a result of the α_{ev} values (Figure 3). At constant N₂O production rates and δ values, the source and emitted δ values can be quantified since the δ values of emitted N₂O must be identical to those of the source and can be calculated from dissolved values (Figure 3; equations 5 and 6).

Our modelling results identified the limitations associated with simple interpretation of dissolved N₂O isotope data since the δ values of dissolved and emitted data are synchronous but rarely offset by a constant value. If N₂O saturation changes with time, the N₂O production rate must also have changed with time,

provided k had been constant (compare model scenarios #1 and #2 in Figures 4 and 5). In contrast, changes in δ values of dissolved N₂O do not require a change in the source δ values (model scenario #1 and Figure 4), while constant δ values of dissolved N₂O do not require constant source δ values (for example model scenario #5 in Figure 8).

When δ values of the source N₂O are variable, the relationship between emitted and source N₂O becomes complicated. The δ values of emitted N₂O will lag behind those of the source and the amplitude of the diel range of δ values will be dampened relative to the source. The amount of lag and dampening is a function of k, N₂O production rate and timing, and the proximity of the source δ values to those of the atmosphere (compare Figures 2 with 3 and Figures 4 with 6). Qualitatively, the δ values of emitted N₂O will be similar to the source if the equilibration time of dissolved N₂O is small relative to the period of source variability (e.g., 24 h period due to diel changes in N cycling [36,45]). Assuming homogeneous N₂O release upstream, the equilibration time can be approximated from a decay curve as $\frac{3z}{k}$, where z is mean depth [49]. If z is small and/or k is high, the equilibration time will be short and the δ values of the emitted N₂O will be close to the source. With decreasing k (or increasing equilibration time), the δ values of emitted N₂O will lag farther behind and will always have a smaller range of δ values than the source. At the most extreme case, the variability in the δ values of emitted N₂O will be reduced to nearly

Table 4. Summary of the results of the SIDNO modelling as the predictive relationship between observations and implications.

Observed Parameter	Implications	Examples
Dissolved N_2O concentration is constant with time	The N_2O production rate is constant with time (if k and temperature are also constant). The N_2O flux to the atmosphere is equal to the production rate. This may not be true if the concentration is close to atmospheric equilibrium.	Scenarios #2 and #3
Dissolved N_2O concentration is variable with time in a sinusoidal pattern	The N_2O production rate is variable with time (if concentration change cannot be explained by change in k or temperature). The average N_2O flux to the atmosphere is equal to the average production rate.	Scenario #1
$\delta^{15}N$ and $\delta^{18}O$ of dissolved N_2O is constant with time	The observation is inconclusive. At concentrations near atmospheric equilibrium, isotopic composition of dissolved N_2O will approximate tropospheric N_2O, egardless of source values. A constant isotopic signature of dissolved N_2O that is different from tropospheric N_2O can indicate either a constant source (if production rate is constant), or a variable source.	Scenario #5
$\delta^{15}N$ and $\delta^{18}O$ of dissolved N_2O is variable with time (slope of data on a $\delta^{18}O$–$\delta^{15}N$ cross-plot tends toward tropospheric N_2O)	The change in $\delta^{15}N$ and $\delta^{18}O$ of dissolved N_2O is likely a result of a change in concentration, but it is possible that the source is variable with time, if the $\delta^{15}N$ and $\delta^{18}O$ values of the source also trend through the value for tropospheric N_2O.	Scenarios #1, #2, #4, and #6
Calculated $\delta^{15}N$ and $\delta^{18}O$ of emitted N_2O is constant with time	The isotopic composition of the source is constant with time, and equal to the calculated value for emitted N_2O	Scenario #1
Calculated $\delta^{15}N$ and $\delta^{18}O$ of emitted N_2O is variable with time	The isotopic composition of the source is variable with time. The range in $\delta^{15}N$ and $\delta^{18}O$ of emitted N_2O is the minimum for the range in that of the source. The flux weighted average $\delta^{15}N$ and $\delta^{18}O$ of emitted N_2O is equal to the production weighted average source values.	Scenarios #2–#6
Long residence time relative to variability of source (need to independently determine k)	The changes in N_2O concentration and $\delta^{15}N$ and $\delta^{18}O$ of emitted N_2O will be dampened relative to, and lag behind, that of the source	Scenarios #3 and #6
Short residence time relative to variability of source (need to independently determine k)	The changes in N_2O concentration and $\delta^{15}N$ and $\delta^{18}O$ of emitted N_2O will be indicative of changes in the source	Scenarios #1, #2, #3 and #4

zero and δ values of the emitted N_2O would be equal to the average production-weighted source δ values. At very long equilibration times, the probability of N_2O consumption increases, a process not explicitly included in SIDNO where the δ value of the source N_2O is simply that which is released to the water column.

Separating N_2O production into nitrification and denitrification requires independent knowledge about the δ values of the source N and O in aquatic ecosystems. It is therefore not possible to state a single $\delta^{15}N$ value for nitrification–N_2O and one for denitrification–N_2O applicable to all aquatic ecosystems. The $\delta^{15}N$ value of the N_2O precursors NH_4^+ and NO_3^- vary across ecosystem as a result of human impact and N loading (agricultural and WWTP) as well as the source of N, and additional N transformations in the aquatic ecosystem. For example, along the length of the Grand River, $\delta^{15}N$ values of NH_4^+ and NO_3^- exhibit systematic trends resulting from the confluence of agricultural tributaries and large urban waste-water treatment plants (Schiff et al., unpublished results, [53]). Nevertheless, these values can be measured and biogeochemical relationships between N species, redox, and N_2O can be used as supporting information for process separation (e.g., [5,12,36,45]). The $\delta^{18}O$ value of N_2O will also vary across ecosystems as a result of its close relationship with $\delta^{18}O$-H_2O and to a lesser extent $\delta^{18}O$-O_2 [10,11,54,56]. Fortunately, $\delta^{18}O$-H_2O values can be easily predicted and measured [57]. Thus, once δ values of N_2O precursors have been identified, biogechemical data can provide an indication about the diel pattern of N_2O

production processes, and ranges of potential end-member δ values can be calculated (e.g., [5] summarize isotopic fractionation for $\delta^{15}N$ and [10,11,54,56] for $\delta^{18}O$) and the model used to fit the field data.

Conclusions

In aquatic ecosystems, the instantaneous δ values of N_2O emitted to the atmosphere are easily calculated if the water temperature and dissolved N_2O concentration and δ values are known. Our modelling efforts illustrate that complex relationships exist between dissolved and source N_2O and that the δ values of dissolved N_2O are not always representative of either the N_2O produced or emitted to the atmosphere. Thus, calculated δ values of the emitted N_2O are the values that should be used to draw conclusions about N_2O emission from aquatic systems and the global N_2O cycle rather than the more commonly used instantaneous values (Table 4). The flux-weighted δ values of emitted N_2O can provide average production-weighted δ values of the N_2O source under dynamic steady-state in aquatic ecosystems.

If the δ values of emitted N_2O are constant with time, either the δ values of the source must also be constant or the N_2O equilibration time is very long. However, if the calculated δ values of emitted N_2O vary with time then the δ values of the source must also vary with time producing a diagnostic pattern. These findings are more robust than using dissolved δ values alone since dissolved

δ values can change simply with a change in N_2O production rate, changes in source δ values, and changes in k. N_2O residence time, dependent on production rate, k, and z, will determine the lag time between the δ values of emitted and source N_2O. The difference in timing between maxima and minima δ values of emitted N_2O and the maxima and minima of dissolved N_2O is indicative of how the δ values of the source change. For all these reasons, we urge caution when using single samples of N_2O concentration and δ to calculate fluxes of N_2O to the atmosphere and inferring N_2O production pathways.

Ultimately, the dynamic model SIDNO may be used to estimate k, N_2O production rate and δ values of the N_2O source, an indication of the production pathway and N cycling, in aquatic ecosystems via inverse modelling. If physical properties, such as depth and temperature are known, SIDNO may be used to fit the measured field data (N_2O concentration and δ values) by adjusting the N_2O source parameters. SIDNO can also be used to explore the dynamics between dissolved, source, and emitted N_2O to query production scenarios and design field campaigns for studies of N cycling processes.

Acknowledgments

We thank MS Rosamond, HM Baulch, the reviewers, and the academic editor for their helpful comments.

Author Contributions

Conceived and designed the experiments: SJT JJV SLS. Performed the experiments: SJT JJV. Analyzed the data: SJT JJV SLS. Contributed reagents/materials/analysis tools: SJT JJV SLS. Wrote the paper: JJV SJT SLS.

References

1. Forster P, Ramaswamy V, Artaxo P, Berntsen T, Betts R, et al. (2007) Changes in atmospheric constituents and in radiative forcing. In: Solomon S, Qin D, Manning M, Chen Z, Marquis M, et al., editors, Climate Change 2007: The Physical Science Basis. Contribution of Working Group I to the Fourth Assessment Report of the Intergovernmental Panel on Climate Change. Cambridge: Cambridge University Press.
2. Denman KL, Brasseur G, Chidthaisong A, Ciais P, Cox PM, et al. (2007) Couplings between changes in the climate system and biogeochemistry. In: Solomon S, Qin D, Manning M, Chen Z, Marquis M, et al., editors, Climate Change 2007: The Physical Science Basis. Contribution of Working Group I to the Fourth Assessment Report of the Intergovernmental Panel on Climate Change. Cambridge: Cambridge University Press.
3. Wada E, Ueda S (1996) Carbon nitrogen and oxygen isotope ratios of CH_4 and N_2O on soil ecosystems. In: Boutton TW, Yamasaki SI, editors.Mass Spectrometry of Soils.New York: Marcel Dekker. pp. 177–204.
4. Aravena R, Mayer B (2010) Isotopes and processes in the nitrogen and sulfur cycles. In: Environmental isotopes in biodegradation and bioremediation. Boca Raton, FL: Lewis Publishers. pp. 203–246.
5. Baulch HM, Schiff SL, Thuss SJ, Dillon PJ (2011) Isotopic character of nitrous oxide emitted from streams. Environ Sci Technol 45: 4682–4688.
6. Park S, Pérez T, Boering KA, Trumbore SE, Gil J, et al. (2011) Can N_2O stable isotopes and isotopomers be useful tools to characterize sources and microbial pathways of N_2O production and consumption in tropical soils? Global Biogeochem Cycles 25: GB1001.
7. Bol R, Toyoda S, Yamulki S, Hawkins JMB, Cardenas LM, et al. (2003) Dual isotope and isotopomer ratios of N_2O emitted from a temperate grassland soil after fertiliser application. Rapid Commun Mass Spectrom 17: 2550–2556.
8. Mandernack KW, Rahn T, Kinney C, Wahlen M (2000) The biogeochemical controls of the $\delta^{15}N$ and $\delta^{18}O$ of N_2O produced in landfill cover soils. J Geophys Res Atmos 105: 17709–17720.
9. Pérez T, Trumbore SE, Tyler SC, Matson PA, Ortiz-Monasterio I, et al. (2001) Identifying the agricultural imprint on the global N_2O budget using stable isotopes. J Geophys Res Atmos 106: 9869–9878.
10. Snider DM, Schiff SL, Spoelstra J (2009) $^{15}N/^{14}N$ and $^{18}O/^{16}O$ stable isotope ratios of nitrous oxide produced during denitrification in temperate forest soils. Geochim Cosmochim Acta 73: 877–888.
11. Snider DM, Venkiteswaran JJ, Schiff SL, Spoelstra J (2012) Deciphering the oxygen isotope composition of nitrous oxide produced by nitrification. Glob Chang Biol 18: 356–370.
12. Beaulieu JJ, Shuster WD, Rebholz JA (2010) Nitrous oxide emissions from a large, impounded river: The Ohio River. Environ Sci Technol 44: 7527–7533.
13. Beaulieu JJ, Tank JL, Hamilton SK, Wollheim WM, Hall RO Jr, et al. (2011) Nitrous oxide emission from denitrification in stream and river networks. Proc Natl Acad Sci USA 108: 214–219.
14. Dore JE, Popp BN, Karl DM, Sansone FJ (1998) A large source of atmospheric nitrous oxide from subtropical north pacific surface waters. Nature 396: 63–66.
15. Kim KR, Craig H (1990) Two-isotope characterization of N_2O in the Pacific Ocean and constraints on its origin in deep water. Nature 347: 58–61.
16. Naqvi S, Yoshinari T, Jayakumar D, Altabet M, Narvekar P, et al. (1998) Budgetary and biogeochemical implications of N_2O isotope signatures in the Arabian Sea. Nature 394: 462–464.
17. Priscu JC, Christner BC, Dore JE, Westley MB, Popp BN, et al. (2008) Supersaturated N_2O in a perenniallyice-covered Antarctic lake: Molecular and stable isotopic evidence for a biogeochemical relict. Limnol Oceanogr 53: 2439–2450.
18. Yoshinari T, Altabet M, Naqvi S, Codispoti L, Jayakumar A, et al. (1997) Nitrogen and oxygen isotopic composition of N_2O from suboxic waters of the eastern tropical north pacific and the arabian sea–measurement by continuous-ow isotope-ratio monitoring. Mar Chem 56: 253–264.
19. Kroeze C, Dumont E, Seitzinger SP (2005) New estimates of global emissions of N_2O from rivers and estuaries. Environ Sci 2: 159–165.
20. Boontanon N, Ueda S, Kanatharana P, Wada E (2000) Intramolecular stable isotope ratios of N_2O in the tropical swamp forest in Thailand. Naturwissenschaften 87: 188–192.
21. Toyoda S, Iwai H, Koba K, Yoshida N (2009) Isotopomeric analysis of N_2O dissolved in a river in the tokyo metropolitan area. Rapid Commun Mass Spectrom 23: 809–821.
22. Inoue HY, Mook WG (1994) Equilibrium and kinetic nitrogen and oxygen isotope fractionations between dissolved and gaseous N_2O. ChemGeol 113: 135–148.
23. Brenninkmeijer CAM, Rckmann T (1999) Mass spectrometry of the intramolecular nitrogen isotope distribution of environmental nitrous oxide using fragmention analysis. Rapid Commun Mass Spectrom 13: 2028–2033.
24. Sutka RL, Ostrom N, Ostrom P, Breznak J, Gandhi H, et al. (2006) Distinguishing nitrous oxide production from nitrification and denitrification on the basis of isotopomer abundances. Appl Environ Microbiol 72: 638–644.
25. Toyoda S, Yoshida N (1999) Determination of nitrogen isotopomers of nitrous oxide on a modified isotope ratio mass spectrometer. Anal Chem 71: 4711–4718.
26. Röckmann T, Kaiser J, Brenninkmeijer CAM, Brand WA (2003) Gas chromatography/isotoperatio mass spectrometry method for high-precision position-dependent ^{15}N and ^{18}O measurements of atmospheric nitrous oxide. Rapid Commun Mass Spectrom 17: 1897–1908.
27. Venkiteswaran JJ, Wassenaar LI, Schiff SL (2007) Dynamics of dissolved oxygen isotopic ratios: a transient model to quantify primary production, community respiration, and air–water exchange in aquatic ecosystems. Oecologia 153: 385–398.
28. Wahlen M, Yoshinari T (1985) Oxygen isotope ratios in N_2O from different environments. Nature 313: 780–782.
29. Zafiriou OC (1990) Laughing gas from leaky pipes. Nature 347: 15–16.
30. Westley MB, Yamagishi H, Popp BN, Yoshida N (2006) Nitrous oxide cycling in the black sea inferred from stable isotope and isotopomer distributions. Deep Sea Res Part II 53: 1802–1816.
31. Well R, Eschenbach W, Flessa H, von der Heide C, Weymann D (2012) Are dual isotope and isotopomer ratios of N_2O useful indicators for N_2O turnover during denitrification in nitratecontaminated aquifers? Geochim Cosmochim Acta 90: 265–282.
32. An S, Joye SB (2001) Enhancement of coupled nitrification-denitrification by benthic photosynthesis in shallow estuarine sediments. Limnol Oceanogr 46: 62–74.
33. Clough TJ, Buckthought LE, Kelliher FM, Sherlock RR (2007) Diurnal uctuations of dissolved nitrous oxide (N_2O) concentrations and estimates of N_2O emissions from a spring-fed river: implications for IPCC methodology. Glob Chang Biol 13: 1016–1027.
34. Laursen AE, Seitzinger SP (2004) Diurnal patterns of denitrification, oxygen consumption and nitrous oxide production in rivers measured at the whole-reach scale. Freshwat Biol 49: 1448–1458.
35. Lorenzen J, Larsen LH, Kjr T, Revsbech NP (1998) Biosensor determination of the microscale distribution of nitrate, nitrate assimilation, nitrification, and denitrification in a diatom-inhabited freshwater sediment. Appl Environ Microbiol 64: 3264–3269.
36. Rosamond MS, Thuss SJ, Schiff SL (2011) Coupled cycles of dissolved oxygen and nitrous oxide in rivers along a trophic gradient in Southern Ontario, Canada. J Environ Qual 40: 256–270.
37. Rosamond MS, Thuss SJ, Schiff SL (2012) Dependence of riverine nitrous oxide emissions on dissolved oxygen levels. Nat Geosci 5: 715–718.
38. Kaiser J, Röckmann T, Brenninkmeijer CAM (2003) Complete and accurate mass spectrometric isotope analysis of tropospheric nitrous oxide. J Geophys Res Atmos 108: 4476.

39. Prinn R, Cunnold D, Rasmussen R, Simmonds P, Alyea F, et al. (1990) Atmospheric emissions and trends of nitrous oxide deduced from 10 years of ALE–GAGE data. J Geophys Res Atmos 95: 18369–18385.

40. Prinn RG, Weiss RF, Fraser PJ, Simmonds PG, Cunnold DM, et al. (2000) A history of chemically and radiatively important gases in air deduced from ALE/GAGE/AGAGE. J Geophys Res Atmos 105: 17751–17792.

41. Lide DR, Frederikse HPR (1995) CRC Handbook of Chemistry and Physics, 76th edition. Boca Raton: CRC Press.

42. McElroy MB, Jones DBA (1996) Evidence for an additional source of atmospheric N_2O. Global Biogeochem Cycles 10: 651–659.

43. Rahn T, Wahlen M (2000) A reassessment of the global isotopic budget of atmospheric nitrous oxide. Global Biogeochem Cycles 14: 537–543.

44. Stein LY, Yung YL (2003) Production, isotopic composition, and atmospheric fate of biologically produced nitrous oxide. Annu Rev Earth Planet Sci 31: 329–356.

45. Baulch HM, Dillon PJ, Maranger R, Venkiteswaran JJ, Wilson HF, et al. (2012) Night and day: short-term variation in nitrogen chemistry and nitrous oxide emissions from streams. Freshwat Biol 57: 509–525.

46. Harrison JA, Matson PA, Fendorf SE (2005) Effects of a diel oxygen cycle on nitrogen transformations and greenhouse gas emissions in a eutrophied subtropical stream. Aquat Sci 67: 308–315.

47. Rock L, Ellert BH, Mayer B, Norman AL (2007) Isotopic composition of tropospheric and soil N_2O from successive depths of agricultural plots with contrasting crops and nitrogen amendments. J Geophys Res Atmos 112: D18303.

48. Kool DM, Wrage N, Oenema O, Dolfifing J, Van Groenigen JW (2007) Oxygen exchange between (de)nitrification intermediates and H_2O and its implications for source determination of NO_3 and N_2O: a review. Rapid Commun Mass Spectrom 21: 356–3578.

49. Chapra SC, Di Toro DM (1991) Delta method for estimating primary production, respiration, and reaeration in streams. J Environ Eng 117: 640–655.

50. Mengis M, Schif SL, Harris M, English MC, Aravena R, et al. (1999) Multiple geochemical and isotopic approaches for assessing ground water no3 elimination in a riparian zone. Ground Water 37: 448–457.

51. Venkiteswaran JJ, Rosamond MS, Schiff SL (2014) Nonlinear response of riverine N_2O uxes to oxygen and temperature. Environ Sci Technol 48: 1566–1573.

52. Spoelstra J, Schiff SL, Brown SJ (2013) Artificial sweeteners in a large canadian river reect human consumption in the watershed. PLoS ONE 8: e82706.

53. Hood JLA, Taylor WD, Schiff SL (2013) Examining the fate of WWTP effluent nitrogen using $\delta^{15}N$ -NH_4^+, $\delta^{15}N$ -NO_3^-, and $\delta^{15}N$ of submersed macrophytes. Aquat Sci. doi: 10.1007/s00027-013-0333-4.

54. Snider DM, Venkiteswaran JJ, Schiff SL, Spoelstra J (2013) A new mechanistic model of $\delta^{18}O$-N_2O formation by denitrification. Geochim Cosmochim Acta 112: 102–115.

55. Well R, Flessa H (2009) Isotopologue enrichment factors of N_2O reduction in soils. Rapid Commun Mass Spectrom 23: 2996–3002.

56. Snider DM, Spoelstra J, Schiff SL, Venkiteswaran JJ (2010) Stable oxygen isotope ratios of nitrate produced from nitrification: ^{18}O-labeled water incubations of agricultural and temperate forest soils. Environ Sci Technol 44: 5358–5364.

57. Bowen GJ, Wassenaar LI, Hobson KA (2005) Global application of stable hydrogen and oxygen isotopes to wildlife forensics. Oecologia 143: 337–348.

58. Clough TJ, Bertram JE, Sherlock RR, Leonard RL, Nowicki BL (2006) Comparison of measured and ef5-r-derived N_2O uxes from a spring-fed river. Glob Chang Biol 12: 352–363.

59. Garnier J, Billen G, Cébron A (2007) Modelling nitrogen transformations in the lower Seine river and estuary (France): impact of wastewater release on oxygenation and N_2O emission. Hydrobiologia 588: 291–302.

60. Reay DS, Smith KA, Edwards AC (2003) Nitrous oxide emission from agricultural drainage waters. Glob Chang Biol 9: 195–203.

5

The Relation between Polyaromatic Hydrocarbon Concentration in Sewage Sludge and Its Uptake by Plants: *Phragmites communis, Polygonum persicaria* and *Bidens tripartita*

Barbara Gworek[1], Katarzyna Klimczak[2], Marta Kijeńska[1]*

1 Institute of Environmental Protection – National Research Institute, Warsaw, Poland, **2** Warsaw University of Life Sciences – SGGW, Department of Soil Environment Sciences, Warsaw, Poland

Abstract

The aim of the study was to define the relationship between the concentration of PAHs in sewage sludge at a particular location and their amount in various plant materials growing on it. The credibility of the results is enhanced by the fact that sewage sludge from two separate sewage-treatment plants were selected for their influence on the content of PAHs in three plant species growing on them. The investigations were carried out for a period of three years. The results demonstrated unequivocally that the uptake of PAHs by a plant depended on polyaromatic hydrocarbon concentration in the sewage sludge. The correlation between accumulation coefficient of PAH in a plant and the content of the same PAH in the sewage sludge had for three-, four- and five-ring hydrocarbons an exponential character and for six-ring hydrocarbons was of a linear character. The accumulation coefficients calculated for three-ring aromatics were several times higher than for four-ring PAHs; further the coefficient values calculated for five-ring PAHs were several times lower than for four-ring hydrocarbons. Finally, the accumulation coefficient values of six-ring PAHs were the lowest in the series of studied polyaromatic hydrocarbons.

Editor: Claude Wicker-Thomas, CNRS, France

Funding: The work was supported by Ministry of Science and Higher Education. The funders had no role in study design, data collection and analysis, decision to publish, or preparation of the manuscript.

Competing Interests: The authors have declared that no competing interests exist.

* Email: marta.kijenska@ios.edu.pl

Introduction

Polycyclic aromatic hydrocarbons (PAHs) are environmental contaminants existing in nature as a complex mixture. They may originate from natural sources, as non-anthropogenic burning of biomass, high temperature thermolysis of organic matter and diagenesis of sedimentary materials. They would be also formed in biosynthetic processes occuring in microorganisms or plants [1]. The concentration of PAHs derived from natural processes is insignificant in comparison with that produced as a result of human activity such as coke and steel manufacturing, waste incineration, domestic heating or transport [2–7]. PAHs constitute the largest class of potential carcinogens, to which mutagenic activity is also ascribed [8]. The most studied and most known compound of the group is benzo[a]pyrene [9] Numerous data illustrate the distribution of concentration of PAHs in the atmosphere in areas of smoke emission and in the vicinity of the large urban centers [10–12]. Because of their mutagenic and/or carcinogenic properties PAHs are a risk to human health.

It has been confirmed [13–14] that the respiratory and dietary exposure to PAHs, placed in the 9th position on the 2011 ATSDR (Agency for Toxic Substances & Disease Registry) [15] substance priority list, may lead to different forms of cancer. PAHs are adsorbed from the atmosphere by aerosol particles, through which they find their way to water, soil and plant foliage during rains. This route is therefore considered appropriate for biomonitoring [16–19]. PAHs may accumulate in the organisms due to their low solubility and high octanol-water partition coefficient and undergo long-range transport [20]. PAHs accumulated in vegetables could make them possibly carcinogenic for humans [21]. In order to control this problem, a proper understanding of the mechanism of PAHs uptake by specific plants is important. On the other hand PAHs found in plants could also originate from intrinsic biochemical transformations. Therefore much research has to be done in order to accurately specify the main source of PAHs in plant material. In an attempt to understand the role of the listed causes of PAHs contamination in plant material (mass transfer between atmosphere and plants, ground and roots as well as biosynthesis) studies have been made to evaluate precisely the relation between the amount of PAHs and its content in plants growing on particular bed. Two possible mechanisms for the uptake of PAHs by plants have been suggested. According to the first, the transfer of PAHs from polluted atmosphere to plants occurs via a particle-phase deposition on the waxy leaf or from the gas phase through the stomata [21–23]. In the second mechanism, an uptake by the root from the soil solution and then translocation to shoots by liquid phase transfer in the transpiration streams is proposed [21–27]. Since the mid 1980s the legislation of the European Union has been focused on the utilization of sewage sludge in agriculture as a natural fertilizer containing especially substantial amount of phosphorus (up to over 12 wt%) and over 7 wt% of nitrogen (e.g. the Directive 86/278/EEC). Due to the accumulation in the plant's leaves (including vegetables) of polyaromatics, showing harmful effects on human health, the determination of the mechanism of their uptake by plants is of the highest importance.

The aim of this study was to determine the correlation between PAHs content in municipal sewage sludge and PAHs uptake by plants growing on them as well as the relationship between PAHs molecular structure and its ability of accumulation in plant.

Sewage sludge from two separate sewage-treatment plants were selected for their influence on the content of PAHs in three plant species growing on them. The investigations were carried out for a period of three years.

Materials and Methods

1. Sampling

Two municipal sewage sludge lagoons were selected for these studies. Samples of the sewage sludge and of the plants growing upon them were collected at each site for a period of 3 years. All plants were self-sown, for the comparison study three plant species (*Phragmites communis*, *Polygonum persicaria* and *Bidens tripartita*) growing in both locations have been chosen. The concentration of thirteen PAHs on the US EPA specification list was determined in the sewage sludge as well as in the plant materials.

2. Characteristics of sampling sites and locations

No specific permissions were required for both locations. We had oral permissions to enter to the area belonging to sewage treatment plant. The permission to access land and take samples of sewage sludge and plants was provided by heads of both sewage treatment plants (Białystok, Wodociągi Białostockie and Ryki, Zakład Wodociągów i Kanalizacji). The field studies did not involve endangered or protected species. Specific location of study:

L1 23°10′E 53°08′N
L2 21°57′E 51°37′N

Location 1 (L1). Municipal sewage plant in a town with a population of about 290 000 inhabitants and textile, metal, electronic, machine-building, building materials, glass, wood and food industries. The sewage plant was used mainly for the treatment of household wastes. Samples were collected over a period of 3 years from 4 sampling fields each with a surface area of ca. 5000 m^2.

Location 2 (L2). Municipal sewage plant in a small town with a population of 11 000, treating household and small-scale industry wastes and sewage.

The sludge from the communal sewage-treatment plant was stored in four lagoons, each of a surface area of 1500 m^2. Samples of sewage sludge and plant materials were collected from each sampling site for a period of 3 years. The sewage sludge from each lagoon was collected separately, thereby avoiding the analysis of mixed samples.

The dominating as well as the accompanying plant species were selected for sampling. The intention in these studies was to collect the same plant species.

3. Methods

The samples for analysis were dried at room temperature (below 20°C), transferred either to a laboratory mill (plant samples) or ground in a mortar and then fractioned in a sieve of 1 mm mesh (sludge samples).

PAHs analyses. The composition of the studied polyaromatic hydrocarbons was determined by high pressure liquid chromatography (HPLC) using Waters chromatograph with photodiode (PDA) and fluorescence (FLSC) detector. The analyses were carried out according to the ISO 13877 method (1998). The identification of the PAHs was based on retention time and UV spectra using capillary column (SUPELCOSIL LC-PAH C18, S-5 μm, 15 cm×4.6 mm) and the quantification of all the investigated PAHs was obtained using signals (pik area) and the calibration curve method. Detection limit of the method was determined at 0.003 ng·g^{-1} for all investigated PAHs analysed together at the same time. Average relative standard deviation for chromatographic analysis was 10% and average expanded uncertainty of method with 95% confidence interval multiplied by the coverage factor k = 2 was less than 30%.

The following procedure was applied for the analysis of sewage sludge and plant material samples: 10–20 g of dried and ground material was mixed with 50 cm^3 of methylene chloride and extracted in the presence of 1 g of metallic copper in Soxtec apparatus for 5 hrs (boiled for 3.5 hrs and subsequently washed for 1.5 hrs); The extracts were evaporated in rotary vacuum evaporator to a dry residue which was dissolved in 1 ml n-hexane portions. The mixtures were introduced into 10 mm-diameter columns containing silica gel and alumina oxide basic powder in layers of 5 cm each in height (layer of column 10 cm in height). The column was eluted with a gradient using 20 cm^3 methylene chloride/n-hexane solutions in ratios 1:3 and 15 cm^3 methylene chloride/n-hexane solutions in ratios 1:1. Then the eluate was concentrated to a dry residue under a flux of nitrogen. The residue was dissolved in 1 ml acetonitrile. The volume of injection for chromatograph analysis was 10 μl.

The relationship between the content of various PAHs in the sewage sludge and in the plants growing on them was determined by the molar accumulation coefficient (m.a.c.) according to the following formula:

$$m.a.c. = n_p/n_s$$

where m.a.c. is molar accumulation coefficient, n_p - content of PAHs in a plant material, n_s - content of PAHs in sewage sludge.

The corresponding concentration of PAHs, their uptake by plants, as well as accumulation coefficient were expressed as molar magnitudes which allowed their direct further use in kinetic deliberations or biological activity description possessing additive character.

Statistical analyses. Relationships between accumulation coefficient of PAH in plant material and the content of PAH in sewage sludge were estimated using simple regression. Linear model ($Y = a+bX$) or multiplicative model ($Y = aX^b$) were selected for the analyses because of good fitting to experimental data. F-test was used for estimation of statistical significance of the examined relationships. The analyses were conducted using Statgraphics 4.1, level of significance was set at 0.05.

Results and Discussion

In both chosen objects the observations were carried for a period of 3 years and samples were collected in both case from 4 different lagoons. The concentrations of 13 PAHs, listed by USEPA as priority pollutants [28] were estimated:

three-ring (fluorene (Frn), phenanthrene (Ph), anthracene (A))

four-ring (fluoranthene (Ftn), pyrene (P), benzo/a/anthracene (BaA), chrysene (Ch))

five-ring (benzo/b/fluoranthene (BbF), benzo/k/fluoranthene (BkF), benzo/a/pyrene (BaP), dibenz/ah/anthracene (DahA))

six-ring (benzo/ghi/perylene (BghiP), indeno/123-cd/pyrene (IP)).

The determination of the concentrations of naphthalene, acenaphthene and acenaphthylene, was considered not necessary due to their negligible impact - these three compounds are highly volatile [29]. The range of PAHs concentration changes over three years and their average values are presented in the table 1. Fluoranthene occured in highest amounts in the sewage sludge in both locations – an average of 4.19 µmol/kg in location 1 (L1) and 3.14 µmol/kg in location 2 (L2). The second most occurring PAH in the sewage sludge of both locations was pyrene – found in average amounts of 3.09 µmol/kg in L1 and 2.10 µmol/kg in L2. The third PAH in L1 is phenanthrene, which was however in the seventh place in sewage sludge of L2. In both locations three PAHs: fluorene, dibenz/ah/anthracene and anthracene were found in lowest amounts.

The sequences of PAH concentrations in the sewage sludge are presented below and in figures 1-2.

L1:   Ftn>P>Ph>BbF>Ch>BaP>BghiP>BaA>IP>BkF> A>Frn>DahA

L2:   Ftn>P>BbF>BaP>Ch>BghiP>Ph>IP>BaA>BkF> Frn>DahA>A

After determining the content of PAHs in sewage sludge, the content of PAHs in the plants growing in them was analyzed. Three plants growing in both locations: common reed (*Phragmites communis*), redshank (*Polygonum persicaria*) and three-lobe beggarticks (*Bidens tripartita*) were analyzed. The shift in the PAH concentration range in each of the plant species during the three years of study and their average values are presented in table 1. Only for Ftn statistically relevant differences between species were stated (between *Phragmites communis* and *Bidens tripartita*). In other cases there were no statistically significant differences observed.

Independently of the content of PAHs in sewage sludge and the species of plant, phenanthrene was a PAH which was present in plant tissue in the highest amount. The content of phenanthrene in the studied plants fluctuated from 0.2055 to 0.3411 µmol/kg (both for *Phragmites communis*). It is worth noting that for each plant species the phenanthrene content was higher in L2, although it occurred in lower amounts in the sewage sludge. The second most abundant PAH in the plants growing in both locations was fluoranthene (0.1073 µmol/kg in *Phragmites communis* collected from L1 to 0.1736 µmol/kg in *Bidens tripartita*, collected in L2). The content of fluoranthene in all plants was also higher in L2, where its amount in the sewage sludge was lower. Fluorene was the third most abundant PAH (although it was fourth after pyrene in *Bidens tripartita* collected in L2) with concentration fluctuating from 0.0729 µmol/kg to 0.1564 µmol/kg. In each case dibenz/ah/anthracene occurred in lowest amounts in the plants (from 0.0014 µmol/kg in *Phragmites communis* collected in L1 to 0.0042 µmol/kg in *Polygonum persicaria* collected in L1).

A comparison of the contents of the studied PAHs in plants did not reveal any clear relationship between the PAH structure and its concentration in plant, probably due to the remarkably varying content of PAHs in the sewage sludge of the various locations. Therefore it was decided to plot the accumulation coefficients which are more objective factors than the content. When the accumulation coefficients for various hydrocarbons are compared, the regularity becomes clear: the uptake efficiency for all plants in

both locations under study was several times higher for three-ring PAHs than for four-ring PAHs and, respectively, for four-rings PAHs than for five-ring and six-ring molecules.

In order to establish the existence of a regularity in the dependence between the concentration of polyaromatics in sewage sludge and the PAH content in plants growing upon the studied sewage sludge, the relationships between the accumulation coefficient and the content of PAH in the sewage sludge was plotted for three-, four-, five- and six-ring hydrocarbons jointly for both studied locations (Fig. 3–6). It was assumed that a general rule can be developed that would confirm the uptake of polyaromatics from the sewage sludge by plants depending on the amount of PAHs. The results presented in Fig. 3–6 indicated the existence of a correlation between the content of PAHs in sewage sludge and the plant material collected from it. The obtained relationships of accumulation coefficient vs. PAH content in sewage sludge exhibited a different character for various classes of PAHs. It should be emphasized that the data depicted on corresponding Figures (Fig. 3–5) possess high convergence level (Tab. 2) in connection with the fact that they were collected for different plants and various locations at various time, made them reliable. As it is shown in the table 2, most of the correlations between accumulation coefficient of PAH in plant material and the content of PAH in sewage sludge were strong (high coefficient of determination – R^2) and statistically significant, because the value $P<0.05$. Only few P values were higher then 0.05.

As shown in Figs. 3–5 for PAHs, containing three, four or five aromatic rings, the relationship between the accumulation coefficient and PAH content in sewage sludge was of an exponential character while in the case of molecules containing six aromatic rings (Fig. 6) a linear character of the discussed relationship was undoubtly observed. The exponential function depicting the accumulation coefficient vs. the PAH content in sewage sludge was of $y = ax^{-b}$ character i.e. the highest accumulation coefficient value corresponding to the PAH lowest content in the sewage sludge rapidly diminished with an increase of PAH concentration and after which the decrease was insignificant after reaching a specific value for each PAH. This specific level corresponded to much higher values of accumulation coefficient for three-ring aromatics than for four-ring hydrocarbons and, respectively, for five-ring molecules. It is noteworthy that for all the discussed PAH groups the profiles depicting anthracene and its derivatives (benz/a/anthracene and dibenz/ah/anthracene, correspondingly) differed significantly from those ascribed to other PAHs, especially for very low accumulation coefficient values.

For the six-ring PAHs the relationship between the accumulation coefficient and PAH content in sewage sludge was of linear character. The accumulation coefficient values, lower than those calculated for the five-ring PAHs, diminished proportionally with the increase of PAHs content in sewage sludge.

In contrast to the anthracene structure to which a limiting effect in the accumulation susceptibility of polyaromatic hydrocarbons can be ascribed, the isomeric phenanthrene structure seemed to facilitate the accumulation of PAHs. This is shown in the examples of phenanthrene, anthracene, benzo(k)fluoranthene and benzo(b)-fluoranthene. Anthracene and phenanthrene are PAHs with a three-ring structure. The linear anthracene molecule should penetrate the plant structure more easily when considering only the physical character of PAH uptake. The phenanthrene molecule possesses a more complex spatial structure, nevertheless, practically in all studied cases its accumulation coefficient was ca. twice higher as compared with those determined for anthracene, and the absolute magnitude of its content was of a range higher

Table 1. The variability and average content of PAHs.

PAH	The range and average content of PAHs in sewage sludge in both locations [μmol/kg]				The range and average content of PAHs in the foliage of Phragmites communis growing on sewage sludge in both locations [μmol/kg]				The range and average content of PAHs in the foliage of Polygonum persicaria growing on sewage sludge in both locations [μmol/kg]				The range and average content of PAHs in the foliage of Bidens tripartita growing on sewage sludge in both locations [μmol/kg]			
	mean		mean±SD	Range (min-max)	mean		mean±SD	Range (min-max)	mean		mean±SD	Range (min-max)	mean		mean±SD	Range (min-max)
	L1	L2	Both locations		L1	L2	Both locations		L1	L2	Both locations		L1	L2	Both locations	
Frn	0.5	0.34	0.417±0.262	0.003–0.952	0.0963	0.0995	0.098±0.051	0.029–0.209	0.1395	0.1023	0.116±0.055	0.064–0.213	0.1564	0.0729	0.123±0.083	0.022–0.251
Ph	2.32	1.43	1.856±1.04	0.372–4.105	0.2055	0.3411	0.299±0.314	0.137–1.458	0.2713	0.2852	0.28±0.077	0.182–0.416	0.2791	0.3191	0.295±0.063	0.212–0.36
A	0.52	0.30	0.404±0.327	0.109–1.362	0.0163	0.0187	0.018±0.010	0.009–0.036	0.0189	0.0164	0.017±0.007	0.008–0.029	0.0164	0.0228	0.019±0.01	0.01–0.032
Ftn	4.01	3.14	3.558±1.745	0.706–7.043	0.1073	0.1206	0.116±0.025[a]	0.061–0.161	0.1525	0.1584	0.156±0.052[ab]	0.084–0.256	0.1674	0.1736	0.170±0.033[b]	0.136–0.217
P	2.94	2.1	2.503±1.457	0.628–5.671	0.0617	0.0686	0.066±0.025	0.031–0.113	0.0971	0.0721	0.081±0.039	0.041–0.179	0.0875	0.1034	0.094±0.019	0.069–0.114
BaA	1.23	1.2	1.215±0.979	0.225–3.053	0.0050	0.0076	0.007±0.004	0.002–0.015	0.0096	0.0097	0.01±0.004	0.006–0.018	0.0119	0.0100	0.011±0.004	0.007–0.017
Ch	1.64	1.59	1.614±1.214	0.23–3.799	0.0126	0.0214	0.019±0.015	0.007–0.073	0.0273	0.0190	0.022±0.011	0.012–0.049	0.0151	0.0254	0.019±0.009	0.009–0.029
BbF	1.85	2.1	1.981±1.034	0.36–3.801	0.0071	0.0187	0.010±0.004	0.004–0.018	0.0202	0.0130	0.016±0.01	0.008–0.041	0.0082	0.0132	0.01±0.003	0.007–0.014
BkF	0.7	0.91	0.811±0.507	0.153–1.937	0.0034	0.0049	0.004±0.002	0.002–0.007	0.0280	0.0064	0.014±0.025	0.002–0.087	0.0038	0.0054	0.004±0.001	0.003–0.006
BaP	1.52	1.83	1.679±0.924	0.266–3.256	0.0051	0.0073	0.007±0.004	0.002–0.012	0.0133	0.0097	0.011±0.006	0.006–0.026	0.0058	0.0103	0.008±0.003	0.003–0.011
DahA	0.34	0.34	0.342±0.314	0.026–1.163	0.0014	0.0027	0.002±0.001	0.001–0.006	0.0042	0.0021	0.003±0.004	0.000–0.014	0.0023	0.0020	0.002±0.002	0.001–0.004
BghiP	1.47	1.49	1.48±0.764	0.228–2.839	0.0025	0.0057	0.005±0.003	0.001–0.012	0.0096	0.0067	0.008±0.005	0.003–0.018	0.0057	0.0039	0.005±0.002	0.003–0.009
IP	1.2	1.43	1.32±0.691	0.217–2.583	0.0025	0.0050	0.004±0.002	0.001–0.009	0.0092	0.0072	0.008±0.005	0.002–0.018	0.0051	0.0054	0.005±0.001	0.003–0.007

[a,b] different letters indicate significant differences between means for species on the basis of Kruskal-Wallis test ($P < 0.05$).

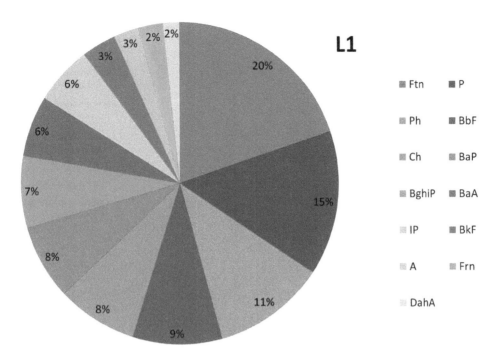

Figure 1. The composition of sewage sludge in L1.

than the value established for anthracene. The same PAH uptake behaviour was observed for the benzo/b/fluoranthene and benzo/k/fluoranthene molecules. The structural differences in the last hydrocarbon pair lie in the existence of the so called "bay region" responsible for the carcinogenic activity in benzo[b]fluoranthene molecule [30].

The remarkable differences in the uptake of very similar PAHs in plant materials could be considered as an indirect proof of a specific biochemical pathway of their collecting by plants.

The existence of dependence between the plant accumulation potential for given PAH and the content of this PAH in a ground suggested the existence of some hydrocarbon molecule transfer from sewage sludge to a plant.

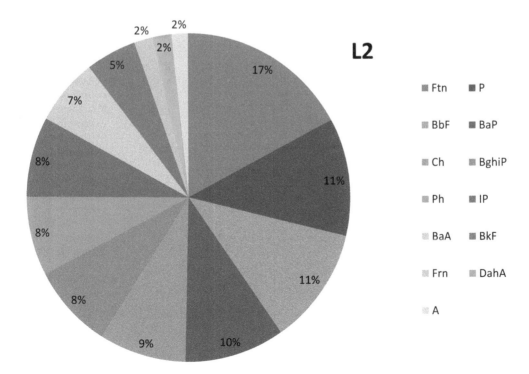

Figure 2. The composition of sewage sludge in L2.

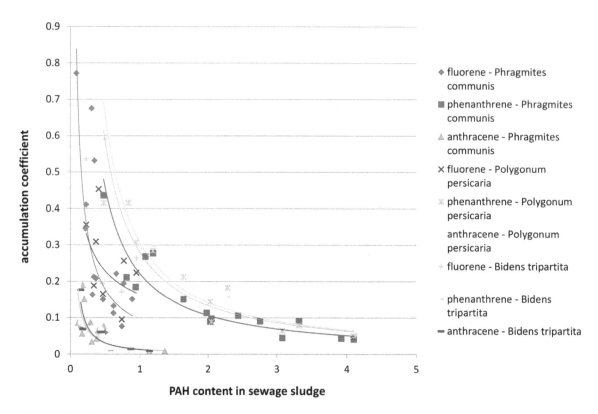

Figure 3. The profiles of dependence: accumulation coefficient of PAH in plant material vs. the content of PAH in sewage sludge for three-ring hydrocarbons.

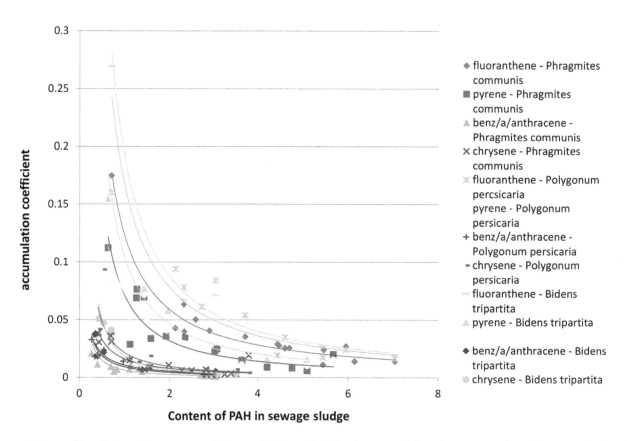

Figure 4. The profiles of dependence: accumulation coefficient of PAH in plant material vs. the content of PAH in sewage sludge for four-ring hydrocarbons.

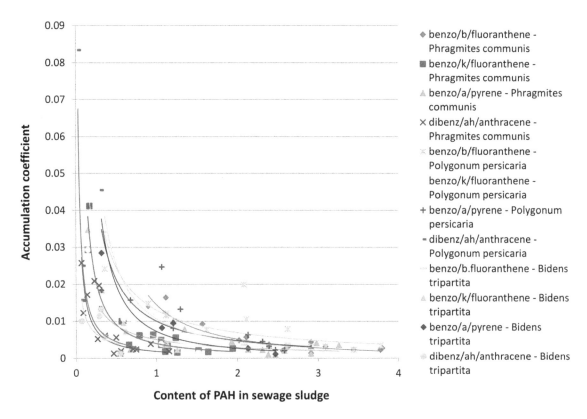

Figure 5. The profiles of dependence: accumulation coefficient of PAH in plant material vs. the content of PAH in sewage sludge for five-ring hydrocarbons.

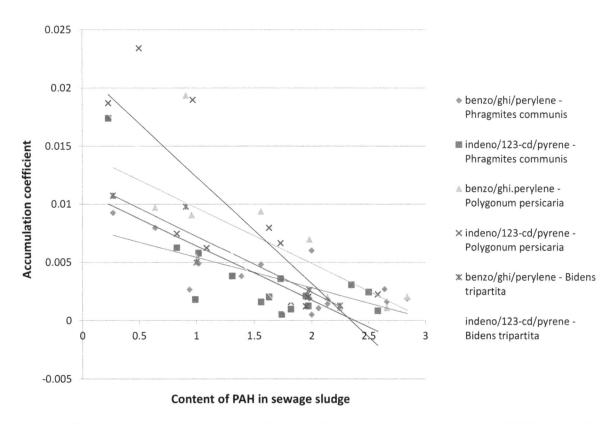

Figure 6. The profiles of dependence: accumulation coefficient of PAH in plant material vs. the content of PAH in sewage sludge for six-ring hydrocarbons.

Table 2. Results of regression analysis between accumulation coefficient of PAH in plant material (Y) vs. the content of PAH in sewage sludge (X).

	Phragmites communis (n=16)*				Polygonum persicaria (n=11)				Bidens tripartita (n=5)			
	a	b	R^2	P	a	b	R^2	P	a	b	R^2	P
three-ring hydrocarbons – parameters of regression model: $Y = aX^b$												
fluorene	0.095	−0.904	52.2	0.002	0.164	−0.478	22.2	0.239	0.153	−0.320	4.5	0.731
phenanthrene	0.221	−1.055	87.3	<0.001	0.280	−1.067	86.4	<0.001	0.307	−1.105	95.7	0.004
anthracene	0.014	−1.127	72.8	<0.001	0.013	−1.186	86.6	<0.001	0.013	−1.270	90.0	0.014
four-ring hydrocarbons – parameters of regression model: $Y = aX^b$												
fluoranthene	0.123	−1.061	86.9	<0.001	0.166	−1.097	81.1	<0.001	0.192	−1.129	97.8	0.001
pyrene	0.071	−1.166	78.5	<0.001	0.072	−0.969	72.2	<0.001	0.104	−1.166	97.5	0.002
benzo/a/anthracene	0.006	−1.244	80.5	<0.001	0.009	−1.036	89.3	<0.001	0.011	−0.953	89.4	0.015
chrysene	0.016	−1.028	69.2	<0.001	0.022	−1.167	88.4	<0.001	0.018	−1.472	97.9	0.001
five-ring hydrocarbons – parameters of regression model: $Y = aX^b$												
benzo/b/fluoranthene	0.015	−1.499	79.4	<0.001	0.013	−0.918	63.8	0.003	0.011	−1.250	95.53	0.004
benzo/k/fluoranthene	0.004	−1.195	82.9	<0.001	0.006	−1.142	75.1	0.001	0.004	−1.208	96.2	0.003
benzo/a/pyrene	0.005	−0.810	40.3	0.008	0.010	−1.089	69.9	0.001	0.007	−1.442	88.0	0.018
dibenz/ah/anthracene	0.002	−1.008	63.1	<0.001	0.002	−0.962	63.9	0.006	0.003	−0.679	30.4	0.335
six-ring hydrocarbons – parameters of regression model: $Y = a+bX$												
Benzo/ghi/perylene	0.008	−0.003	59.3	<0.001	0.014	−0.005	55.7	0.008	0.012	−0.005	86.7	0.021
indeno/123-cd/pyrene	0.011	−0.005	52.7	0.002	0.021	−0.009	69.3	0.002	0.020	−0.011	77.2	0.049

*- in some cases sample size was lower because of missing data.

(a-intercept, b- regression coefficient, R^2- coefficient of determination, P – observed significance level for F-test, value below 0.05 means significant correlation).

In spite of the differences in the character of the studied relationship noted for three-, four- and five-ring PAHs, they all exhibited a diminishing character.

Conclusions

The results of the present study carried out on three different plant species growing on two independent localizations demonstrated unequivocally that:

- a correlation exists between the content of PAHs in the plant and in sewage sludge;

- this correlation is of an exponential character for three-, four- and five-ring hydrocarbons and of a linear character for six-ring PAHs;

- the observed relationship does not depend on the plant species;

- the appearance of such correlation unequivocally indicated the existence of a path of PAHs uptake by plants from the soil bed;

- the accumulation coefficients calculated for three-ring aromatics were several times higher than for four-ring PAHs; moreover, the coefficient values calculated for five-ring PAHs were several times lower than for four-ring hydrocarbons; finally, the accumulation coefficient values of six-ring PAHs were the lowest in the series of studied polyaromatic hydrocarbons;

- the anthracene derived series of hydrocarbons exhibited lower accumulation coefficient values than other hydrocarbons of the same ring number;

- in contrast, the phenanthrene series was characterized by high accumulation coefficient values.

Author Contributions

Conceived and designed the experiments: BG KK. Performed the experiments: BG KK. Analyzed the data: BG KK MK. Contributed reagents/materials/analysis tools: BG KK. Wrote the paper: BG MK.

References

1. Garcia R, Diaz-Somoano M, Calvo M, Lopez-Anton MA, Suarez S, et al. (2012) Impact of the semi-industrial coke processing plant in the surrounding surface soil. Part II: PAH content. Fuel Process Tech 104: 245–252.
2. Ravindra K, Sokhi R, Grieken RV (2008) Atmospheric polycyclic aromatic hydrocarbons source attribution, emission factors and regulation. Atmos Environ 42: 2895–2921.
3. Wild SR, Jones KC (1995) Polynuclear aromatic hydrocarbons in the United Kingdom environment – A preliminary source inventory and budget. Environ Pollut 88: 91–108.
4. Lim L, Harrison R, Harrad S (1999) The contribution of traffic to atmospheric concentrations of polycyclic aromatic hydrocarbons. Environ Sci Technol 33(20): 3538–3542.
5. Benner B, Gordon G (1989) Mobile sources of atmospheric polycyclic aromatic hydrocarbons: a roadway tunnel study. Environ Sci Technol 23(10): 1269–1278.
6. Harvey PJ, Campanella BF, Castro PML, Harms H, Lichtfouse E, et al. (2002) Phytoremediation of polyaromatic hydrocarbons, anilines and phenols. Environ Sci Pollut Res 9(1): 29–47.
7. Weisman D, Alkio M, Colon-Carmona A (2010) Transcriptional responses to polycyclic aromatic hydrocarbon-induced stress in *Arabidopsis thaliana* reveal the involvement of hormone and defense signaling pathways. BMC Plant Biol 10: 59.
8. Van Metre PC, Mahler BJ (2003) The contribution of particles washed from rooftops to contaminant loading to urban streams. Chemosphere 52: 1727–1741.
9. Dutta K, Ghosh D, Nazmi A, Kumawat KL, Basu A (2010) A common carcinogen benzo[a]pyrene causes neuronal death in mouse via microglial activation. PLOS ONE 5 (4): 1–14.
10. Nielsen T (1996) Traffic contribution of polycyclic aromatic hydrocarbons in center of a large city. Atmos Environ 30: 3481–3490.
11. De Nicola F, Maisto G, Prati MV, Alfani A (2005) Temporal variations in PAH concentrations in *Quercus ilex* L (holm oak) leaves in an urban area. Chemosphere 61: 432–440.
12. Arruti A, Fernandez-Olmo I, Irabien A (2012) Evaluation of the urban/rural particle-bond PAH and PCB levels in the Northern Spain (Cantabria region). Environ Monit Assess 184: 6513–6526.
13. Lee BM, Shim GA (2007) Dietary exposure estimation of benzo[a]pyrene and cancer risk assessment. J Tox Environ Health A 70: 1391–1394.
14. Yoon E, Park K, Lee H, Yang JH, Lee C (2007) Estimation of excess cancer risk on time-weighted lifetime average daily intake of PAHs from food ingestion. Human Ecol Risk Assess 13: 669–680.
15. ATSDR, Priority List of Hazardous Substances, http://www.atsdr.cdc.gov/spl/
16. De Nicola F, Lancelotti C, Prati MV, Maisto G, Alfani A (2011) Biomonitoring of PAHs by using *Quercus ilex* leaves: Source diagnostic and toxicity assessment. Atmos Environ 45: 1428–1433.
17. De Souza Pereira M, Heitmann D, Reifenhäuser W, Ornellas Meire R, Silva Santos L, et al. (2007) Persistent organic pollutants in atmospheric deposition and biomonitoring with *Tillandsia usenoides* (L.) in an industrialized area in Rio de Janeiro state, southeast Brazil – Part II: PCB and PAH. Chemosphere 67: 1736–1745.
18. Lehndorff E, Schwark L. (2004) Biomonitoring of air quality in the Cologne Conurbation using pine needles as a passive sampler – Part II: polycyclic aromatic hydrocarbons (PAH). Atmos Environ 38: 3793–3808.
19. Orecchio S (2007) PAHs associated with the leaves of *Quercus ilex* L.: Extraction, GC-MS analysis, distribution and sources - Assessment of air quality in the Palermo (Italy) area. Atmos Environ 41: 8669–8680.
20. Yang Y, Woodward LA, Li QX, Wang J (2014) Concentrations, source and risk assessment of polycyclic aromatic hydrocarbons in soils from Midway Atoll, North Pacific Ocean. PLOS ONE 9 (1): 1–7.
21. Ashraf MW, Taqvi SIH, Solangi AR, Qureshi UA (2013) Distribution and risk assessment of polycyclic aromatic hydrocarbons in vegetables grown in Pakistan. J Chem: 1–5.
22. Meudec A, Dussauze J, Deslandes E, Poupart N (2006) Evidence for bioaccumulation of PAHs within internal shoot tissues by a halophytic plant artificially exposed to petroleum-polluted sediments. Chemosphere 65: 474–481.
23. Duarte-Davidson R, Jones KC (1996) Screening the environmental fate of organic contaminants in sewage sludge applied to agricultural soils: II. The potential for transfers to plants and grazing animals. Sci Total Environ 185: 59–70.
24. Gao Y, Zhu L (2004) Plant uptake, accumulation and translocation of phenanthrene and pyrene in soils. Chemosphere 55: 1169–1178.
25. Yang Z, Zhu L (2007) Performance of the partition-limited model on predicting ryegrass uptake of polycyclic aromatic hydrocarbons. Chemosphere 67: 402–409.
26. Zhu L, Gao Y (2004) Prediction of phenanthrene uptake by plants with a partition-limited model. Environ Pollut 131: 505–508.
27. Gao Y, Li H, Gong S (2012) Ascorbic acid enhances the accumulation of polycyclic aromatic hydrocarbons (PAHs) in roots of tall fescue (*Festuca arundinacea* Scherb.). PLOS ONE 7 (11):1–8.
28. Zhou B, Zhao B (2014) Analysis of intervention strategies for inhalation exposure to polycyclic aromatic hydrocarbons and associated lung cancer risk based on a Monte Carlo population exposure assessment model. PLOS ONE 9 (1): 1–11.
29. Wild SR, Waterhouse KS, McGrath SP, Jones KC (1990) Organic contaminants in an agricultural soil with known history of sewage sludge amendments: Polycyclic aromatic hydrocarbons. Environ Sci Technol 24: 1706–1711.
30. LaVole EJ, Amin S, Hecht SS, Furuya K, Hoffmann D (1982) Tumour initiating activity of dihydrodiols of benzo[b]fluoranthene, benzo[j]fluoranthene and benzo[k]fluoranthene. Carcinogenesis 3(1): 49–52.

6

Predicting Greenhouse Gas Emissions and Soil Carbon from Changing Pasture to an Energy Crop

Benjamin D. Duval[1,2¤a], Kristina J. Anderson-Teixeira[1¤b], Sarah C. Davis[1¤c], Cindy Keogh[3], Stephen P. Long[1,2,4], William J. Parton[3], Evan H. DeLucia[1,2,4]*

1 Energy Biosciences Institute, University of Illinois at Urbana-Champaign, Urbana, Illinois, United States of America, 2 Global Change Solutions, Urbana, Illinois, United States of America, 3 Natural Resource Ecology Laboratory, Fort Collins, Colorado, United States of America, 4 Department of Plant Biology, University of Illinois at Urbana-Champaign, Urbana, Illinois, United States of America

Abstract

Bioenergy related land use change would likely alter biogeochemical cycles and global greenhouse gas budgets. Energy cane (*Saccharum officinarum* L.) is a sugarcane variety and an emerging biofuel feedstock for cellulosic bio-ethanol production. It has potential for high yields and can be grown on marginal land, which minimizes competition with grain and vegetable production. The DayCent biogeochemical model was parameterized to infer potential yields of energy cane and how changing land from grazed pasture to energy cane would affect greenhouse gas (CO_2, CH_4 and N_2O) fluxes and soil C pools. The model was used to simulate energy cane production on two soil types in central Florida, nutrient poor Spodosols and organic Histosols. Energy cane was productive on both soil types (yielding 46–76 Mg dry mass$\cdot$ha^{-1}). Yields were maintained through three annual cropping cycles on Histosols but declined with each harvest on Spodosols. Overall, converting pasture to energy cane created a sink for GHGs on Spodosols and reduced the size of the GHG source on Histosols. This change was driven on both soil types by eliminating CH_4 emissions from cattle and by the large increase in C uptake by greater biomass production in energy cane relative to pasture. However, the change from pasture to energy cane caused Histosols to lose 4493 g CO_2 eq$\cdot$m^{-2} over 15 years of energy cane production. Cultivation of energy cane on former pasture on Spodosol soils in the southeast US has the potential for high biomass yield and the mitigation of GHG emissions.

Editor: Chenyu Du, University of Nottingham, United Kingdom

Funding: This research was funded by the Energy Bioscience Institute. The funders had no role in study design, data collection and analysis, decision to publish, or preparation of the manuscript.

Competing Interests: Please note that Duval, Long and DeLucia are affiliated with a company, Global Change Solutions LLC. This affiliation does not represent a "competing interest".

* E-mail: delucia@illinois.edu

¤a Current address: Dairy Forage Research Center, United States Department of Agriculture, Agricultural Research Service, Madison, Wisconsin, United States of America
¤b Current address: Conservation Ecology Center, Smithsonian Conservation Biology Institute, National Zoological Park, Front Royal, Virginia, United States of America
¤c Current address: Voinovich School for Leadership and Public Affairs, Ohio University, Athens, Ohio, United States of America

Introduction

Land use has a pervasive influence on atmospheric greenhouse gas (GHG) concentrations and thereby on climate [1,2,3]. Carbon emissions from land use change, often to make way for agriculture, have contributed substantially to anthropogenic increases in the atmospheric CO_2 concentration [2]. For example, C emissions from tropical deforestation have been estimated at 10.6 ± 1.8 Pg CO_2 per year between 1990 and 2007, equal to ~40% of global fossil fuel emissions [3]. Likewise, it is estimated that 40–52 Pg CO_2 have been released by plowing high-C native prairie soils [4]. Agricultural practices are important to global GHG budgets, with agroecosystems contributing ~14% of global anthropogenic GHG emissions [1]. Agricultural practices can also reduce GHG emissions and enhance soil carbon, and have the potential to mitigate climate change [4,5].

Land use and land management changes associated with the emerging bioenergy industry are likely to have substantial impacts on global GHG budgets [6,7,8]. A change from fossil fuels to an energy economy more reliant on plant-derived biofuels has the potential to reduce GHG emissions [9]. The prospect of lowering emissions is one factor leading to the United States' mandate to produce 136 billion liters of renewable fuel by 2022 [10]. However, meeting this mandate will require substantial land area [11,12], which implies potentially major changes to regional biogeochemical cycling [12,13].

Corn grain (*Zea mays*) is the dominant crop used for ethanol production in the US [14]. However, the ability of corn ethanol to reduce GHG emissions is questionable [15,16], and corn production exacerbates nitrogen pollution and other environmental problems [17–19]. Of particular concern is the possibility that diversion of corn for ethanol production will increase global grain prices and trigger agricultural expansion and deforestation elsewhere in the world [7]. The emerging commercial technology to convert ligno-cellulose to ethanol could redress the reliance on corn grain as an ethanol feedstock [20]. This could be particularly beneficial if cellulosic biofuel crops are grown on land that is not important for food production, while having lower GHG emissions than traditional row-crop agriculture [21,22]. Therefore, considerable research has focused on understanding the soil C and greenhouse gas consequences of replacing traditional agriculture used for bioenergy with perennial grasses like switchgrass (*Panicum virgatum* L.), Miscanthus (*Miscanthus x giganteus* J. M. Greef & Deuter ex Hodk. & Renvoize), or restored prairie cropping systems in the Midwestern United States [17,23–27].

The Southeastern United States holds particular potential for cultivation of second-generation biofuel crops [28,29]. In comparison with the corn-soy and wheat belts of the Midwestern US, this region's longer growing season, high precipitation and relatively lower land costs make it attractive for biofuel crop production. However, far less is known about the biogeochemical consequences of land-use change to biofuel crop production in this region.

Energy cane, a promising crop for ligno-cellulosic fuel production, is a variety of sugarcane (*Saccharum officinarum*) that is higher yielding, more cold tolerant and has lower sucrose content than commercially produced sugarcane [28]. Because of its lower sugar concentration, it has not been widely cultivated, but has been of interest commercially as a genetic stock for improving cold tolerance in higher sucrose sugarcane strains [28]. With the development of ligno-cellulosic ethanol conversion technologies, sucrose concentration is less important for ethanol production, and energy cane could become an important biofuel feedstock as yields are high, ranging from 25–74 $Mg \cdot ha^{-1} \cdot yr^{-1}$ dry mass (Table 1).

Florida is the largest sugarcane producing state in the US and is therefore a likely location for large-scale energy cane production [30]. Currently, 466,000 hectares of land in Florida are used for low-intensity grazing, and converting some portion of this land could provide an option for growing energy cane [31,32]. However, it is unknown if converting pasture to cultivated land will affect GHG exchange with the atmosphere and soil carbon storage. More frequent soil disturbance and the presence of larger quantities of litter from growing energy cane could increase CO_2 efflux to the atmosphere [33,34], while removing cattle from the landscape will displace methane (CH_4) efflux [35]. If fields are fertilized, nitrous oxide (N_2O) emissions may increase because of greater substrate availability for denitrifying microbes [36], and

indeed, high rates of N_2O efflux have been measured from sugarcane grown on highly fertilized soils in Australia [37]. However, considering the entire suite of greenhouse gasses, there may be an overall reduction in GHG flux due to the offset provided by greater atmospheric carbon uptake into the crop.

The region of Florida where energy cane is likely to be grown has two distinct soil types. The most common soils are Spodosols, which are low nutrient and low organic matter sands requiring significant fertilizer to maintain agricultural productivity [38]. Substantial sugarcane production in Florida also occurs on Histosols, which are high organic matter "mucks" that are not typically fertilized, as production on these soils can be maintained by N mineralization from organic matter [39]. The cultivation of Histosols began by draining swamplands, where organic matter had accumulated under anaerobic conditions. Drainage accelerates decomposition and further cultivation of these organic soils is associated with rapid oxidation of organic matter, resulting in significant soil C loss and emissions of CO_2 and N_2O to the atmosphere [9,40,41].

Theoretical [13,25] and empirical research [42,43] indicate that the conversion of land in the rain-fed Midwest currently used to produce corn for ethanol to perennial biofuel feedstocks such as switchgrass or Miscanthus (a close relative of sugarcane) would greatly reduce or reverse the emission of GHG to the atmosphere and rebuild depleted carbon stocks in the soil. Prior studies with Miscanthus in Europe have measured substantial decreases in nitrogen use, and large increases in soil biomass and organic matter relative to other agricultural land uses [44,45]. There have been no experimental studies that address how changing a landscape to cultivate energy cane will impact GHG emissions and soil C stocks. This is addressed here by using the process-based biogeochemical model DayCent to run *in silico* experiments to ask how land use change from pasture to energy cane production changes ecosystem GHG flux and soil C storage. We test the hypotheses that converting pastures to energy cane will lead to reductions in GHG flux to the atmosphere and increase soil C stocks, and that soil type is an important modulator of that change.

Methods

Plant and Soil Analyses

To parameterize the DayCent model, plants and soils were collected on private land in Highlands County, Florida (27° 21′ 49″ N, 81° 14′ 56″ W) in May 2011. Paired 4-m² plots (n = 3) were randomly located in energy cane fields that had been recently (<2 months) converted from pasture and in adjacent non-cultivated pasture on both Spodosols (hyperthermic Arenic Alaquods) and Histosols (hyperthermic Histic Glossaqualfs). We harvested all aboveground biomass from each plot. Soil samples were taken from the pastures in areas not yet under energy cane cultivation. Three soil cores to a depth of 1 m were extracted from each plot with a 1.75-cm diameter wet sampling tube (JMC product # PN010, Newton, IA). Soil cores were separated by depth (0–30 cm, >30 cm). Plant material and soils were oven dried at 65°C (plant material) and 105°C (soils) until they reached constant mass. Dried soils were coarse ground with a mill (model F-4, Quaker City, Phoenixville, PA), and then fine ground with a coffee grinder (Sunbeam Products Inc., Boca Raton, FL). Total C and N content may have been slightly underestimated from the dried Histosols due to volatilization, but the values we measured (Table 1) fall well within the range reported by NRCS Web Soil Survey [46]. Plant material was ground to pass a 425-µm mesh (Wiley mill, Thomas Scientific, Swedesboro, NJ, USA). Plant and soil subsamples

Table 1. Input parameters (mean and one standard error of the mean; SEM) for carbon and nitrogen concentration of energy cane and soils collected from the Highlands Ethanol farm, Highlands County, Florida.

| | | %C | | %N | |
		Mean	SEM	Mean	SEM
Energy cane	Live leaves	43.68	0.22	1.80	0.18
	Dead leaves	39.77	0.22	0.52	0.03
	Stalks	41.18	0.33	0.87	0.10
Soils	Soil Depth				
Histosols	0–60 cm	7.77	2.48	0.50	0.20
	60–100 cm	7.77	2.48	0.50	0.20
Spodosols	0–30 cm	0.77	0.17	0.04	<0.01
	30–60 cm	0.36	0.03	0.02	0.01
	60–100 cm	0.36	0.03	0.02	0.01

When site-specific data were not available, plant information was used from reference [65], and soil data were collected from the NRCS Web Soil Survey (http://websoilsurvey.nrcs.usda.gov/).

Table 2. Site information for studies used in DAYCENT model validation.

Site	Lit. Yield	Model Yield	Max. Temp.	Min. Temp.	Precipitation	Latitude	Longitude	Reference
Auburn, AL	26.1	25.4	24.2	9.8	1160	32.67	−85.44	Woodard and Prine, 1993
Belle Glade, FL	25.0	28.3	27.8	16.4	1378	26.68	−80.67	Korndorfer, 2009
EREC, FL	51.3	43.5	29.1	17.7	1181	26.65	−80.63	Gilbert et al., 2006
Gainesville, FL	35.6	27.3	27.0	13.7	1123	29.68	−82.27	Woodard and Prine, 1993
Hendry, FL	39.2	55.9	28.4	18.3	1362	27.78	−82.15	USDA, 2011
Hillsboro, FL	60.7	62.5	28.5	18.3	1547	27.90	−82.49	Gilbert et al., 2006
Houma, LA (1st ratoon)	36.6	38.2	25.2	14.8	500	29.57	−90.65	Legendre and Burner, 1995
Houma, LA (2nd ratoon)	34.9	37.6	25.2	14.8	500	29.57	−90.65	Legendre and Burner, 1995
Hundley, FL	73.5	62.0	28.5	18.1	1457	26.30	−80.16	Gilbert et al., 2006
Jay, FL (plant cane)	35.8	33.6	26.9	16.4	1321	28.65	−80.82	Woodard and Prine, 1993
Jay, FL (1st ratoon)	27.8	32.8	26.9	16.4	1321	28.65	−80.82	Woodard and Prine, 1993
Lakeview, FL	71.3	62.5	28.5	17.4	1275	26.30	−80.15	Gilbert et al., 2006
Hidalgo, TX	34.6	42.5	28.7	18.1	576	26.17	−97.93	Weidenfeld, 1995
Ona, FL (1st ratoon)	40.5	38.1	28.6	16.0	1160	27.48	−81.92	Woodard and Prine, 1993
Ona, FL (2nd ratoon)	30.2	31.6	28.6	16.0	1160	27.48	−81.92	Woodard and Prine, 1993
Pahokee, FL	60.5	65.5	28.4	17.5	1269	26.82	−80.66	Glaz and Ulloa, 1993
Palm Beach, FL	32.3	35.4	29.0	16.6	851	26.67	−80.15	USDA, 2011
Quincy, FL (1st ratoon)	26.3	25.1	25.8	12.9	1445	30.59	−84.58	Woodard and Prine, 1993
Quincy, FL (2nd ratoon)	27.8	21.7	25.8	12.9	1445	30.59	−84.58	Woodard and Prine, 1993
Shorter, AL	26.4	25.4	24.9	10.9	1119	32.40	−85.94	Sladden et al., 1991
Sundance, FL	42.1	43.5	28.6	17.5	1303	26.60	−80.87	Gilbert et al., 2006

Yield values from the literature and modeled yields for energy cane and sugarcane represent total aboveground biomass expressed as Mg ha^{-1} on a dry mass basis. Climate variables include mean annual maximum and minimum temperature (°C) and mean annual precipitation (mm).

within each plot were combined, and C and N concentrations were measured for depth-stratified soil samples (Table 1) and total above ground biomass with a flash combustion chromatographic separation elemental analyzer (Costech 4010 CHNSO Analyzer, Costech Analytical Technologies Inc. Valencia, CA). The instrument was calibrated with acetanilide obtained from Costech Analytical Technologies, Inc. Other physical soil attributes, including texture, bulk density and water holding capacity were obtained from the NRCS Web Soil Survey [46] for Highlands County, Florida.

The DayCent Model

The DayCent model [47,48] was developed to simulate ecosystem dynamics for agricultural, forest, grassland and savanna ecosystems [49–51]. The model is a daily time step version of the Century model [52,53], using the same soil carbon and nutrient cycling submodels to simulate soil organic matter dynamics (C and N) and nitrogen mineralization. DayCent uses more mechanistic submodels than Century to simulate daily plant production, plant nutrient uptake, trace gas fluxes (N_2O, CH_4), NO_3 leaching, and soil water and temperature [48,54–58].

The DayCent soil organic matter model is widely used to simulate the impacts of management practices on soil carbon dynamics and nutrient cycling. Specifically, the soil organic matter submodel has been used to simulate the impacts of soil tillage practices; no-tillage, minimum tillage and conventional tillage [59,60], crop rotations [59], and biofuel crops; woody biomass, switchgrass (*Panicum virgatum*), Miscanthus (*Miscanthus* X *giganteus*), and sugarcane [13,61] on soil carbon dynamics for agricultural systems. These studies test model performance against observed

data and demonstrate general success in simulating changes in soil carbon levels associated with management practices.

The soil trace gas submodel has been extensively tested using observed soil CH_4 and N_2O data sets from agricultural and natural ecosystems, and once parameterized with plant production data, provides accurate predictions of trace gas fluxes. Specifically, DayCent has successfully simulated the observed impacts of N fertilizer additions and cropping systems [50,58,59] on soil N_2O and CH_4 fluxes. The model results and observed data sets demonstrate that increasing N fertilizer levels increases soil N_2O fluxes and that soil N_2O fluxes are much lower for perennial crops as compared to annual crops.

The DayCent model has been used extensively to simulate grassland and crop yields [50,58,59,62], and to evaluate the environmental impacts of growing crops. Adler et al. [63] used the DayCent model to simulate net greenhouse gas fluxes (soil C status and soil CH_4 and N_2O fluxes) associated with the use of corn, soybeans, alfalfa, hybrid popular, reed canary grass and switchgrass for biofuel energy production in Pennsylvania. Davis et al. [25] used the DayCent model to simulate the environmental impacts of growing switchgrass and Miscanthus in Illinois and compared simulated plant production for switchgrass and Miscanthus with observed yield data. The authors also compared the net soil greenhouse gas fluxes (soil C changes and soil CH_4 and N_2O fluxes) associated with growing switchgrass and Miscanthus and growing corn and soybeans. Davis et al. [13] recently used the DayCent model to simulate the environmental impact of replacing the corn currently grown for ethanol production in the Corn Belt with perennial grasses (Miscanthus and switchgrass) for second-generation biofuel production. The authors found that the

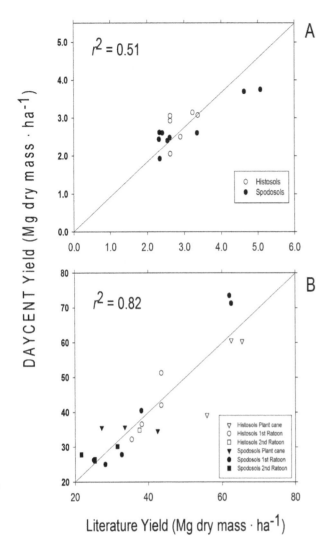

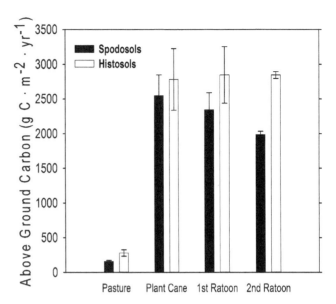

Figure 2. Modeled above ground production of grazed pasture and energy cane in Highlands County, Florida. Values are mean above ground carbon (g C·m^{-2}·yr^{-1}, ± SD) for 15 years in pasture, and for 5 × 3-year ratoon cycles in energy cane (each bar represents the average of 5 values, one for each year for each stage in the planting cycle).

Figure 1. Regression analysis used in DayCent model validation. Model output for dry mass yield was compared to literature values for A) pasture yield values from the USDA-NASS database, B) sugarcane and energy cane dry mass yield. Data points are compared to a 1:1 line.

DayCent model successfully predicted corn, Miscanthus, and switchgrass biomass production for U.S. sites with multiple N fertilizer levels. They also showed that the DayCent model successfully simulated observed annual soil N$_2$O fluxes from corn and switchgrass grown with multiple N fertilizer levels and showed that soil N$_2$O fluxes are much lower for fertilized switchgrass than for corn.

Furthermore, the basis for the DayCent model, Century, has been used to simulate sugarcane production in Brazil [61,64] and Australia [65]. These authors show that the Century/DayCent soil organic matter sub-model can correctly simulate the impacts of fertilizer, and organic matter additions on soil carbon levels and surface litter decay.

Model Parameterization

Energy cane is a variety of sugarcane, and thus parameterizing DayCent for this crop required only minor changes to the previously published input data used for sugarcane [61,65]. Energy cane differs from sugarcane in that it has increased cold

tolerance, decreased sucrose content, and higher cellulose content. We adjusted parameters based on direct measurement of energy cane tissue traits described above. The principal changes from sugarcane to energy cane were reducing the minimum C:N ratio of leaves from 28.6 to 22.1 and changing the C:N of stems from 160 to 30.5. Because of this change in C:N, the parameter for C allocation to stems in DayCent also was modified (from 60% to 40%), to reflect the lower C content of stems relative to N for energy cane versus the previously modeled sugarcane parameters [65]. Bahiagrass (*Paspalum notatum* Flueggé) pasture was simulated using the existing DayCent model parameters for warm season grasses [66,67].

Histosols are challenging to model as organic matter and C content typically are uniform throughout the soil profile [68], but they also are known to subside because of oxidation of the highly labile organic C pools characteristic of these soils [69]. This subsidence was calculated from the modeled rate of organic matter loss and bulk density. DayCent simulates soil C flux to a depth of 30 cm [47], so as soil was lost with subsidence new soil and organic matter became part of this upper 30 cm column from below. This assumes that loss only occurred in the upper 30 cm, which is reasonable since this is the disturbed and aerated part of the soil. The C and N added from low in the soil profile was calculated from the rate of subsidence and the measured elemental contents and bulk density of the soil that was below 30 cm, when sampled, which is at time zero in our model. However, model output calculates GHG and soil C to soil depths to 30 cm.

Model Validation

Literature values of aboveground production (dry mass) for grazed pasture, sugarcane and energy cane (Table 2) were used to validate DayCent. Validation focused on aboveground biomass production because this variable has been measured widely across a range of sites. While there were insufficient data on trace gas flux or changes in soil C in sugarcane or energy cane for validation of

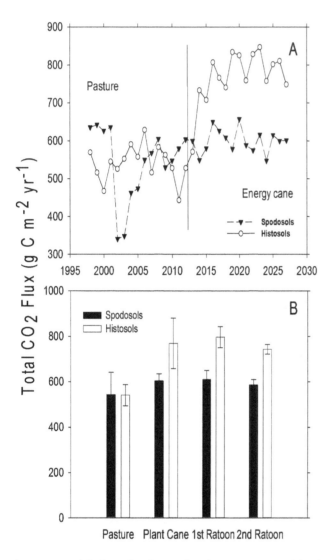

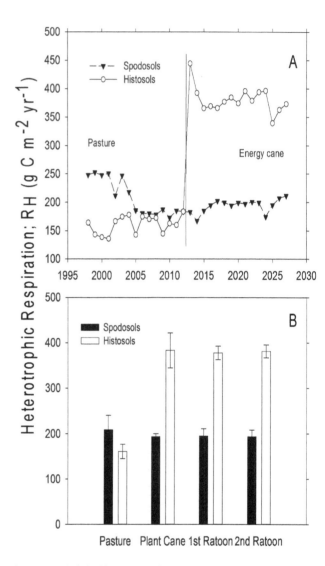

Figure 3. Modeled total soil CO₂ flux from pasture and land converted to energy cane in Highlands County, Florida. A) Total annual soil CO₂ flux (expressed as g C·m⁻²). Dashed line represents year of land use conversion from pasture to energy cane. B) Mean total soil CO₂ flux (g C·m⁻²·yr⁻¹, ± SD) for 15 years in pasture, and for 5, 3-year ratoon cycles in energy cane (each bar represents the average of 5 values, one for each year for each stage in the planting cycle).

Figure 4. Modeled heterotrophic respiration (R_H) from pasture and land converted to energy cane in Highlands County, Florida. A) Total annual heterotrophic respiration (g C·m⁻²). Dashed line represents year of land use conversion from pasture to energy cane. B) Mean heterotrophic respiration (g C·m⁻²·yr⁻¹, ± SD) for 15 years in pasture, and for 5, 3-year ratoon cycles in energy cane (each bar represents the average of 5 values, one for each year for each stage in the planting cycle).

these variables, validation based on productivity for other crops reliably predicts trace gas flux [59,60,63,70–72].

We compiled a literature database of 17 sites that had reliable data on sugarcane and energy cane yield. There were also pasture productivity data for 15 of those sites [73]. In some instances we were able to contact researchers directly to access unpublished data (Table 2). The geographic range of sites represents the breadth of sugarcane production in the continental United States, and the potential range of energy cane production on currently grazed pastures. For all sites, daily weather data inputs (minimum and maximum temperature, daily precipitation) from 1980 to 2002 were obtained from the DayMet database [74]. The model was run using the DayCent growing degree-day subroutine to determine plant emergence, senescence and death, based on plant phenological characters and daily weather data. Soil data for the validation sites were obtained from the NRCS Web Soil Survey [75]. Using the same schedule of management events used for the

in silico experiments (described below), DayCent was run with site-specific soil and weather data for each sites. The fit of modeled to measured above ground dry mass production (Mg dry matter ha⁻¹) of our simulations of grazed pasture and energy cane were separately tested via linear regression, using the linear model function in R [76].

Initial Simulation Conditions

A "spin-up" period in DayCent based on historical land use and vegetation type was used to set initial soil conditions. The dominant, historic vegetation type for this area of south-central Florida was savanna, with a mixture of grasses and several species of scrub-oak, or sawgrass for the swamp areas [77]. A mix of perennial C₃ grasses species and symbiotic N₂ fixing plants, were used as initial conditions for the savanna simulation (initial vegetation type "savanna" in DayCent). A period of 2000 years was simulated to obtain an initial soil C and N conditions prior to

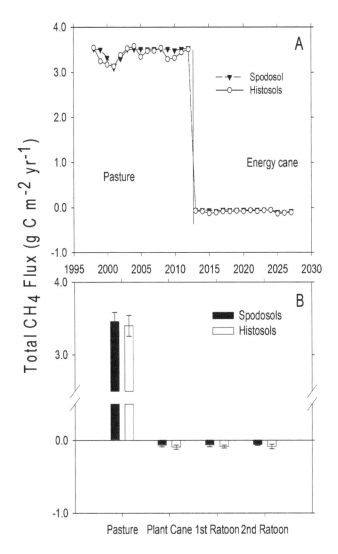

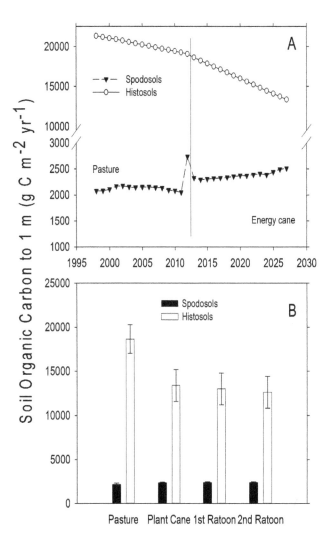

Figure 5. Modeled CH₄ flux from pasture and land converted to energy cane in Highlands County, Florida. A) Total annual CH₄ flux (g C·m⁻²). The solid vertical line represents year of land use conversion from pasture to energy cane, positive values indicate CH₄ efflux and negative values indicate CH₄ uptake. B) Mean CH₄ flux (g C·m⁻²·yr⁻¹, ± SD) for 15 years in pasture, and for 5, 3-year ratoon cycles in energy cane (each bar represents the average of 5 values, one for each year for each stage in the planting cycle).

Figure 6. Changes in total soil organic C from pasture and land converted to energy cane in Highlands County, Florida. A) Total annual SOC flux (g C·m⁻²). The solid vertical line represents year of land use conversion from pasture to energy cane. B) Mean SOC flux (g C·m⁻²·yr⁻¹, ± SEM) for 15 years in pasture, and for 5, 3-year ratoon cycles in energy cane (each bar represents the average of 5 values, one for each year for each stage in the planting cycle).

our *in silico* experiments. The model was run for spin ups and all subsequent experiments using the growing degree-day sub-routine.

In silico Experiments

Model simulations were then run to determine the GHG soil-atmosphere exchange and change in soil C predicted for conversion of pasture to energy cane on the two dominant soil types, Spodosols and Histosols. We used daily weather data inputs (minimum and maximum temperature, daily precipitation) from 1951 to 2002, which was the longest time period available for Highlands County, Florida obtained from the DayMet database [74]. This weather file is used by DayCent to create a mean and standard deviation of weather parameters, thus the more weather data available for a given site, the more accurately the variability of a site will be captured by the model.

To initiate the experimental simulations, in 1998 we converted the savanna by removing all above ground biomass and plowing to a depth of 30 cm. A landscape conversion to a grazed Bahiagrass (*Paspalum notatum* Flueggé) ecosystem was then simulated. Bahia-grass is a common forage grass for this part of Florida that would be considered "improved pasture", although usually not fertilized or irrigated [78]. We simulated grazing in our modeling experiment by annually removing 10% of live shoot and 1.0% of standing dead shoots. Prior to planting energy cane, another plow event to 30 cm was initiated to remove the pasture vegetation and simulate the physical land use change.

The simulated cycle of energy cane planting and harvest was based on the sugarcane literature [65,79,80] and discussions with University of Florida and USDA sugarcane agronomists [81,82]. In the simulations, energy cane was planted in January of the first year (2013), followed by a two-year ratoon (crop regenerated from remaining biomass) from which 80% of the above ground biomass was harvested in December. At the end of the second ratoon, the crop was removed and the land plowed before planting a new

Table 3. Modeled ecosystem carbon, nitrogen and greenhouse gas fluxes after converting pasture to energy cane on nutrient poor Spodosols and organic matter rich Histosols.

	Spodosols			Histosols		
	Pasture	Energy cane	Δ	Pasture	Energy cane	Δ
SOC (g C·m^{-2})	2736	2513	−224	16087	10373	−5715
Nitrogen Mineralization (g N·m^{-2})	134	203	69	216	293	77
Heterotrophic Respiration (g C·m^{-2})	3130	2913	−218	2413	5715	3302
Total Soil CO$_2$ Efflux (g C·m^{-2})	8148	8993	845	8111	11540	3429
CH$_4$ (g CO$_2$eq·m^{-2})	2980	−33	−3013	2958	−46	−3004
N$_2$O (g CO$_2$eq·m^{-2})	214	649	435	6713	1742	−4970
Total System C Flux (g CO$_2$eq·m^{-2})	−1159	−2812	−1653	−1367	924	2291
Total Greenhouse Gas Flux (g CO$_2$eq·m^{-2})	2035	−2196	−4231	8304	2620	−5684

Greenhouse gas and N mineralization values are the sum of values from pasture 15 years prior to conversion to energy cane and the sum values for 15 years following the conversion to energy cane. Positive values indicate a flux to the atmosphere and negative values indicate uptake from the atmosphere by the ecosystem. Soil organic matter values are the differences between the last year of energy cane production and the last year of pasture. Total GHG values are the sums of CH$_4$, N$_2$O and total system C flux (calculated in DayCent as the difference between all C uptake and storage versus efflux from respiration) expressed as CO$_2$e. Differences (Δ) represent the values for energy cane minus pasture.

plant crop. This cycle of ratooning and planting was repeated in the simulation for fifteen years following conversion from pasture; i.e. five cycles of three years each. This three-year planting cycle is typical for sugarcane production in Florida [83,84].

Irrigation events were scheduled every month throughout the dry season, and every two months during the rainy season to maintain soil water at field capacity. Fertilizer $(NH_4^+ - NO_3^-)$ was applied in mid February and mid June of each year of the simulation, at a rate of 102 kg N· ha^{-1} per fertilization event for Spodosols. No fertilizer was added to the organic rich Histosols. This fertilization schedule was based on studies that suggest that a split fertilization regime at this rate maximizes sugarcane yield, and that fertilizing above this level does not increase yield but increases N$_2$O efflux [38,65,85]. The input files used to drive DayCent (e.g. schedule files, plant input parameters, and soil input files) are available online [86].

Calculations and Statistical Analyses

We summed daily GHG and soil C fluxes from DayCent to calculate yearly fluxes and report those in g C or N·m^{-2} yr^{-1}, with the exception of total GHG values which are reported as CO$_2$ equivalents [87] and factored by warming potential (CO$_2$ = 1, CH$_4$ = 23, N$_2$O = 296; ref. 85). Total ecosystem C flux was calculated as the annual change in total ecosystem C storage between the beginning and end of a year and represents the net ecosystem carbon balance expressed in CO$_2$eq [88,89].

Because the model experiments were performed using the same site with the same weather data, but controlled for soil type, the simulations had the structure of a paired design where each year was a replicate [90]. We therefore used paired t-tests to determine differences between soil types within a plant type (n = 15) and between plant types within a soil type (n = 15). The variation reported with mean annual values represents inter-annual variation in the predicted variables. Heteroscedasticity was examined with the Fligner-Killeen test, and output data distributions, which did not meet variance assumptions, were compared with the Wilcoxon rank-sum test. The routines t.test (paired = TRUE) and wilcox.test were performed using R [76,90].

Because of the large number of pair-wise comparisons of our model results, the False Discovery Rate (FDR) test was used to account for multiple comparisons. The FDR test is less conser-

vative than a P-value adjustment such as the Bonferroni correction, and determines the probability of a Type I error. We calculated a FDR of 0.024 for our matrix of tests, and therefore justified the use of multiple paired t-tests without P-value adjustment [91].

Results

Predicted harvested yields for both pasture and energy cane in our validation sites agreed well with measured values from the literature (Pasture: $r^2 = 0.52$, Energy cane: $r^2 = 0.82$, Figure 1A & 1B), indicating that our modeled predictions provided a good representation of the productivity that drives the biogeochemical dynamics of DayCent.

For our modeled site, DayCent estimated a large increase in aboveground plant biomass production after conversion of pasture to energy cane (Figure 2); annual aboveground biomass production increased by a factor of 14 on Spodosols and by a factor of 10 on Histosols, relative to pasture. Energy cane production ranged from 1911–3153 g C m^{-2} yr^{-1} (46–76 Mg dry biomass·ha^{-1}). Predicted energy cane production remained high through the three harvests on Histosols, but declined through the modeled ratoon cycle on Spodosols (Figure 2).

There was considerable temporal variation in predicted soil CO$_2$ efflux from pasture in the 15 years simulated prior to the conversion to energy cane (Figure 3a). This variation was particularly evident for pasture on Spodosols and was driven primarily by variation in precipitation. Total soil CO$_2$ efflux was similar for pasture on both soil types, but significantly increased when averaged over 15 years after conversion to energy cane on the Histosols (Figure 3a; t = 10.65, d.f. = 14, $P<0.001$). Land use conversion did not increase CO$_2$ efflux on Spodosols (t = 0.58, d.f. = 14, P = 0.57). Following conversion to energy cane CO$_2$ efflux from Histosols was significantly higher than energy cane on Spodosols (Figure 3b; t = 9.56, d.f. = 14, $P<0.001$).

The conversion of land from pasture to energy cane had no significant effect on the predicted heterotrophic component of soil respiration (R_H) on Spodosols (Figure 4a), but caused a large increase in R_H from the Histosols (Figure 4a; t = 31.86, d.f. = 14, $P<0.001$) and resulted in higher R_H on Histosols than Spodosols following the conversion to energy cane (Figure 4b; t = 23.68, d.f.

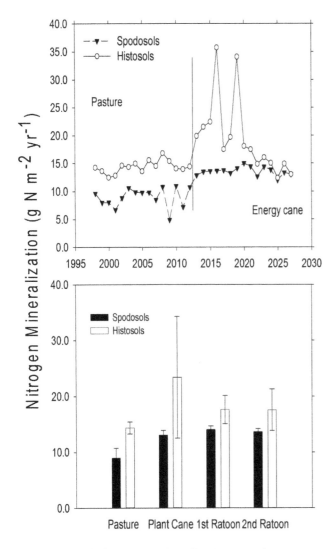

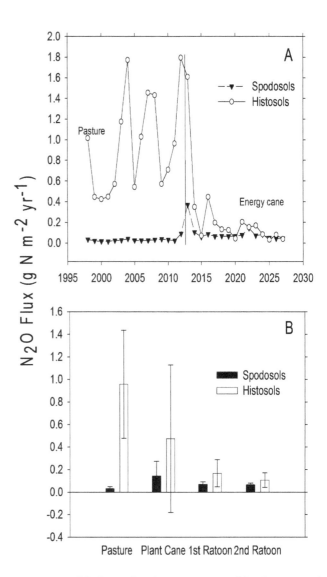

Figure 7. Modeled nitrogen mineralization rates from pasture and land converted to energy cane in Highlands County, Florida. A) Total annual N mineralization rate (g $N \cdot m^{-2}$). The solid vertical line represents year of land use conversion from pasture to energy cane. B) Mean N mineralization rate (g $N \cdot m^{-2} \cdot yr^{-1}$, ± SD) for 15 years in pasture, and for 5, 3-year ratoon cycles in energy cane (each bar represents the average of 5 values, one for each year for each stage in the planting cycle).

Figure 8. Modeled N_2O flux from pasture and land converted to energy cane in Highlands County, Florida. A) Total annual N_2O flux (g $N \cdot m^{-2}$). The solid vertical line represents year of land use conversion from pasture to energy cane, positive values indicate N_2O efflux and negative values indicate N_2O uptake. B) Mean N_2O flux (g $N \cdot m^{-2} \cdot yr^{-1}$, ± SD) for 15 years in pasture, and for 5, 3-year ratoon cycles in energy cane (each bar represents the average of 5 values, one for each year for each stage in the planting cycle).

= 14, $P<0.001$). Prior to the conversion to energy cane, modeled (R_H) was slightly higher in pasture on Spododols than on Histosols (Figure 4; $t = 31.86$, d.f. = 14, $P<0.001$).

On both soil types, the removal of cattle associated with the conversion of pasture to energy cane caused a substantial change in predicted CH_4 flux ($t = 185$, d.f. = 14, $P<0.001$ on Spodosols; $t = 167$, d.f. = 14, $P<0.001$ on Histosols; Figure 5a). Without cattle, pastures were a small CH_4 sink (0.16–0.60 g $C \cdot m^{-2} \cdot yr^{-1}$ uptake in Spodosols, 15 year sum = 112 g $CO_2eq \cdot m^{-2}$, 0.12–0.57 $gC \cdot m^{-2} \cdot yr^{-1}$ uptake for Histosols, 15 year sum = 135 g $CO_2eq \cdot m^{-2}$). Introducing cattle at stocking rates and grazing intensity typical for this region (1 head $cattle \cdot ha^{-1}$: ref. 31), caused pasture on both soil types to be a substantial source of CH_4 to the atmosphere (Figure 5).

Changes in vegetation and management practices altered soil organic carbon (SOC), and these changes were particularly evident on the Histosols (Table 2; Figure 6). Histosols had a larger pool of active C (weekly to monthly turnover) than Spodosols under both pasture and energy cane (pasture, $t = 19.25$, d.f. = 14, $P<0.001$; energy cane, $t = 14.21$, d.f. = 14, $P<0.001$). Comparing the remaining total SOC pools between the end of pasture and the last year of the energy cane simulation, Histosols lost a large amount of soil organic C; 5714 g $C \cdot m^{-2}$ to 1 m depth (Figure 6; $t = 296$, d.f. = 14, $P<0.001$), compared to the SOC loss from Spodosols of 224 g $C \cdot m^{-2}$ to 1 m (Table 3).

Nitrogen mineralization increased after pasture was converted to energy cane on both the fertilized Spodosols ($t = 9.02$, d.f. = 14, $P<0.001$) and the non-fertilized Histosols ($t = 2.72$, d.f. = 14, $P = 0.02$). After conversion to energy cane, Histosols had higher rates of N mineralization than Spodosols (Figure 7; $t = 3.43$, d.f. = 14, $P = 0.004$), and this increase in available N likely accounted for the continued high yields on Histosols.

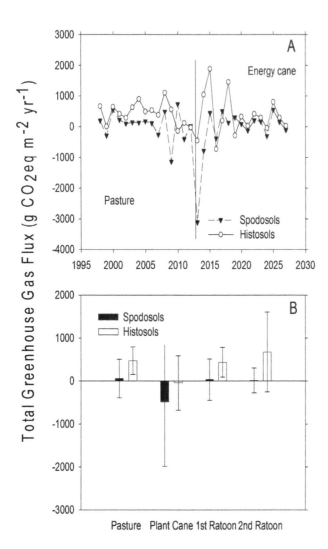

Figure 9. Changes in total greenhouse gas (GHG) from pasture and land converted to energy cane in Highlands County, Florida. Positive values indicate GHG efflux and negative values indicate GHG uptake. A) Total annual GHG flux, reported as CO_2 equivalents converted to account for differences in warming potential (g $CO_2e \cdot m^{-2}$). The solid vertical line represents year of land use conversion from pasture to energy cane. B) Mean greenhouse gas flux in CO_2 equivalents converted to account for differences in warming potential (g $CO_2e \cdot m^{-2} \cdot yr^{-1}$, ± SD) for 15 years in pasture, and for 5, 3-year ratoon cycles in energy cane (each bar represents the average of 5 values, one for each year for each stage in the planting cycle).

Prior to conversion to energy cane, N_2O efflux was higher in pastures on Histosols compared to Spodosols (Figure 8; Wilcoxon rank sum, W = 225, $P<0.001$). After conversion to energy cane, Histosols remained greater sources of N_2O than Spodosols ($t = 12.15$, d.f. = 14, $P<0.001$). Conversion of pasture to energy cane decreased N_2O efflux on Histosols (Figure 8a; $t = 4.30$, d.f. = 14, $P<0.001$), but increased N_2O efflux on Spodosols (Figure 8b; $t = 2.87$, d.f. = 14, $P = 0.01$). It is likely that N_2O emission from Histosols decreased following conversion because the increase in productivity resulted in a higher uptake of nitrate that would otherwise be available for denitrification.

Total GHG exchange (global warming potential) was calculated by converting the fluxes of CH_4 and N_2O to CO_2 equivalents based on their warming potential relative to CO_2 [87] and

summing these with total system C flux (Table 2). Variation in weather caused substantial inter-annual variation in total GHG flux, with both pasture and energy cane varying between net GHG sinks and sources (Figure 9); no significant differences in annual GHG flux were resolved on either soil type (Spodosols: $t = 1.15$, d.f. = 14, $P = 0.27$; Histosols: $t = 0.13$, d.f. = 14, $P = 0.90$). When the cumulative GHG emission were calculated for the fifteen years prior to conversion, pasture was a net source to GHGs to the atmosphere on both soil types, and pasture was a stronger source on Histosols (8304 $gCO_2eq \cdot m^{-2}$) than on Spodosols (2035 $gCO_2eq \cdot m^2$; Table 3). Conversion of pasture to energy cane caused the Spodosols to transition from a source to a sink for GHGs and reduced the flux of GHGs to the atmosphere on Histosols. On both soil types, the reduction GHG emission to the atmosphere was associated with a large decrease in CH_4 emissions caused by the elimination of cattle grazing. On the Histosols, the reduction in N_2O emissions to the atmosphere also contributed to reduced emission of GHGs. This analysis of GHG emissions and their corresponding global warming potentials did not account for the displacement of fossil fuel emissions by the biofuel product.

Discussion

Parameterization of the DayCent model for energy cane, an emerging bioenergy crop, successfully simulated biomass production across the southeast United States (Figure 1). Our simulations suggested high yields for energy cane on former pastureland in a subtropical climate when Spodosols are highly fertilized (200 kg $N \cdot ha^{-1} \cdot yr^{-1}$), and when microbial activity in Histosols leads to high rates of N mineralization (rates were 44% higher on Histosols). When integrated over 15 years (Table 3), conversion of pasture to energy cane on Spodosols converted a net source of GHG (due to cattle CH_4 emissions) to a sink driven by the removal of cattle and the increase in C uptake by energy cane. While Histosols were a net GHG source under both pasture and energy cane, the source was reduced by the land use conversion (Table 3). The GHG improvement resulting from this conversion from pasture to energy cane would be even greater if fossil fuel displacement by cellulosic ethanol had been included.

The range of our simulated energy cane yields was 46–76 Mg ha^{-1} dry mass per year on fertilized Spodosols and unfertilized Histosols. Using published values for the conversion efficiency for the production of cellulusoic ethanol [22], a hypothetical energy cane farm of 10,000 ha could therefore produce between 142–236 million liters of ethanol [92]. In comparison, equal areas of land devoted to corn grain and Miscanthus in the Midwest would yield between 25 and 73% this amount of ethanol, respectively, assuming the maximum yields reported by other authors [11,22].

Typically, sugarcane yield declines with ratooning, the repeated harvests of aboveground material generated by vegetative growth [93]. The model reproduced the yield decline for energy cane on Spodosols but not on Histosols, but the model in its current configuration probably failed to capture the mechanisms that would normally cause a decline in yield. Various factors ranging from increases in nematode populations and ratoon stunting disease, to mechanical compaction of the soil have been implicated in ratoon decline [94,95], and these were not accounted for in the model. Although sugarcane in Florida typically is grown for three years and three annual harvests before re-planting, if it were grown for more years between re-planting, we would expect a continuing yield decline on the Spodosols. In contrast, continued mineralization of organic matter on Histosols may sustain high yields beyond the 3-year period simulated in the model. On both soils the GHG benefits would be improved with longer ratoon cycles because of less soil disturbance due to decreased frequency of soil disturbance for replanting.

Organic matter (OM) content of soils is important for sustaining high yields of sugarcane, in part because OM mineralization provides the labile N necessary to sustain plant growth. Spodosols had much less OM than Histosols (Appendix I; Figure 6; [65,96]). The high CO_2 efflux rates from Histosols (Figures 3–4) and the patterns of SOC loss following land use change (Figure 6) correspond to higher rates of OM mineralization. The associated higher rates of N mineralization (Table 3; Figure 7) on Histosols provided additional N to energy cane and improved crop yield (Figure 2). Although energy cane on Spodosols was fertilized to offset the low N content of these soils, rates of nitrification (the process by which NH_4^+ is converted into the highly mobile NO_3^- anion) were higher on these soils. The fertilizer applied to energy cane crops on Spodosols in the simulations was NH_4^+ - NO_3^-, a labile substrate for nitrification [66]. Spodosols had consistently higher nitrification rates than Histosols, and therefore higher NO_3^- content because of fertilization, and it is possible that some fraction of fertilizer was lost before plant uptake [97]. We hypothesize that a combination of NO_3^- leaching from fertilizer before plant uptake, lower initial N content, and lower mineralization rates may have created a stronger N limitation to yield on Spodosols but not on Histosols.

Before land use change, pasture on both soil types was a net source of GHGs to the atmosphere (Table 3). This is consistent with both direct measurements [98] and modeling efforts [99] that have found grazed pastures to be net sources of GHGs, but this is also a function of grass species present and animal stocking density [100]. The model estimated that pastures were sinks for CO_2, with total C uptake of 1159 g CO_2 m^{-2} and 1367 g CO_2 m^{-2} over 15 years on Spodosols and Histosols, respectively (Table 3). In the absence of cattle, both soil types were CH_4 sinks (112 and 135 g CO_2eq, respectively), but including reasonable estimates of CH_4 efflux from cattle (Figure 5) and N_2O efflux from soils (Figure 8) resulted in net GHG emission to the atmosphere on both pasture soils (Table 3). Following conversion to energy cane, the production of N_2O on Spodosols increased (Figure 8) within the range of N_2O flux rates previously reported for Australian sugarcane fertilized at similar rates to this study [37]. The increase in N_2O was offset by uptake of CO_2 and the change from a source to a sink for CH_4 (Figure 5), with the net effect that Spodosols became a net GHG sink (Table 3). Indeed, over 15 years energy cane on Spodosol was a GHG sink of >40 Mg CO_2eq per hectare (Table 3). On Histosols, eliminating grazing following the conversion of pasture to energy cane caused a similar decrease in CH_4 efflux to the atmosphere (Figure 5) and this land use change also reduced N_2O emissions (Figure 8; Table 3). However, following land conversion this system switched from a net CO_2 sink to a source, and this change in total system C prevented energy cane on Histosols from becoming a net sink for GHGs. The driver for GHG production on Histosols was higher R_H, and significant losses of soil organic matter [69] that resulted in total C efflux from these soils (Table 3).

The model successfully simulated energy cane biomass production across a range of sites across the southern United States (Figure 1). Previous studies have shown that DayCent reliably predicts soil biogeochemistry and GHG exchange when parameterized for net primary production [51], suggesting that the estimates of GHG flux and soil C dynamics were reasonable. Eddy-flux measurements of GHG exchange that are now being initiated at this site will provide an independent test of the predictions of GHG effects of conversion made here.

Indirect land use change (ILUC) – the stimulation of deforestation or increased agriculture in other parts of the world driven by diversion of current agricultural land to bioenergy production – potentially poses an environmental risk of bioenergy production [7,101]. Growing energy cane on land converted from low stocking density pasture would be unlikely to trigger significant increases in food price or ILUC in the way that large-scale shifts from corn or soy production in the Midwestern United States would motivate greater production of those crops elsewhere [13]. Indeed, the recommended stocking density for Bahiagrass pasture in this region is ~1 animal·ha^{-1} [31], and cattle and calf operations in Florida account for less than 6% of the state's annual agricultural revenue [102]. The loss in meat production could be redressed with minimal increases in current stocking rates, and would be unlikely to trigger the type of large-scale landscape changes that may occur through the diversion of midwestern agricultural land [7]. However, displacing cattle for energy cane production may potentially increase methane emissions elsewhere, which would negate the local benefit of reduced methane flux to the atmosphere.

The environmental impacts of changing land use from pasture to energy cane were highly dependent on the soil type. Whereas the cultivation of Histosols results in high CO_2 efflux and the reduction of soil carbon (Figures 3, 4, and 6), the model predicted that energy cane crops on Spodosols would act as a net C and GHG sink (Figure 6, Table 3). From both a biofuel and biogeochemical perspective, these results suggest that energy cane grown on nutrient poor soils, as opposed to organic soils, has the potential to be a high-yielding bio-ethanol feedstock that creates a GHG sink in the Southeastern United States.

Acknowledgments

We thank Michael Masters for lab analysis of plant and soil C and N. Lykes Brothers Inc. graciously provided access to energy cane plantations and pastures, which greatly helped in parameterizing our model. We also thank Dr. Robert Gilbert, Dr. Barry Glaz, and Mr. Pedro Korndorfer for sharing their knowledge on sugar and energy cane agronomy and data that aided our modeling effort.

Author Contributions

Conceived and designed the experiments: BDD KJAT WJP EHD. Performed the experiments: BDD SCD CK WJP. Analyzed the data: BDD SCD KJAT WJP SPL EHD. Contributed reagents/materials/analysis tools: CK WJP. Wrote the paper: BDD KJAT WJP SPL EHD.

References

1. IPCC (2007) Climate Change 2007: The Physical Science Basis. Contribution of Working Group I to the Fourth Assessment Report of the Intergovernmental Panel on Climate Change. Solomon S, Qin D, Manning M, Chen Z, Marquis M et al. Eds. Cambridge, UK, Cambridge, UK.

2. Le Quéré C, Raupach MR, Canadell JG, Marland G, Bopp L, et al. (2009) Trends in the sources and sinks of carbon dioxide. Nat Geosci 2: 831–836.

3. Pan Y, Birdsey RA, Fang J, Houghton R, Kauppi PE, et al. (2011) A large and persistent carbon sink in the world's forests. Science 333: 988–993.

4. Lal R (2004) Soil carbon sequestration impacts on global climate change and food security. Science 304: 1623–1627. doi: 10.1126/science.1097396.

5. Tilman D (1998) The greening of the green revolution. Nature 396: 211–212.

6. Fargione JE, Hill JD, Tilman D, Polasky S, Hawthorne P (2008) Land clearing and the biofuel carbon debt. Science 319: 1235–1238.

7. Searchinger T, Heimlich R, Houghton RA, Dong F, Elobeid A, et al. (2008) Use of U.S. croplands for biofuels increases greenhouse gases through emissions from land-use change. Science 319: 1238–1240.

8. Melillo JM, Reilly JM, Kicklighter DW, Gurgel AC, Cronin TW (2009) Indirect emissions from biofuels: How important? Science 326: 1397–1399.

9. Börjesson P (2009) Good or bad bioethanol from a greenhouse gas perspective – What determines this? Applied Energy 86: 589–594.

10. United States Congress (2007) The Energy Independence and Security Act of 2007 (H.R. 6). Available: http://frwebgate.access.gpo.gov/cgibin/getdoc.cgi?dbname=110_cong_bills&docid=f: h6enr.txt.pdf. Accessed 2010 Dec 19.

11. Heaton EA, Dohleman FG, Long SP (2008) Meeting US biofuel goals with less land: the potential of Miscanthus. Glob Change Biol 14: 2000–2014.

12. Fargione JE, Plevin RJ, Hill JD (2010) The ecological impact of biofuels. Ann Rev Ecol Evol Syst 41: 351–377.

13. Davis SC, Parton WJ, Del Grosso SJ, Keough C, Marx E, et al. (2012) Impacts of second-generation biofuel agriculture on greenhouse gas emissions in the corn-growing regions of the US. Front Ecol Environ 10: 69–74. doi:10.1890/110003.

14. Dien BS, Bothast RJ, Nichols NN, Cotta MA (2002) The U.S. corn ethanol industry: an overview of current technology and future prospects. Int Sugar J 103: 204–211.

15. Davis SC, Anderson-Teixeira KJ, DeLucia EH (2009) Life-cycle analysis and the ecology of biofuels. Trends Plant Sci 14: 140–146.

16. O'Hare M, Plevin RJ, Martin JI, Jones AD, Kendall A, et al. (2009) Proper accounting for time increases crop-based biofuels' greenhouse gas deficit versus petroleum. Environ Res Lett 4: doi:10.1088/1748-9326/4/2/024001.

17. Donner SD, Kucharik CJ (2008) Corn-based ethanol production compromises goal of reducing nitrogen export by the Mississippi River. Proc Natl Acad Sci USA 105: 4513–4518.

18. Hill J, Polasky S, Nelson E, Tilman D, Huo H (2009) Climate change and health costs of air emissions from biofuels and gasoline. Proc Nat Acad Sci USA 106: 2077–2082.

19. Smeets EMW, Bouwman LF, Stehfest E, van Vuuren DP, Posthuma A (2009) Contribution of N_2O to the greenhouse gas balance of first-generation biofuels. Glob Change Biol 15: 1–23.

20. Solomon BD, Barnes JR, Halvorsen KE (2007) Grain and cellulosic ethanol: history, economics, and energy policy. Biomass Bioenergy 31: 416–425.

21. Tilman D, Socolow R, Foley JA, Hill J, Larson E (2009) Beneficial biofuels-the food, energy and environment trilemma. Science 325: 270–271.

22. Somerville C, Youngs H, Taylor C, Davis SC, Long SP (2010) Feedstocks for lignocellulosic biofuels. Science 329: 790–792.

23. Tilman D, Hill J, Lehman C (2006) Carbon-negative biofuels from low-input high-diversity grassland biomass. Science 314: 1598–1600.

24. Anderson-Teixeira KJ, Davis SC, Masters MD, DeLucia EH (2009) Changes in soil organic carbon under biofuel crops. Glob Change Biol Bioenergy 1: 75–96.

25. Davis SC, Parton WJ, Dohleman FG, Smith CM, Del Grosso S, et al. (2010) Comparative biogeochemical cycles of bioenergy crops reveal nitrogen fixation and low greenhouse gas emissions in a Miscanthus x giganteus agro-ecosystem. Ecosystems 13: 144–156.

26. Robertson GP, Hamilton SK, Del Grosso SJ, Parton WP (2011) The biogeochemistry of bioenergy landscapes: carbon, nitrogen, and water considerations. Ecol App 21: 1055–1067. doi: 10.1890/09-0456.1.

27. Zeri M, Anderson-Teixeira KJ, Masters MD, Hickman G, DeLucia EH, et al. (2011) Carbon exchange by establishing biofuel crops in central Illinois. Agric Ecosyst Environ 144: 319–329.

28. Sladden SE, Bransby DI, Aiken GE, Prine GM (1991) Biomass yield and composition, and winter survival of tall grasses in Alabama. Biomass Bioenergy 1: 123–127.

29. Mark T, Darby P, Salassi M (2009) Energy cane usage for cellulosic ethanol: estimation of feedstock costs. Southern Agricultural Economics Association Annual Meeting, Atlanta, Georgia, January 31-February 3, 2009.

30. Baucum LE, Rice RW (2009) An overview of Florida sugarcane. University of Florida IFAS Extension document SS-AGR-232.

31. Hersom M (2005) Pasture stocking density and the relationship to animal performance. Animal Science Department, Florida Cooperative Extension Service, Institute of Food and Agricultural Sciences, University of Florida, Document number AN155.

32. Steiner J (2012) personal communication.

33. Bowden RD, Nadelhoffer KJ, Boone RD, Melillo JM, Garrison JB (1993) Contributions of aboveground litter, belowground litter, and root respiration to soil respiration in a temperate mixed hardwood forest. Can J For Res 23: 1402–1407.

34. Paustian K, Six J, Elliott ET, Hunt HW (2000) Management options for reducing CO_2 emissions from agricultural soils. Biogeochemistry 48: 147–163.

35. DeRamus HA, Clement TC, Giampola DD, Dickison PC (2003) Methane emissions of beef cattle on forages: efficiency of grazing management systems. J Environ Qual 32: 269–277.

36. Mosier A, Kroeze C, Nevison C, Oenema O, Seitzinger S, et al. (1998) Closing the global N_2O budget: nitrous oxide emissions through the agricultural nitrogen cycle. Nutr Cycl Agroecosys 52: 225–248.

37. Thorburn PJ, Biggs JS, Collins K, Probert ME (2010) Nitrous oxide emissions from Australian sugarcane production systems – are they greater than from other cropping systems? Agric Ecosyst Environ 136: 343–350.

38. Rice RW, Gilbert RA, Lentini RS (2002) Nutritional requirements for Florida sugarcane. Florida Cooperative Extension Service. UF/IFAS, Document SS-ARG-228. University of Florida Institute of Food Agricultural Science.

39. Morgan KT, McCray JM, Rice RW, Gilbert RA, Baucum LE (2009) Review of current sugarcane fertilizer recommendations: a report from the UF/IFAS sugarcane fertilizer standards task force. Document SL 295, Soil and Water

40. Morris DR, Gilbert RA, Reicosky DC, Gesch RW (2004) Oxidation potentials of soil organic matter in Histosols under different tillage methods. Soil Sci Soc Am J 68: 817–826.

41. Stehfest E, Bouwman LF (2006) N_2O and NO emission from agricultural fields and soils under natural vegetation: summarizing available measurement data and modeling of global annual emissions. Nutr Cycl Agroecosys 74: 207–228.

42. Anderson-Teixeira KJ, Masters MD, Black CK, Zeri M, Hussain MZ, et al. (2012) Altered belowground carbon cycling following land use change to perennial bioenergy crops. Ecosystems, in press.

43. Smith CM, David MB, Mitchell CA, Masters MD, Anderson-Teixeira KJ, et al. (2013) Reduced nitrogen losses following conversion of row crop agriculture to perennial biofuel crops. J Env Qual 42: 219–228, doi: 10.2134/jeq2012.0210.

44. Beale CV, Long SP (1997) Seasonal dynamics of nutrient accumulation and partitioning in the perennial C-4-grasses Miscanthus x giganteus and Spartina cynosuroides. Biomass Bioenergy 12: 419–428.

45. Hansen EM, Christensen BT, Jensen LS, Kristensen K (2004) Carbon sequestration in soil beneath long-term Miscanthus plantations as determined by ^{13}C abundance. Biomass Bioenergy 26: 97–105.

46. NRCS Web Soil Survey website. Available: http://websoilsurvey.nrcs.usda.gov/app/HomePage.htm. Accessed 2010 Oct 7.

47. Parton WJ. Hartman MD, Ojima DS, Schimel DS (1998) DAYCENT and its land surface submodel: description and testing. Glob Planet Change 19: 35–48.

48. Parton WJ, Holland EA, Del Grosso SJ, Hartmann MD, Martin RE, et al. (2001) Generalized model for NO_x and N_2O emissions from soils. J Geophys Res-Atmos 106: 17403–17420.

49. DayCent: Daily Century Model website. Available: http://www.nrel.colostate.edu/projects/daycent/. Accessed 2010 Jun 4.

50. Del Grosso SJ, Halvorson AD, Parton WJ (2008a) Testing DayCent model simulations of corn yields and nitrous oxide emissions in irrigated tillage systems in Colorado. J Environ Qual 37: 1383–1389, doi:10.2134/jeq2007.0292.

51. Parton WJ, Hanson PJ, Swanston C, Torn M, Trumbore SE, et al. (2010) ForCent model development and testing using the Enriched Background Isotope Study experiment. J Geophys Res 115: G04001.

52. Parton WJ, Schimel DS, Cole CV, Ojima DS (1987) Analysis of factors controlling soil organic levels of grasslands in the Great Plains. Soil Sci Soc Am J 51: 1173–1179.

53. Parton WJ, Ojima DS, Cole CV, Schimel DS (1994) A general model for soil organic matter dynamics: sensitivity to litter chemistry, texture and management, R.B. Bryant,R.W. Arnoldm, Editors, Quantitative Modeling of Soil Forming Processes, Soil Science Society of America, Madison, WI. Pp. 147–167.

54. Eitzinger J, Parton WJ, Hartman M (2000) Improvement and validation of a daily soil temperature submodel for freezing/thawing periods. Soil Sci 165: 525–534.

55. David MB, Del Grosso SJ, Hu X, McIsaac GF, Parton WJ, et al. (2009) Modeling denitrification in a tile-drained, corn and soybean agroecosystem of Illinois, USA. Biogeochemistry 93: 7–30.

56. Del Grosso SJ, Parton WJ, Mosier AR, Ojima DS, Hartmann MD (2000a) Interaction of soil carbon sequestration and N2O flux with different land use practices. In: van Ham J, Baede APM, Meyer LA, Ybema R (eds.), Non-CO_2 Greenhouse Gases: Scientific Understanding, Control and Implementation. Kluwer Academic Publishers, The Netherlands. 303–311.

57. Del Grosso SJ, Parton WJ, Mosier AR, Ojima DS, Kulmala AE, et al. (2000b) General model for N_2O and N_2 gas emissions from soils due to denitrification. Global Biogeochem Cycles 14: 1045–1060.

58. Del Grosso SJ, Mosier AR, Parton WJ, Ojima DS (2005) DayCent model analysis of past and contemporary soil N_2O and net greenhouse gas flux for major crops in the USA. Soil Tillage Res 83: 9–24.

59. Del Grosso SJ, Ojima DS, Parton WJ, Mosier AR, Peterson GA, et al. (2002) Simulated effects of dryland cropping intensification on soil organic matter and GHG exchanges using the DAYCENT ecosystem model. Environ Pollution 116, S75–S83.

60. Del Grosso SJ, Parton WJ, Ojima DS, Keough CA, Riley TH, et al. (2008b) DAYCENT simulated effects of land use and climate on county level N loss vectors in the USA. Pages 571–595 in: R.F. Follett, and J.L. Hatfield (eds.) Nitrogen in the Environment: Sources, Problems, and Management, 2nd ed. Elsevier Science Publishers, The Netherlands.

61. Galdos MV, Cerri CC, Cerri CEP, Paustian K, Van Antwerpen R (2010) Simulation of sugarcane residue decomposition and aboveground growth, Plant Soil 326: 243–259. DOI 10.1007/s11104–009–004–3.

62. Hartmann MD, Merchant EK, Parton WJ, Gutmann MP, Lutz SM, et al. (2011) Impact of historical land use changes in the U.S. Great Plains, 1883 to 2003. Ecol App 21: 1105–1119.

63. Adler PR, Del Grosso SJ, Parton WJ (2007) Life-Cycle assessment of net greenhouse-gas flux for bioenergy cropping systems. Ecol Appl 17: 675–691.

64. Galdos MV, Cerri CC, Cerri CEP, Paustian K, Van Antwerpen R (2009) Simulation of soil carbon dynamics under sugarcane with the CENTURY Model, Soil Sci Soc Am J 73: 802–811.

65. Vallis I, Parton WJ, Keating BA, Wood AW (1996) Simulation of the effects of trash and N fertilizer management on soil organic matter levels and yields of sugarcane. Soil Tillage Res 38: 115–132.

66. Pepper DA, Del Grosso SJ, McMurtrie RE, Parton WJ (2005) Simulated carbon sink response of shortgrass steppe, tallgrass prairie and forest ecosystems to rising [CO_2], temperature and nitrogen input, Global Biogeochem Cycles 19: GB1004 doi:10.1029/2004GB002226.

67. Kelly RH, Parton WJ, Hartman MD, Stretch LK, Ojima DS, et al. (2000), Intra-annual and interannual variability of ecosystem processes in shortgrass steppe, J Geophys Res 105(D15) 20093–20100 doi:10.1029/2000JD900259.

68. Brady NC, Weil RR (2002) The nature and properties of soils. Prentice Hall, Upper Saddle River, New Jersey. 960 p.

69. Morris DR, Gilbert RA (2005) Inventory, crop use, and soil subsidence of Histosols in Florida. J Food Agr Environ 3: 190–193.

70. Newman Y, Vendramini J, Blount A (2010) Bahiagrass (*Paspalum notatum*): overview and management. University of Florida IFAS Extension. Publication #SS-AGR-332. http://edis.ifas.ufl.edu/ag342. Accessed 2011 Aug 7.

71. Valentine DW, Holland EA, Schimel DS (1994) Ecosystem and physiological controls over methane production in northern wetlands. J Geophys Res 99: 1563–1571.

72. Del Grosso SJ, Parton WJ, Mosier AR, Walsh MK, Ojima DS, et al. (2006) DayCent national scale simulations of N_2O emissions from cropped soils in the USA. J Environ Qual 35: 1451–1460.

73. US Department of Agriculture, National Agricultural Statistics Service website. Available: http://www.nass.usda.gov/Quick_Stats/. Accessed 2010 Feb 25.

74. DAYMET United States Data Center-A source for daily surface weather data and climatological summaries website. Available: www.daymet.org. Accessed 2010 Jun 1.

75. US Department of Agriculture Natural Resources Conservation Service, Web Soil Survey website. Available: http://websoilsurvey.sc.egov.usda.gov/app/HomePage.htm. Accessed 2010 Dec 12.

76. R Development Core Team (2007) R: A language and environment for statistical computing. R Foundation for Statistical Computing, Vienna, Austria.

77. Barbour MG, Billings WD (1988) North American terrestrial vegetation. Press Syndicate of the University of Cambridge. Melbourne, Australia.

78. Pitman WD, Portier KM, Chambliss CG, Kretschmer AE (1992) Performance of yearling steers grazing bahia grass pastures with summer annual legumes or nitrogen fertilizer in subtropical Florida. Trop Grasslands 26: 206–211.

79. Glaz B, Ulloa MF (1993) Sugarcane yields from plant and ratoon sources of seed cane. J Am Soc Sugar Cane Tech 13: 7–13.

80. Wiedenfeld RP, Encisco J (2008) Sugarcane responses to irrigation and nitrogen in semiarid south Texas. Agron J 100: 665–671.

81. Gilbert RA, Shine JM, Miller JD, Rice RW, Rainbolt CR (2006) The effect of genotype, environment and time of harvest on sugarcane yields in Florida, USA. Field Crop Res 95: 156–170.

82. Glaz B (2012) Personal communication.

83. Glaz B, Morris DR (2010) Sugarcane Responses to water-table depth and periodic flood. Agron J 102: 372–380.

84. US Environmental Protection Agency (2011) Florida Sugarcane Metadata. Environmental Protection Agency, Washington DC. Available: http://www.epa.gov/oppefed1/models/water/met_fl_sugarcane.htm. Accessed 2011 August 1.

85. Muchovej RM, Newman PR (2004) Nitrogen fertilization of sugarcane on a sandy soil: II soil and groundwater analysis. J Am Soc Sugar Cane Tech 24: 225–240.

86. University of Illinois, DeLucia Laboratory Public Data Archive website. Available: http://www.life.illinois.edu/delucia/Public%20Data%20Archive/. Accessed 2013 Jun 4.

87. Department of Energy and Climate Change (DECC) and the Department for Environment, Food and Rural Affairs website. Available: https://www.gov.uk/government/publications/2012-guidelines-to-defra-decc-s-ghg-conversion-factors-for-company-reporting-methodology-paper-for-emission-factors. Accessed: 2013 Jul 23.

88. Forster P, Ramaswamy V, Artaxo P, Berntsen T, Betts R, et al. (2007) Climate Change 2007: The Physical Science Basis. Contribution of Working Group I to the Fourth Assessment Report of the Intergovernmental Panel on Climate Change. Cambridge University Press, Cambridge, United Kingdom and New York, NY, USA.

89. Chapin FS, Woodwell G, Randerson J, Rastetter E, Lovett G, et al. (2006) Reconciling carbon-cycle concepts, terminology, and methods. Ecosystems 9: 1041–1050.

90. Crawley MJ (2007) The R Book. John Wiley and Sons, West Sussex, England. 942 p.

91. Storey JD (2003) The positive false discovery rate: a Bayesian interpretation and the q-value. Ann Stat 31: 2013–2035.

92. Graham-Rowe D (2011) Agriculture: beyond food versus fuel. Nature 474: S6–S8. doi: 10.1038/474S06a.

93. Ball-Coelho B, Sampaio EVSB, Tiessen H, Stewart JWB (1992) Root dynamics in plant and ratoon crops of sugar cane. Plant Soil 142: 297–305.

94. Hoy JW, Grisham MP, Damann KE (1999) Spread and increase of ratoon stunting disease of sugarcane and comparison of disease detection methods. Plant Disease 83: 1170–1175.

95. Stirling GR, Blair BL, Pattemore JA, Garside AL, Bell MJ (2001) Changes in nematode populations on sugarcane following fallow, fumigation and crop rotation, and implications for the role of nematodes in yield decline. Australas Plant Pathol 30: 323–335.

96. Yadav RL, Prasad SR (1992) Conserving the organic matter content of the soil to sustain sugarcane yield. Exp Ag 28: 57–62.

97. Chapin FS, Matson PA, Mooney HA (2002) Principals of Terrestrial Ecosystem Ecology. Springer Science, New York, New York, USA.

98. Rowlings D, Grace P, Kiese R, Scheer C (2010) Quantifying N_2O and CO_2 emissions from a subtropical pasture. 19th World Congress of Soil Science, Soil Solutions for a changing World. 1–6 August 2010, Brisbane, Australia.

99. Howden SM, White DH, Mckeon GM, Scanlan JC, Carter JO (1994) Methods for exploring management options to reduce greenhouse gas emissions from tropical grazing systems. Clim Change 27: 49–70.

100. Liebig MA, Gross JR, Kronberg SL, Phillips RL (2010) Grazing management contributions to net global warming potential: A long-term evaluation in the Northern Great Plains. J Environ Qual 39: 799–809.

101. Plevin RJ, O'Hare M, Jones AD, Torn MS, Gibbs HK (2010) Greenhouse gas emissions from biofuels: Indirect land use change are uncertain but may be much greater than previously estimated. Env Sci Tech 44: 8015–8021.

102. Florida Department of Agriculture and Consumer Services website. Available: http://www.fl-ag.com/agfacts.htm. Accessed 2011 Oct 1.

7

Combining XCO$_2$ Measurements Derived from SCIAMACHY and GOSAT for Potentially Generating Global CO$_2$ Maps with High Spatiotemporal Resolution

Tianxing Wang*, Jiancheng Shi, Yingying Jing, Tianjie Zhao, Dabin Ji, Chuan Xiong

State Key Laboratory of Remote Sensing Science, Institute of Remote Sensing and Digital Earth, Chinese Academy of Sciences. Beijing, China

Abstract

Global warming induced by atmospheric CO$_2$ has attracted increasing attention of researchers all over the world. Although space-based technology provides the ability to map atmospheric CO$_2$ globally, the number of valid CO$_2$ measurements is generally limited for certain instruments owing to the presence of clouds, which in turn constrain the studies of global CO$_2$ sources and sinks. Thus, it is a potentially promising work to combine the currently available CO$_2$ measurements. In this study, a strategy for fusing SCIAMACHY and GOSAT CO$_2$ measurements is proposed by fully considering the CO$_2$ global bias, averaging kernel, and spatiotemporal variations as well as the CO$_2$ retrieval errors. Based on this method, a global CO$_2$ map with certain UTC time can also be generated by employing the pattern of the CO$_2$ daily cycle reflected by Carbon Tracker (CT) data. The results reveal that relative to GOSAT, the global spatial coverage of the combined CO$_2$ map increased by 41.3% and 47.7% on a daily and monthly scale, respectively, and even higher when compared with that relative to SCIAMACHY. The findings in this paper prove the effectiveness of the combination method in supporting the generation of global full-coverage XCO$_2$ maps with higher temporal and spatial sampling by jointly using these two space-based XCO$_2$ datasets.

Editor: Juan A. Añel, University of Oxford, United Kingdom

Funding: The work described in this paper has been jointly supported by project of "Climate Change: Carbon Budget and Related Issues" (Grant nr. XDA05040402) from Chinese Academy of Sciences (CAS), the CAS/SAFEA International Partnership Program for Creative Research Teams (Grant nr. KZZD-EW-TZ-09) and National Natural Science foundation of China (Grant nr. 41301177). The funders had no role in study design, data collection and analysis, decision to publish, or preparation of the manuscript.

Competing Interests: The authors have declared that no competing interests exist.

* Email: wangtx@radi.ac.cn

Introduction

In recent years, global warming caused by emission of CO$_2$ has attracted considerable attention from the public. During the past decade, although tremendous efforts have been made toward improving the understandings of the mechanism between CO$_2$ increase in the atmosphere and global warming, some uncertainties still exist in the spatiotemporal characteristics of CO$_2$ sinks/sources on regional and global scales due to the lack of high-density measurements of such variables with good accuracy [1,2]. To date, the estimates of CO$_2$ flux from inverse methods rely mainly on ground-based measurements [3,4]. Although providing highly accurate atmospheric CO$_2$ records, the traditional ground-based networks intrinsically suffer from sparse spatial coverage [2,5]. Satellite-based measurements with various spatial and temporal resolutions provide a unique opportunity to accurately map atmospheric CO$_2$ in both daytime and nighttime over large areas, thus having the potential to bridge this gap. As a result, various satellite-based platforms have been equipped in recent years for deriving the CO$_2$ concentrations.

Generally, methods for retrieving CO$_2$ from space can be grouped into two categories: (1) inferring CO$_2$ concentrations by measuring shortwave infrared (SWIR) reflected solar radiation around 1.6 and 2.0 μm with sufficient spectral resolution. This includes the Greenhouse gases Observing SATellite (GOSAT),

operating since 2009 [6], the Scanning Imaging Absorption spectrometer for Atmospheric CartograpHY (SCIAMACHY), in orbit since 2002 [7], and the second Orbiting Carbon Observatory (OCO-2), which, as a rebuild of OCO [8,9], is planned to be launched in July 2014. In addition, CarbonSat will also be scheduled to be launched in 2018 (http://www.iup.uni-bremen.de/carbonsat/). These measurements have a nearly uniform sensitivity to CO$_2$ from the surface up through the middle troposphere, and thus are frequently used to derive the column-average dry air mole fraction of atmosphere CO$_2$ (XCO$_2$) during the daytime; (2) retrieving CO$_2$ concentrations by interpreting the recorded spectra of the Earth-atmosphere system in thermal infrared (TIR) bands (around 15 μm). Instruments that work in such a way include AIRS [10,11], IASI [12,13], and FTS (Band 4) of GOSAT [6]. These measurements bring the advantage that they can detect CO$_2$ during both day and night time, while the lack of sensitivity in the lower troposphere makes them inappropriate to estimate CO$_2$ near the surface where the largest signals of CO$_2$ sources and sinks occur [1]. The complementarities of these platforms allow us to combine the SWIR and TIR measurements for obtaining enhanced understanding of CO$_2$ spatiotemporal variations globally. Since XCO$_2$ is much less affected by vertical transport of CO$_2$, it is particularly useful for investigation of CO$_2$ sources and sinks using inversion modeling [14,15]. On the other hand, the spatial and temporal variations in XCO$_2$ are even smaller than that in the surface CO_2; therefore, unprecedented measurement precision and accuracy are highly required for such column measurements [16–19]. SCIAMACHY (operation stopped in April 2012) and GOSAT are two typical instruments that can be used to derive XCO_2 from space, and a variety of retrieval algorithms have been developed for SCIAMACHY [1,20–27] and GOSAT [2,4,5,28–30] with eyes on improving XCO_2 retrieval accuracy to a great extent. At present, a number of XCO_2 products have been released. These will definitely enhance our understanding of the global carbon cycle.

Unfortunately, almost all typical instruments currently used to derive atmospheric CO_2 concentration are working in the infrared spectral range (less than 16 μm). Thus, except for the instrument's observation mode (for example, GOSAT observes in lattice points), the spatial coverage of the derived CO_2 is severely restricted by the presence of clouds. In addition, the lower signal-to-noise level over ice/snow covered surfaces and ocean for SWIR instruments (e.g., SCIAMACHY) also contributes to the CO_2 sparse coverage. For instance, it has been pointed out that only about 10% of GOSAT data can be used for retrieval of XCO_2 due to the cloud contaminations [4]. The amount of CO_2 measurements will be even smaller if additional screening criteria such as quality of spectral fit, aerosol loadings, etc. are further applied. Although the amount of remaining CO_2 measurements from certain space-based instruments may largely surpass that of ground-based sites, it is still not sufficient enough for accurately quantifying the spatiotemporal distribution of CO_2 over the global scale. As a result, it is greatly desired to jointly use these available CO_2 measurements derived from various space-based data. Recently, a novel method has been proposed for combining CO_2 values from seven different algorithms, and a new Level-2 CO_2 database (EMMA) from one algorithm is composed according to the median of monthly average of seven CO_2 products in each $10° \times 10°$ latitude/longitude grid box [31]. In fact, this method cannot increase the number of CO_2 observations but chooses a product with moderate oscillation among the available products. Despite the usefulness of the XCO_2 measurements (Level 2) in their own right, further spatiotemporal analysis for interpreting their scientific merit is essentially necessary due to the retrieval uncertainties and sparse coverage of such Level-2 observations [32]. For this point, many works have attempted to generate global full-coverage (i.e., Level 3) maps from XCO_2 values derived from single satellite observations using a geospatial statistics approach [32–34]. However, as reflected in these studies (for instance, Fig. 1 in the work of [33]), a compromise has to be made between the interpolated accuracy and the spatiotemporal resolution of Level-3 product because of the limited amount of Level-2 XCO_2 observations being used. For this point, instead of using Level-2 XCO_2 from a single dataset (e.g., GOSAT or OCO-2) as performed in the existing literature, we attempt to explore the potential of combining two CO_2 datasets (GOSAT and SCIA-MACHY) in assisting in global Level-3 generation, aiming to: (1) propose a general strategy for combining (fusing) various CO_2 datasets with different instruments, algorithms, averaging kernels, etc.; and 2) increase the number of daily CO_2 points (utilized in Level-3 map interpolations) through the combination of two datasets, so that potentially improved Level-3 maps with higher accuracy and shorter time scale can be generated. The better the interpretation of the satellite-based CO_2 observations one can make, the higher the resolution (both temporal and spatial) of the generated global CO_2 maps.

Datasets

For GOSAT, the Fourier transform spectrometer (FTS) on GOSAT is the fundamental unit to retrieve atmospheric CO_2 and CH_4. It observes sunlight reflected from the earth's surface, and light emitted from the atmosphere and the surface. It is composed of three narrow bands in the SWIR region (0.76, 1.6, and 2.0 μm) and a wide TIR band (5.5–14.3 μm) at a spectral and spatial resolution of 0.2 cm^{-1} and 10.5 km, respectively [35]. Specifically, four CO_2 products from GOSAT have currently been released to the public: University of Leicester product [9,36], the RemoTeC product [28], NIES GOSAT product [35] and the product generated by NASA's Atmospheric CO_2 Observations from Space (ACOS) team (hereafter called ACOS product) [2,30]. The difference between some of the above mentioned products with various versions have been investigated in a recent study [37]. In the present paper, the ACOS product of 2009–2010 with version v2.9 has been employed.

SCIAMACHY was successfully launched on board Environmental Satellite (ENVISAT) in 2002 (unfortunately ceased in April 2012), which is a detector elements satellite spectrometer covering the spectral range 0.24–2.38 μm with a moderate spectral resolution of about 0.2–1.6 nm, and spatial resolution at nadir of 60×30 km [7]. It has eight spectral channels, with 1024 individual detector diodes for each band, observing the spectral regions 0.24–1.75 μm (band 1–6), 1.94–2.04 μm (band 7), and 2.26–2.38 μm (band 8) simultaneously in nadir and limb and solar and lunar occultation viewing geometries [22]. As mentioned in Section 1, till today, a number of CO_2-retrieval algorithms have been developed for SCIAMACHY. The IUP/IFE of University of Bremen has released two XCO_2 products, i.e., WFM-DOAS product [21,22] and the Bremen Optimal Estimation DOAS (BESD) product [1,26]. In this study, the BESD product with the versions of v02.00.08 for 2009–2010 is used.

In addition, CO_2 profiles of CT [38] are also collected here to allow the data mentioned above to be properly fused. CT is a NOAA data assimilation system, which provides the 3D profiles of CO_2 mole fractions in the atmosphere over the globe. For this study, CT data with version CT2011 is collected. This dataset provides global CO_2 profiles with $3° \times 2°$ latitude/longitude grid and 3 hours temporal resolution (a total 8 times from 01 to 22 in UTC) spanning the time period from January 2000 to December 2010. The CT dataset is used here mainly to assist in adjusting and time-shifting of the two CO_2 products being combined.

Methodologies

For combining the different space-based CO_2 measurements, three steps are adapted in this study. First, taking the global ground measurements of CO_2 as reference, remove the bias of the individual CO_2 retrievals for ensuring the accuracy of the fused CO_2 product; then make some adjustment for both the ACOS and BESD products, so that they can be physically comparable and thus combined; finally fuse the ACOS and BESD CO_2 products considering their retrieval uncertainties, spatial scales, differences in averaging kernels and overpass times, etc.

3.1 Global bias corrections

Removal of any global bias of the retrieved CO_2 when compared with the ground *in situ* measurements is essential before performing joint use. Many researches [4,25] frequently pointed out that CO_2 retrievals from GOSAT are low biased with different levels due to the uncertainties in pressure, radiometric calibration, line shape model, cloud and aerosol scattering, etc.

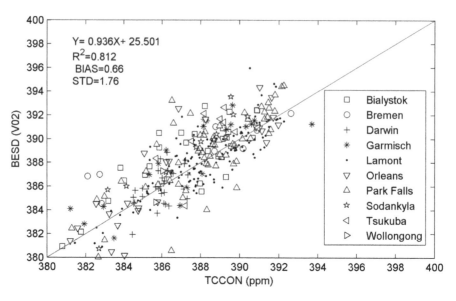

Figure 1. Validation of the BESD products against *in situ* **TCCON CO_2 measurements over globe for 2009–2010.**

Fortunately, a recent study has proposed a method for evaluating systematic errors in CO_2 and showed that the new version of ACOS product (v2.9) has a low global bias (<0.5 ppm) [39]. Thus, there is no global bias correction for the ACOS product being conducted here, but only the ACOS retrievals that pass the filter of table B1 in the work of [38] and marked as "good" in the quality flag are used. For the BESD product, we select Total Carbon Column Observing Network (TCCON) [15] measurements for 2009–2010 as the ground truth to determine its global bias. Specifically, BESD retrievals within ±2.5° and ±2.5° latitude/longitude box centered at each TCCON site and the mean FTS value (within ±1 h time window of satellite overpass time) are

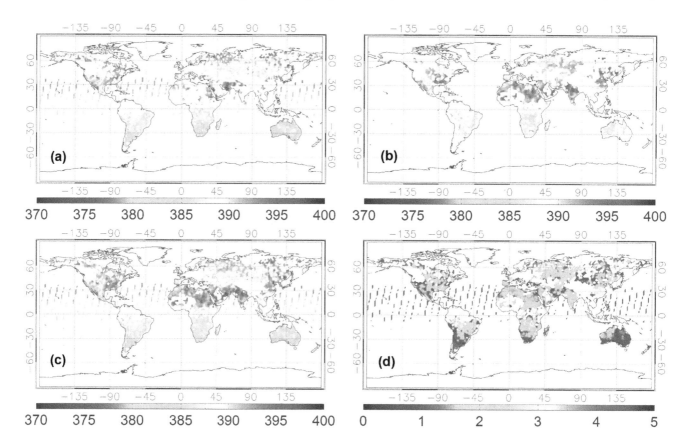

Figure 2. XCO₂ monthly mean maps in May of 2010. ((a) ACOS XCO₂, (b) BESD XCO₂, (c) combined product, and (d) XCO₂ uncertainties of the combined product).

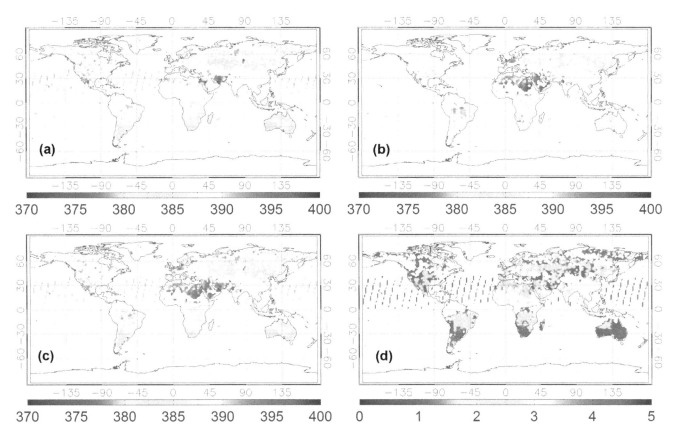

Figure 3. XCO$_2$ monthly mean maps in June of 2010. ((a) ACOS XCO$_2$, (b) BESD XCO$_2$, (c) combined product, and (d) XCO$_2$ uncertainties of the combined product).

extracted and compared (totally ten TCCON sites are utilized). The coincidence criteria mentioned above ultimately yield a total of 338 pairs of CO$_2$ measurements. The comparison result is shown in Fig. 1.

3.2 Retrieval adjustments

As pointed out by most researchers, it is not reasonable to directly compare or use two XCO$_2$ measurements. A suitable way to do that is to take the a priori profiles and variations in averaging kernel into account during the comparison [26,40]. To tackle the a priori issue, after correcting their global biases, both BESD and ACOS products are adjusted for a common a priori profile, which we assume to be the CT profile interpolated at the middle of the two overpass times (Equation (1)). Specifically, the a priori CO$_2$ profile of both the ACOS and BESD are first interpolated or extrapolated to the level of the CT CO$_2$ profile according to their pressure layers. After interpolation, the a priori profiles for both ACOS and BESD have the same dimension as the CT profile. Here the reason we take the CT profile at the middle of the two overpass times is that the time difference for GOSAT (1:00 pm) and SCIAMACHY (10:00 am) is relative large (3 hours), if we take one satellite time as reference, the induced error would be large for the other satellite measurements considering the CO$_2$ natural diurnal variation. So a middle time between these two satellite overpass times is selected for minimizing the CO$_2$ uncertainties during the adjustment.

$$XCO_2\_adj = XCO_2\_ret + (h^T I - a)(xCT - xa) \qquad (1)$$

Here, XCO_2_adj is the adjusted XCO$_2$ for ACOS or BESD; XCO_2_ret corresponds to retrieved XCO$_2$ of ACOS or BESD; a is the column-averaging kernel (row vector) of ACOS or BESD; h is pressure-weighting function (column vector); I is an identity matrix; xCT and xa (column vectors) are the common CT CO$_2$ profile and the corresponding a priori CO$_2$ profile for ACOS or BESD, respectively.

While it is not trivial to accurately consider the smoothing error without an estimate of the true atmospheric variability which is generally not readily available for most cases [39]. Fortunately, some works revealed that the smoothing error is generally small [26,39]. Consequently, for the remainder of this paper, only the adjustment in Equation (1) is applied for both the ACOS and BESD CO$_2$ products (after bias corrections).

3.3 Combination and time shifting

Based on the processes described above, the world is divided into a number of $0.5° \times 0.5°$ latitude/longitude grid box (totally 720×360). For each grid cell, Equation (2) is used to combine the corresponding CO$_2$ measurements within that grid.

$$XCO_2\_Fued = \sum_{i=1}^{m} \left(XCO_2\_i \times \frac{1 - Uncert\_ratio^i}{\sum\limits_{i=1}^{m}(1 - Uncert\_ratio^i)} \right) \qquad (2)$$

where XCO_2_Fued is the combined XCO$_2$; m is the total number of space-based CO$_2$ retrievals (ACOS and/or BESD) within a certain grid; XCO_2_i is the ith XCO$_2$ retrieval in a grid for which

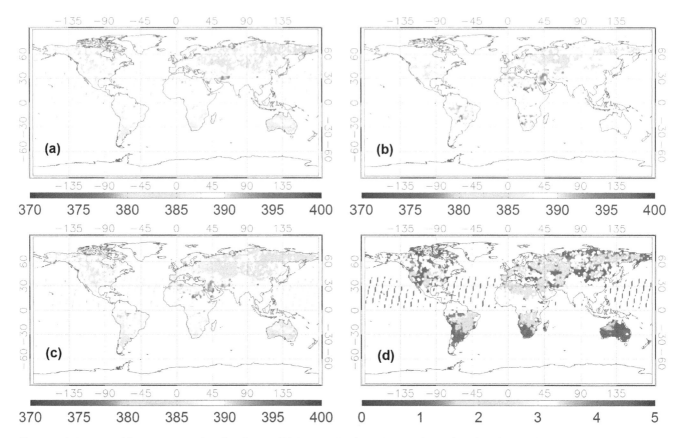

Figure 4. XCO$_2$ monthly mean maps in July of 2010. ((a) ACOS XCO$_2$, (b) BESD XCO$_2$, (c) combined product, and (d) XCO$_2$ uncertainties of the combined product).

the global bias and Equation (1) are supposed to be applied; *Uncert_ratio^i* is the ratio of uncertainty of the ith XCO$_2$ retrieval to its XCO$_2$ value.

Please note that since different CO$_2$ retrievals have distinct overpass times, it is necessary to unify them to avoid uncertainties induced from the time discrepancy before fusion. To this end, a method for considering the CO$_2$ shifting along time has been developed (Equation (3)). First, designate a specific time or select one overpass time as reference, then transfer CO$_2$ measurements at various overpass times to that of the reference time by interpolating the CT CO$_2$ at temporal scale. Here, it should be pointed out that despite the CO$_2$ absolute values of CT not being accurate enough, the daily cycle pattern of atmospheric CO$_2$ it reflects is assumed to be correct.

$$XCO_2\_ref = \frac{\omega^T X\_ref^{CT}}{\omega^T X\_t^{CT}} \times XCO_2\_t \qquad (3)$$

Here, XCO_2_ref is the transformed XCO$_2$ (ACOS or BESD) at the reference time; XCO_2_t is the retrieved XCO$_2$ from ACOS or BESD at overpass time t; X_ref^{CT} and X_t^{CT} are CO$_2$ profiles of CT at times of reference and t, respectively; ω is the pressure-weighting vector (column vector).

Based on the time-shifting strategy proposed here, a global CO$_2$ map at any specific time can be theoretically produced by employing the pattern of the CO$_2$ daily cycle reflected by CT data. For instance, we can unify all XCO$_2$ retrievals being combined with various overpass times to that of UTC = 1.

Results

Evaluation analysis showed that the global bias for the BESD product is generally small. In this study, the bias of the BESD product is corrected by subtracting 0.6 ppm from all XCO$_2$ values according to the results in Fig. 1. Although the systematic bias of the XCO$_2$ retrievals is removed, it is supposed that the error characteristics (random error) within the data are still unchanged. The bias-corrected XCO$_2$ retrievals of both ACOS and BEDS are used as fundamental data for the combination algorithm.

By applying the series of processes shown in Section 3, daily, weekly, as well as monthly maps of combined XCO$_2$ for 2009 and 2010 are generated. Here, as an example, only four maps (from May to August) of monthly mean XCO$_2$ of 2010 are shown here (Fig. 2–Fig. 5). In addition, the total XCO$_2$ uncertainties of the combined product which mainly depend on the uncertainties of the original ACOS or BESD XCO$_2$ retrievals are also illustrated.

From Fig. 2–Fig. 5, it is not difficult to observe that the combined data realize the physical complementary of the two products in terms of spatial coverage. The number of valid CO$_2$ measurements in the fused product is the union of the CO$_2$ data from both the ACOS and BESD at the same geographical location. In addition, the combined XCO$_2$ demonstrates similar spatiotemporal characteristics with that of ACOS and BESD over the globe, which implies that all processes associated with the combination do not distort the essential information of the original XCO$_2$ products (ACOS or BESD). Similar findings can also be observed in the daily mean and weekly mean XCO$_2$ maps. To quantitatively investigate the improvement of fused XCO$_2$ in spatial coverage, the fractional coverage of all three variables

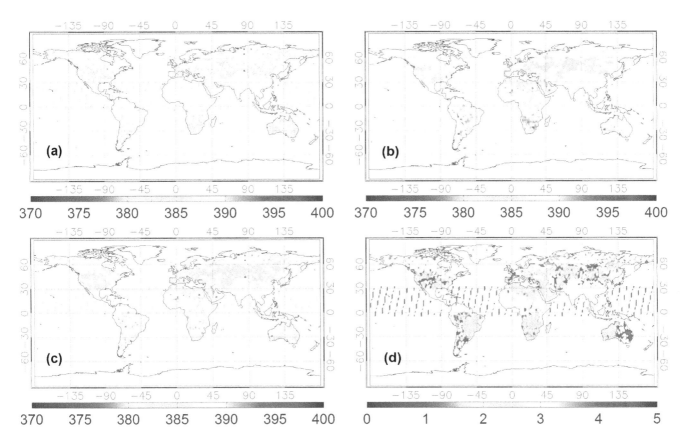

Figure 5. XCO$_2$ monthly mean maps in August of 2010. ((a) ACOS XCO$_2$, (b) BESD XCO$_2$, (c) combined product, and (d) XCO$_2$ uncertainties of the combined product).

(ACOS, BESD, and combined XCO$_2$) on both daily and monthly scales is calculated (Fig. 6). From Fig. 6, it can be seen that the average global coverage of ACOS and BESD is around 0.46% and 0.21%, respectively, on a daily scale. The monthly mean coverage of such products accounts for about 5.70% and 3.75%, respectively. While spatial coverage of combined XCO$_2$ can reach up to 0.65% and 8.42% on daily and monthly scales, respectively, it accounts for increments of 41.3% and 47.7% on the daily and

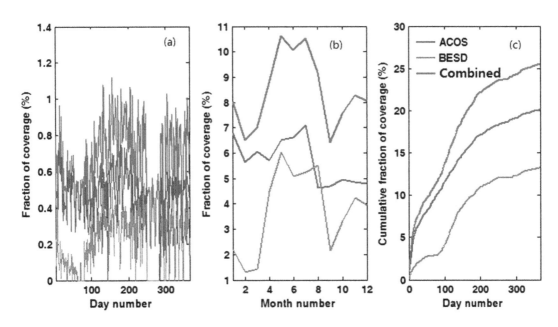

Figure 6. XCO$_2$ fraction of coverage of ACOS, BESD, and combined products. (a) Daily coverage. (b) Monthly coverage. (c) Cumulative coverage.

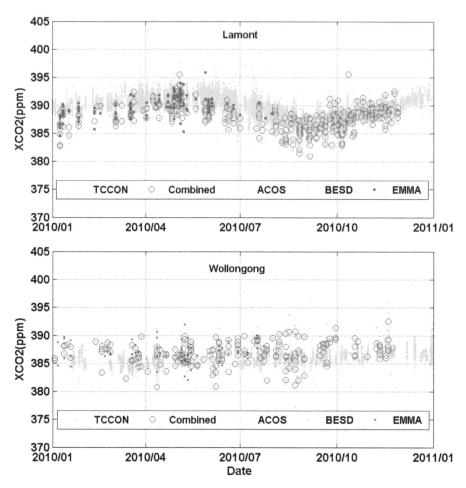

Figure 7. Comparison of XCO₂ measurements from TCCON, ACOS, BESD, EMMA, and our new combination method over Wollongong and Lamont sites (distance<0.25 degree, temporal difference<1 hour).

monthly scales with respect to that of GOSAT and it is even higher relative to the coverage of SCIAMACHY. Likewise, the cumulative fraction of coverage of the combined XCO_2 has risen to 25% when compared with 20% and 13% for ACOS and BESD, respectively. The increase in the XCO_2 spatial coverage indicates the potential advantage of the combined XCO_2 observations in generating global Level-3 XCO_2 maps when compared with any single dataset by providing more satellite-based XCO_2 retrievals used for optimal interpolating.

For evaluating the performance of our combination strategy, the combined XCO_2 values are compared with that retrieved from ACOS and BESD as well as XCO_2 in the EMMA database at two TCCON sites (Fig. 7). The results reveal that the XCO_2 values from the combination method show generally consistent variation in time with TCCON measurements except for a small overall bias (especially for the Lamont site). On the whole, the new combined XCO_2 product shows good consistency with the EMMA data, and they are comparable in terms of CO_2 magnitude, while the combined XCO_2 are shown with a longer time period, which is in line with the satellite observations, and possess more data points even over the same period.

Discussions and Conclusions

Despite the fact that space-based measurements can provide a unique opportunity to map atmospheric CO_2 over large areas, the number of valid CO_2 measurements from a single space-based instrument is generally limited for a certain day over a specific region due to the presence of clouds. In addition, although these Level-2 XCO_2 retrievals themselves are very important for inversion modeling of surface carbon sources/sinks, further comprehensive analysis by investigating the spatiotemporal full-coverage XCO_2 (Level 3) distribution is needed for interpreting their significant scientific merit [32]. While the limited satellite observations restrict the generation of Level-3 XCO_2 maps with high spatial and temporal resolutions when only a single satellite-based XCO_2 dataset is considered. This is our main motivation in this paper.

In this study, a strategy for combining SCIAMACHY and GOSAT CO_2 measurements has been proposed by fully accounting for the CO_2 global bias, differences in averaging kernels and overpass times, and the Level-2 retrieval errors of the CO_2 measurements being used. The results indicated that the average global coverage of both ACOS and BESD is less than 0.5% on a daily scale, and less than 6% on a monthly scale. While spatial coverage of combined XCO_2 can reach up to 0.65% and 8.42% on daily and monthly scales, respectively, the comparison analysis reveals that the combined XCO_2 product is consistent with TCCON and EMMA in both temporal variation and magnitude except for a small bias when compared with the TCCON measurements. All these findings herein prove the effectiveness of the combination method in supporting generation global full-coverage XCO_2 maps with higher temporal and spatial sampling by jointly using two space-based XCO_2 datasets. Similar to the existing studies (e.g. [32–34]), although these combined XCO_2 are not intended to be used in inverse modeling studies, they deliver a key complement for such research, and can be deemed as an independent dataset for comparison with model predictions. Similar to the existing study [31], an improved fusion approach (based on multiple XCO_2 datasets) to create Level-2 XCO_2 measurements that can be directly used for inverse modeling is also attempted and will be presented in another paper.

A last point that needs to be addressed is that although we employed CO_2 data of GOSAT and SCIAMACHY in this study, the proposed strategies are not restricted to such data. As a general strategy, it can be refined and adapted to further combine other XCO_2 products, such as OCO-2, CarbonSat, etc. in the future,

and even to be applied to the fusion of other trace gases, such as O_3, CH_4.

Acknowledgments

The authors would like to thank the SCIAMACHY team at University of Bremen IUP/IFE as well as the ACOS scientific teams for providing us the CO_2 products. The authors also thank the anonymous reviewers for their helpful and valuable comments to improve this work.

Author Contributions

Conceived and designed the experiments: TW JS YJ TZ DJ CX. Performed the experiments: TW JS YJ TZ DJ CX. Analyzed the data: TW JS YJ TZ DJ CX. Contributed reagents/materials/analysis tools: TW JS YJ TZ DJ CX. Wrote the paper: TW JS YJ TZ DJ CX.

References

1. Reuter M, Bovensmann H, Buchwitz M, Burrows JP (2010) A method for improved SCIAMACHY CO_2 retrieval in the presence of optically thin clouds. Atmos. Meas. Tech 3: 209–232.

2. O'Dell CW, Connor B, Bösch H, O'Brien D, Frankenberg C, et al. (2012) The ACOS CO_2 retrieval algorithm – Part 1: Description and validation against synthetic observations. Atmos Meas Tech 5(1): 99–121.

3. Baker DF, Law RM, Gurney KR, Rayner P, Peylin P, et al (2006) TransCom 3 inversion intercomparison: Impact of transport model errors on the interannual variability of regional CO_2 fluxes, 1988–2003. Global Biogeochem Cy 20: GB1002, doi:10.1029/2004GB002439.

4. Morino I, Uchino O, Inoue M, Yoshida Y, Yokota T, et al. (2011) Preliminary validation of column-averaged volume mixing ratios of carbon dioxide and methane retrieved from GOSAT short-wavelength infrared spectra. Atmos Meas Tech 4: 1061–1076.

5. Butz A, Hasekamp OP, Frankenberg C, Aben I (2009) Retrievals of atmospheric CO_2 from simulated space-borne measurements of backscattered near-infrared sunlight: accounting for aerosol effects. Appl Opt 18: 3322–3336.

6. Kuze A, Suto H, Nakajima M, Hamazaki T (2009) Thermal and near infrared sensor for carbon observation Fourier-transform spectrometer on the Greenhouse Gases Observing Satellite for greenhouse gases monitoring, Appl Opt 35: 6716–6733.

7. Bovensmann H, Burrows JP, Buchwitz M, Frerick J, Noel S, et al (1999) SCIAMACHY –Mission objectives and measurement modes. J Atmos Sci 56: 127–150.

8. Crisp D, Atlas RM, Breon FM, Brown LR, Burrows JP, et al (2004) The Orbiting Carbon Observatory (OCO) mission. Adv Space Res 34: 700–709.

9. Boesch H, Baker D, Connor B, Crisp D, Miller C (2011) Global Characterization of CO_2 Column Retrievals from Shortwave-Infrared Satellite Observations of the Orbiting Carbon Observatory-2 Mission. Remote Sens 3: 270–304.

10. Aumann HH, Chahine MT, Gautier C, Goldberg MD, Kalnay E, et al. (2003) AIRS/AMSU/HSB on the Aqua Mission: Design, science objectives, data products, and processing systems. IEEE Trans Geosci Remote Sens 41: 253–264.

11. Chahine M, Barnet C, Olsen ET, Chen L, Maddy E (2005) On the determination of atmospheric minor gases by the method of vanishing partial derivatives with application to CO_2. Geophys Res Lett 32: L22803, doi:10.1029/2005GL024165.

12. Phulpin T, Cayla F, Chalon G, Diebel D, Schlussel P (2002) IASI on board Metop: Project status and scientific preparation. Proceedings of the 12th International TOVS Study Conference. Lorne, Vi ctoria, Australia.

13. Turquety S, Hadji-Lazaro J, Clerbaux C, Hauglustaine DA, Clough SA, et al. (2004) Operational trace gas retrieval algorithm for the Infrared Atmospheric Sounding Interferometer. J Geophys Res 109: D21301, doi:10.1029/2004JD004821.

14. Yang Z, Washenfelder RA, Keppel-Aleks G, Krakauer NY, Randerson JT, et al. (2007) New constraints on Northern Hemisphere growing season net flux. Geophys Res Lett 34: L12807. doi:10.1029/2007GL029742.

15. Wunch D, Toon GC, Blavier JFL, Washenfelder R, Notholt J, et al. (2011) The Total Carbon Column Observing Network. Philos T Roy Soc A 369: 2087–2112.

16. Rayner PJ, O'Brien DM (2011) The utility of remotely sensed CO_2 concentration data in surface source inversions. Geophys Res Lett 28: 175–178.

17. Houweling S, Breon FM, Aben I, Rodenbeck C, Gloor M, et al. (2004) Inverse modeling of CO_2 sources and sinks using satellite data: a synthetic inter-comparison of measurement techniques and their performance as a function of space and time. Atmos Chem Phys 4: 523–538.

18. Olsen SC, Randerson JT (2004) Differences between surface and column atmospheric CO_2 and implications for carbon cycle research. J Geophys Res 109; D02301. doi:10.1029/2003JD003968.

19. Miller CE, Crisp D, DeCola PL, Olsen SC, Randerson JT, et al. (2007) Precision requirements for space-based XCO_2 data. J Geophys Res 112: D10314, doi:10.1029/2006JD007659.

20. Buchwitz M, Beek R, Burrows JP, Bovensmann H, Warneke T, et al. (2005) Atmospheric methane and carbon dioxide from SCIAMACHY satellite data: initial comparison with chemistry and transport models, Atmos Chem Phys 5: 941–962.

21. Buchwitz M, Beek R, Nol S, Burrows JP, Bovensmann H, et al. (2005) Carbon monoxide, methane and carbon dioxide columns retrieved from SCIAMACHY by WFM-DOAS: year 2003 initial data set. Atmos Chem Phys 5: 3313–3329.

22. Buchwitz M, Rozanov VV, Burrows JP (2000) A near-infrared optimized DOAS method for the fast global retrieval of atmospheric CH_4, CO, CO_2, H_2O, and N_2O total column amounts from SCIAMACHY Envisat-1 nadir radiances. J Geophys Res 105: 15231–15245, doi:10.1029/2000JD900191.

23. Houweling S, Hartmann W, Aben I, Schrijver H, Skidmore J, et al. (2005) Evidence of systematic errors in SCIAMACHY-observed CO_2 due to aerosols. Atmos Chem Phys 5: 3003–3013.

24. Barkley MP, Frieβ U, Monks PS (2006) Measuring atmospheric CO_2 from space using Full Spectral Initiation (FSI) WFM-DOAS, Atmos Chem Phys 6: 3517–3534.

25. Schneising O, Buchwitz M, Burrows JP, Bovensmann H, Reuter M, et al. (2008) Three years of greenhouse gas column-averaged dry air mole fractions retrieved from satellite – Part 1: Carbon dioxide. Atmos Chem Phys 8: 3827–3853.

26. Reuter M, Bovensmann H, Buchwitz M, Burrows JP, Connor BJ, et al. (2011) Retrieval of atmospheric CO_2 with enhanced accuracy and precision from SCIAMACHY: Validation with FTS measurements and com-parison with model results. J Geophys Res 116: D04301, doi:10.1029/2010JD015047.

27. Bösch H, Toon GC, Sen B, Washenfelder RA, Wennberg PO, et al. (2006) Space-based near-infrared CO_2 measurements: Testing the Orbiting Carbon Observatory retrieval algorithm and validation concept using SCIAMACHY observations over Park Falls, Wisconsin. J Geophys Res 111: D23302, doi:10.1029/2006JD007080.

28. Butz A, Guerlet S, Hasekamp O, Schepers D, Galli A, et al. (2011) Toward accurate CO_2 and CH_4 observations from GOSAT. Geophys Res Lett 14: L14812. DOI:10.1029/2011GL047888.

29. Yoshida Y, Ota Y, Eguchi N, Kikuchi N, Nobuta K, et al. (2011) Retrieval algorithm for CO_2 and CH_4 column abundances from short-wavelength infrared spectral observations by the Greenhouse Gases Observing Satellite. Atmos Meas Tech 4: 717–734.

30. Crisp D, Fisher B, O'Dell C, Frankenberg C, Basilio R, et al. (2012)The ACOS CO_2 retrieval algorithm - Part II: Global XCO_2 data characterization. Atmos Meas Tech 5: 687–707.

31. Reuter M, Bösch H, Bovensmann H, Bril A, Buchwitz M, et al. (2013) A joint effort to deliver satellite retrieved atmospheric CO_2 concentrations for surface flux inversions: the ensemble median algorithm EMMA. Atmos Chem Phys 13: 1771–1780.

32. Hammerling DM, Michalak AM, O'Dell C, Kawa SR (2012) Global CO_2 distributions over land from the Greenhouse Gases Observing Satellite(GO-SAT). Geophys Res Lett 39: L08804, doi:10.1029/2012GL051203.

33. Hammerling DM, Michalak AM, Kawa SR (2012) Mapping of CO_2 at high spatiotemporal resolution using satellite observations: Global distributions from OCO-2. J Geophys Res 117: D06306, doi:10.1029/2011JD017015.

34. Zeng Z, Lei L, Hou S, Ru F, Guan X, et al. (2014) A Regional Gap-Filling Method Based on Spatiotemporal Variogram Model of CO_2 Columns. IEEE Trans Geosci Remote Sens 6: 3594–3603.

35. Yokota T, Yoshida Y, Eguchi N, Ota Y, Tanaka T, et al. (2009) Global Concentrations of CO_2 and CH_4 Retrieved from GOSAT: First Preliminary Results. SOLA 5: 160–163.

36. Parker R, Boesch H, Cogan A, Fraser A, Feng L, et al. (2011) Methane observations from the Greenhouse Gases Observing SATellite: Comparison to

ground-based TCCON data and model calculations. Geophys Res Lett 15: L15807, doi:10.1029/2011GL047871.

37. Wang TX, Shi JC, Jing YY, Xie YH (2013) Investigation of the consistency of atmospheric CO_2 retrievals from different space-based sensors: Intercomparison and spatio-temporal analysis, Chin Sci Bull, 33: 4161–4170.

38. Peters W, Jacobson AR, Sweeney C, Andrews AE, Conway TJ, et al. (2007) An atmospheric perspective on North American carbon dioxide exchange: Carbon Tracker. Proc Natl Acad Sci U.S.A. 48: 18925–18930.

39. Wunch D, Wennberg PO, Toon GC, Connor BJ, Fisher B, et al. (2011) A method for evaluating bias in global measurements of CO_2 total columns from space. Atmos Chem Phys 11: 12317–12337.

40. Rodgers CD (2000) Inverse Methods for Atmospheric Sounding: Theory and Practice. World Scientific Publishing Co. Ltd.193 p.

8

Carbon Dioxide Flux from Rice Paddy Soils in Central China: Effects of Intermittent Flooding and Draining Cycles

Yi Liu[1], Kai-yuan Wan[1], Yong Tao[1], Zhi-guo Li[1], Guo-shi Zhang[1], Shuang-lai Li[2], Fang Chen[1]*

1 Laboratory of Aquatic Botany and Watershed Ecology, Wuhan Botanical Garden, Chinese Academy of Sciences China, Wuhan, China, 2 Institute of Plant Protection and Soil Fertilizer, Hubei Academy of Agricultural Sciences, Wuhan, China

Abstract

A field experiment was conducted to (i) examine the diurnal and seasonal soil carbon dioxide (CO_2) fluxes pattern in rice paddy fields in central China and (ii) assess the role of floodwater in controlling the emissions of CO_2 from soil and floodwater in intermittently draining rice paddy soil. The soil CO_2 flux rates ranged from -0.45 to 8.62 $\mu mol.m^{-2}.s^{-1}$ during the rice-growing season. The net effluxes of CO_2 from the paddy soil were lower when the paddy was flooded than when it was drained. The CO_2 emissions for the drained conditions showed distinct diurnal variation with a maximum efflux observed in the afternoon. When the paddy was flooded, daytime soil CO_2 fluxes reversed with a peak negative efflux just after midday. In draining/flooding alternating periods, a sudden pulse-like event of rapidly increasing CO_2 efflux occured in response to re-flooding after draining. Correlation analysis showed a negative relation between soil CO_2 flux and temperature under flooded conditions, but a positive relation was found under drained conditions. The results showed that draining and flooding cycles play a vital role in controlling CO_2 emissions from paddy soils.

Editor: Dorian Q. Fuller, University College London, United Kingdom

Funding: The study was supported by the National Natural Science Foundation of China (31100386), and the Cooperated Program with International Plant Nutrition Institute (IPNI-HB-33). The funders had no role in study design, data collection and analysis, decision to publish, or preparation of the manuscript.

Competing Interests: The authors have declared that no competing interests exist.

* E-mail: fchenipni@126.com

Introduction

Increases in the emission of greenhouse gases such as carbon dioxide (CO_2), methane (CH_4), and nitrous oxide (N_2O) from soil surface to the atmosphere have been a worldwide concern for several decades [1–3]. CO_2 is recognized as a significant contributor to global warming and climatic change, accounting for 60% of global warming or total greenhouse effect [4]. Measuring the soil CO_2 efflux is crucial for accurately evaluating the effects of soil management practices on global warming and carbon cycling. Temporal variations in soil CO_2 flux have been observed in almost all ecosystems [5,6]. Soil CO_2 fluxes are usually higher during warm seasons and lower during cold seasons [7,8]. The seasonal variation is driven largely by changes in temperature, moisture, and photosynthate production [5,9,10]. The main factors controlling seasonal variations in soil CO_2 flux may depend on the type of ecosystems and the climate.

The increase in population in areas where rice is the main cultivated crop has led to the increase in worldwide area under rice cultivation by approximately 40% over the last 50 years [11]. In particular, Asian countries (China, India, Indonesia, etc.) have accounted for approximately 90% of the total global area under rice cultivation for the last 50 years [11]. Rice paddies in monsoonal Asia play an important role in the global budget of greenhouse gases such as CH_4 and CO_2 [12,13]. Carbon emissions (esp. CH_4) from rice paddies are expected to be a long-term contributor to greenhouse gases, perhaps

increasingly over the past 5000 years [14]. Efforts have been made recently to model carbon emissions based on the history and archaeology of rice cultivation in Asia. However, since these emissions from rice cultivation vary a great deal, this poses a major challenge in modeling this phenomenon [15]. As a result, experimental research from rice paddies assumes greater importance. Many of the factors controlling gas exchange between rice paddies and the atmosphere are different from those in dryland agriculture and other ecosystems because rice is flooded during most of its cultivation period. The dynamics of soil CO_2 fluxes in a paddy field differs significantly from that in fields with upland crop cultivation in which aerobic decomposition process is dominant [6,16,17]. Field studies designed to measure soil CO_2 fluxes and improve our understanding of the factors controlling the fluxes are thus needed.

Intermittent draining and flooding, which is one of the most important water management practices in rice production, was found to be the most promising option for CH_4 mitigation also [18,19]. Mid-season aeration was also found to be one of the basic techniques for raising rice yields in China [20] and was widely adopted in rice cultivation where irrigation/drainage system was well managed. The management induced change of anaerobic and aerobic conditions results in temporal and spatial (vertical, horizontal) variations in reduction and oxidation (redox) reactions affecting the dynamics of organic and mineral soil constituents [21,22]. Thus, intermittent drainage with increased impacts can strongly affect soil CO_2 emissions [6,16]. However, the mechanism of CO_2 exchange between rice paddies and the atmosphere is not fully understood. For example, using eddy covariance measurements, Miyata et al. [16] found a significantly larger net CO_2 flux from the rice paddy soil to atmosphere when the field was drained compared to when it was flooded. These differences in the CO_2 flux were mainly due to increased CO_2 emissions from the soil surface under drained conditions resulting from the removal of diffusion barrier caused by the floodwater. The existence of floodwater, anaerobic soil, or changes in the micrometeorological environment with flooding influences root activity, photosynthesis, and respiration of rice plants [23]. Activity of aquatic plants such as algae in the floodwater may also affect CO_2 exchange between rice paddies and the atmosphere [22]. Most of the data obtained so far were not sufficiently detailed to examine the influence of these factors on the CO_2 exchange in rice paddies.

The scale and dynamics of growing-season CO_2 emissions from paddy fields have been documented mostly through flux measurements made with low time resolution using manual chambers [6,16,17]. In this study, we report a data set that extends hourly CO_2 flux measurements during the rice-growing season in 2011 to improve the understanding of the process controlling CO_2 exchanges in rice paddy soils. The measurements were used to assess the role of floodwater in controlling the exchanges of CO_2 from the paddy soil. The objectives of this study were to: (i) analyze seasonal and diurnal variation of soil CO_2 fluxes in rice paddy fields in the Yangtze River valley; and (ii) determine the effects of related environmental factors associated with flooding and draining cycles in paddy soils on CO_2 flux from the soil surface.

Materials and Methods

Site Description

Field experiments were conducted over one rice growing season, i.e. from June to October 2011, at Nanhu Agricultural Research Station (30°28′N, 114°25′E, altitude 20 m). The research site is owned by Hubei Academy of Agricultural Sciences. The field studies did not involve endangered or protected species and no specific permits were required for the described field studies. The site lies in a typical area of the humid mid-subtropical monsoon climate in the Yangtze River valley of China. The mean annual temperature of the site is 17°C, the cumulative temperature above 10°C is 5,190°C, and the average annual frost-free period is 276 d. The average annual precipitation is 1,300 mm, with most of the rainfall occurring between April and August. The paddy field soil is a Hydromorphic paddy soil, which is a silty clay loam derived from Quaternary yellow sediment. Some physical and chemical properties of the experimental soil (0–20 cm depth) were: pH, 6.3; organic matter, 30.23 g.kg^{-1}; total N, 2.05 g.kg^{-1}; available P, 5 mg.kg^{-1}; available K, 101 mg.kg^{-1}; soil bulk density, 1.26 g.cm^{-3}. The experimental site has been under rice-wheat cultivation since last 30 years, where rice is planted from June to October each year and wheat is planted from November to May the following year. Daily meteorological information (including rainfall and temperature) during the 2011 rice-growing season is presented in Fig. 1.

Field Management

In 2011, rice was transplanted to the paddy field on 15 June with a plant to plant spacing of 20 cm and a row spacing of 27 cm. Irrigation started on 13 June and the field was flooded continuously until 17 July. This was followed by five intermittent flooding and draining cycles, with 3–7 days of flooding and 2–8

days of draining. The field was not irrigated and drained about a month before harvesting. The number of flooded days were 55, while the number of drained days were 53 during the 2011 rice-growing seasons. The depth of standing water during flooding periods was, on average, 10 cm. Before transplanting, base fertilizer – consisting of 36 kg N ha^{-1} in the form of urea (N 46%), 45 kg P_2O_5 ha^{-1} in the form of calcium superphosphate (P_2O_5 12%), and 90 kg K_2O ha^{-1} in the form of potassium sulfate (K_2O 45%) – was broadcast over the soil, which was then turned over by plowing to transfer the fertilizer to the subsurface (i.e., beyond 20 cm soil depth). Additional nitrogen, in the form of urea, was applied at tillering and heading stages of rice growth at rates of 36 and 18 kg N ha^{-1}, respectively. Rice grain was harvested from 1 to 3 October, 2011.

Measurement of Soil CO_2 Flux

The soil CO_2 flux was measured using the soil respiration method, where a cylinder static chamber of 22.5 cm diameter and 30 cm height was placed on the soil. The rate of increase in CO_2 concentration within the chamber was monitored with an ACE (ADC BioScientific Ltd) automated soil CO_2 flux system. The automated design means that during analysis cycles, the soil can be exposed to ambient conditions before the chamber closes to take measurements. This means the ACE will continue to collect data without any human intervention for as long as permitted by its battery life. This makes the ACE an ideal research instrument for continuous assessment of below-ground respiration and carbon stores in on-going experiments. Static chambers were inserted to a depth of approximately 7 cm, extending 23 cm above the soil surface to allow placing of the chamber. During the flooding period, the water remained in situ. The time span between chamber contact with the soil and the start of measurements (the deadband) was 20 s; this has previously been determined to be sufficient for pressure equilibration. The measurement time was set to 180 s. The ACE has a highly accurate CO_2 infrared gas analyzer housed directly inside the soil chamber, with no long gas tubing connecting the soil chamber and no separate analyzer. This ensures accurate and robust measurements, and the fastest possible response times to fluxes in gas exchange. During the soil CO_2 flux measurements, air temperature within the canopy and soil temperature at 2 cm depth were also recorded by the ACE analyzer unit. And the measurements were made at 1-hour intervals during the rice-growing season. During a 24-hour period, the values were averaged to give the mean daily soil CO_2 flux. Survey sites of three replications were taken from the experiment plot. Survey sites were located in the space between two rows, and the two sites were located 5–7 m apart. Three ACE stations were connected via an ACE Master control unit. Each CO_2 flux measurement from the experiment plot was thus an average of three individual measurements.

In order to examine the diurnal soil CO_2 flux pattern in a paddy field, soil CO_2 flux as well as canopy air temperature, soil temperature and PAR were also measured simultaneously at 1 hour intervals for 24 hours under both flooded (6/28~6/29 and 8/14~8/15) and drained (7/20~7/21 and 9/4~9/5) conditions. During these 24 hour periods, the sky was clear and with no clouds.

To study the soil CO_2 emissions in relation to draining and flooding cycle system, two draining/flooding alternation and circulation periods (7/23~7/28 and 8/29~9/4) were tested. We continuously monitored soil CO_2 fluxes along with air temperature within the canopy and soil temperature before, during, and after each flooding and draining cycle in the experiment paddy soil. Clear days continued during the experiment, but temperature

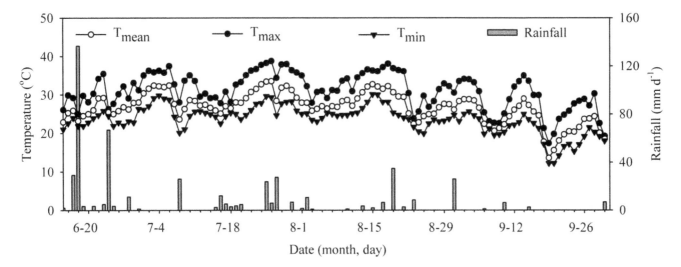

Figure 1. Air temperature records and rainfall events at the study site during the experimental period. (T_{mean}: mean temperature; T_{max}: maximum temperature; T_{min}: minimum temperature).

conditions were a little different from day to day. Flooding started at 9 am (09:00 h) and water depth reached 10 cm around midday. The water level was gradually decreased with cessation of irrigation.

Results

Seasonal Variations in Soil CO_2 Fluxes from Paddy Fields

The daily course of soil CO_2 flux rate is shown in Fig. 2A, while Fig. 2B shows the air temperature within the canopy and soil temperature (2 cm). The soil CO_2 flux rates ranged from -0.45 to 8.62 $\mu mol.m^{-2}.s^{-1}$, exhibiting a wide seasonal fluctuation during the rice-growing season. The soil CO_2 fluxes were generally low at the rice seedling stage, when it remained at about $0\sim1 \mu mol.m^{-2}.s^{-1}$ until the first mid-summer drainage. Then the fluxes increased gradually until the tillering stage, with a midway peak near the end of the first mid-summer drainage. From the tillering stage to the physiological maturity stage (i.e, from July to September), the daily average soil CO_2 flux rates had a magnitude ranging between 0 and 9 $\mu mol.m^{-2}.s^{-1}$, which then settled at around $1\sim3$ $\mu mol.m^{-2}.s^{-1}$ until the end of the season. The differences in the rates of soil CO_2 fluxes between drained and flooded conditions are also shown in Fig. 2a. Mean soil CO_2 fluxes under flooded conditions was 0.72 (with standard deviation of 0.48) $\mu mol.m^{-2}.s^{-1}$ (n = 55), whereas under drained conditions, the corresponding value was 2.79 (with standard deviation of 1.73) $\mu mol.m^{-2}.s^{-1}$ (n = 53). It is likely that floodwater decreased topsoil diffusivity, and may thus have decreased soil CO_2 effluxes [24]. Reduction of biological activity under anoxic condition may be another reason for low soil CO_2 fluxes during the flooding period [22].

The air temperature within the canopy and soil temperature (2 cm) exhibited seasonal patterns similar to soil CO_2 fluxes. The temperature varied from 15 to 33°C during the whole growing period of rice in 2011. From June to September, the temperature ranged from 21 to 33°C, and several peaks occurred. From the mid of September (9/18) to the day before harvesting (about 15 days), the average temperature of 19.7°C for air temperature within the canopy and 19.8°C for soil temperature (0–2 cm) are shown in Fig. 2.

Diurnal Patterns of Soil CO_2 Fluxes in Paddy Fields

The diurnal variations in soil CO_2 fluxes and incident PAR, air temperature within the canopy, and soil temperature under both flooding (6/28~6/29 and 8/14~8/15) and draining (7/20~7/21 and 9/4~9/5) conditions are shown in Fig. 3. These experiments began in the early evening, running for just under 24 h. Under flooding conditions, fluxes of CO_2 were, as expected, lower because the diffusivity and biological activity of the topsoil was substantially reduced by floodwater. Initially, there was a slow release of CO_2 into the atmosphere as a positive efflux settled at around 0–1 $\mu mol.m^{-2}.s^{-1}$ throughout the night. At sunrise the fluxes decreased, even negatively peaked at around 16:00 (negative values indicate carbon sequestration). This may have been because some aquatic plants, such as algae, inside the floodwater began to photosynthesize again. In contrast, CO_2 flux under draining conditions was positive and settled around 2~4 $\mu mol.m^{-2}.s^{-1}$ throughout the night, despite falling temperatures (Fig. 3). After sunrise, CO_2 fluxes remained positive and increased with temperature, reaching a peak at 2 pm (14:00 h) before falling again as temperatures declined.

Soil CO_2 Fluxes Related to Conversion Processes of Draining and Flooding Cycles

Fig. 4 shows soil CO_2 fluxes, canopy air temperature, and soil temperature before, during, and after the flooding and draining cycle. Soil CO_2 fluxes increased immediately after flooding, and exceeded pre-flooding values by two-thirds. This increase was abrupt and pulselike. Replacement of soil air by water should thus cause an enriched CO_2 pulse. And then, the soil CO_2 flux rate subsequently decreased by 70~90% within only one hour after the water pulse. Within the following days, the CO_2 fluxes remained at minimum levels (about $-2\sim2$ $\mu mol.m^{-2}.s^{-1}$) during flooding. As standing water declined and eventually disappeared, the CO_2 fluxes gradually increased and finally reached to maximum levels (about 6~8 $\mu mol.m^{-2}.s^{-1}$). This indicates that draining and flooding cycles play vital roles in controlling CO_2 emissions in a paddy soil.

Variability of Soil CO_2 Fluxes Related to Temperature

Temperature has a marked effect on CO_2 emissions from the soil surface. To study the relationship between soil CO_2 flux rates

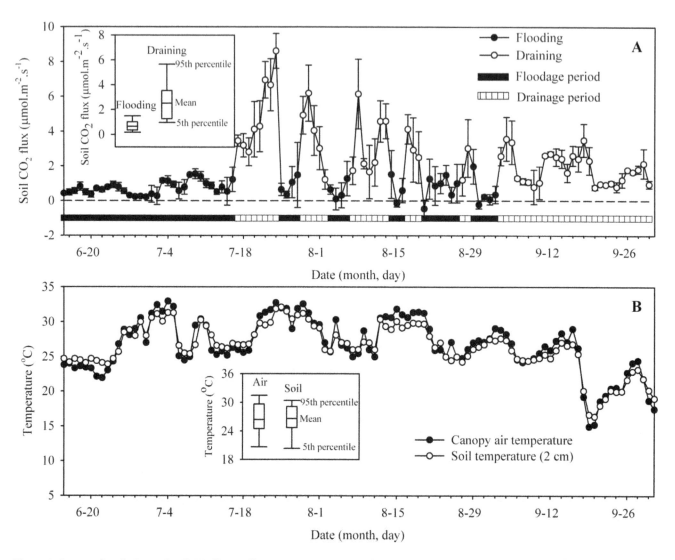

Figure 2. Seasonal variations of soil CO₂ flux, soil temperature (2 cm) and crop canopy temperature in an intermittent draining and flooding rice paddy. The insets indicate box-plot of soil CO_2 fluxes under flooded and drained conditions, and box-plot of canopy air and soil temperature during the 2011 rice-growing season.

and temperature, two environmental temperatures (air temperature within the canopy and soil temperature) were tested in this study (Fig. 5). Linear and exponential regression analysis were used to model the influence of temperature on soil CO_2 flux rates under both flooded and drained conditions. Negative linear correlations between temperature and soil CO_2 fluxes were found under flooded conditions ($R^2 = 0.1524$, P<0.001 and $R^2 = 0.0535$, P<0.001 for canopy air and soil temperatures, respectively), presumably because standing water limited soil CO_2 emissions. On the contrary, soil CO_2 flux rates increased as an exponential function of temperature under drained conditions ($R^2 = 0.1963$, P<0.001 and $R^2 = 0.2382$, P<0.001 for canopy air and soil temperatures, respectively).

Discussion

Previous research had revealed that water management systems show the highest potential in controlling CH_4 emissions [25]. CH_4 emissions were higher under continuous flooding than intermittent draining practices [26,27], while they declined during the drainage period to near zero and increased after re-flooding [28]. Drainage

during the rice cultivation period significantly increased CO_2 emissions in our study, while CH_4 emissions were clearly reduced and has been shown by other research [18,29]. Miyata et al. [16] also found that flooded or drainage conditions of paddy soils had strong effects not only on CH_4 emissions but also on CO_2 emissions. Lower CH_4 emissions due to water drainage may increase CO_2 emission. However, during the submerged period of paddy rice cultivation, CO_2 production in the soil is severely restricted under flooding condition. This effect can be explained with two basic mechanisms [8], which could be observed in a paddy soil (Fig. 6). First, flooding a field for subsequent rice cultivation cuts off the oxygen supply from the atmosphere and the microbial activities switch from aerobic (i.e. oxic condition) to facultative (i.e. hypoxic condition) and to anaerobic (i.e. anoxic condition) conditions [22]. As a consequence, biological activity reduction under anoxic condition, rather than completely, inhibits CO_2 production. At the same time, water replaces the gaseous phase in the soil pores. Since CO_2 diffusion rates in water are four orders of magnitude lower than those in air, a part of the produced CO_2 is stored in the soil. Hence, the soil CO_2 fluxes can be dramatically reduced by flooding during the paddy rice cultivation

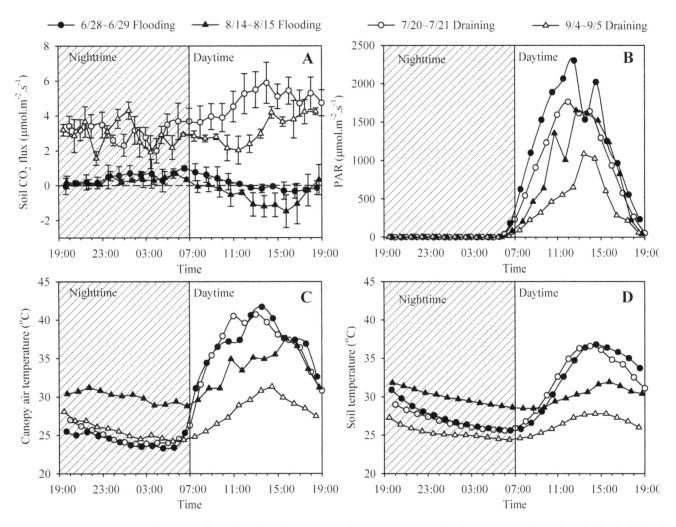

Figure 3. Diurnal patterns of soil CO_2 flux related to PAR, canopy air temperature and soil temperature (2 cm) under both drained and flooded condtions in a rice paddy.

[6,16,23]. Results from the present study provide indirect support for this conclusion, since the soil CO_2 flux rates under flooded conditions were significantly lower than those observed under drained conditions (Fig. 2).

Our study also demonstrated that, in rice fields exposed to intermittent flooding and draining cycles, environmental factors regulating diurnal fluctuations in CO_2 flux are quite different from those governing seasonal variations. Under drainage conditions, soil CO_2 flux showed a single peak at 2 pm (14:00 h), and was lowest in the wee hours. This is in agreement with patterns recorded in forests [5], grassland [30] and dryland areas [31]. Furthermore, correlation analysis revealed that canopy air temperature and soil temperature explained most of the diurnal fluctuations in soil CO_2 flux. In contrast, soil CO_2 flux during the flooding period fluctuated within ± 2 $\mu mol.m^{-2}.s^{-1}$ and soil CO_2 flux rates had small negative values in the daytime (i.e., the paddy soil was obviously a net CO_2 sink.), although soil CO_2 fluxes were positive throughout the night. This occurred primarily because of the layer of standing water, which is the habitat of bacteria, phytoplankton, macrophytes and small fauna. The photosynthesis process of these aquatic organisms affects ecosystem respiration [22].

Sudden pulse-like events of rapidly increasing CO_2 efflux occur in soils under paddy fields in response to re-flooding after draining.

Similarly, an abrupt rise in near-surface soil moisture due to precipitation can cause an instantaneous soil respiration pulse [24,32]. Soil respiration is shown to respond rapidly and instantaneously to the onset of rain and return to the pre-rain rate shortly after the rain stops [32]. The likely reason for this is that CO_2 is heavier than air and accumulates by gravitation within the air spaces of the soil. Replacement of this gaseous carbon by dilution will not occur without water and, unstirred by turbulent mixing, accumulation of CO_2 within the soil will increase. A sudden flooding might simply seal the soil pores, replace the captured CO_2 by water, and release it back into the air [33]. These occurrences, termed "Birch effect", can have a marked influence on the ecosystem carbon balance [34,35]. Indeed, this transient effect was observed in several studies at the ecosystem [36] and soil [37] scales. **On the other hand**, our analysis indicates that soil CO_2 flux was gradually increased during flooding to draining conversion processes. Response of soil CO_2 flux rates to these processes can be viewed in terms of increased diffusivity due to decrease in water filled pore space. Besides this general effect of soil aeration on soil CO_2 flux, the higher soil respiration rates during the drainage periods may have resulted from the higher physiological activity of microorganismsin not limiting soil oxic conditions [22].

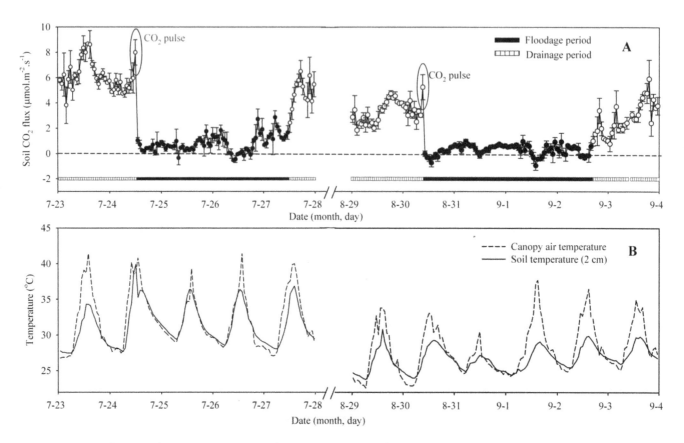

Figure 4. Soil CO_2 fluxes, soil temperature (2 cm) and canopy temperature before, during, and after the flooding and draining cycle in a rice paddy field.

We examined possible seasonal effects of temperature on soil CO_2 flux and found significant relation between the two under both flooded and drained conditions, but with widely differing mechanisms. In the present study, we found a negative relation between temperature and soil CO_2 flux, as long as soil CO_2 diffusivity is limiting as is the case during flooding period. An alternative explanation is based on the photosynthetic activity of the aquatic botany. The periods with the high photosynthetic

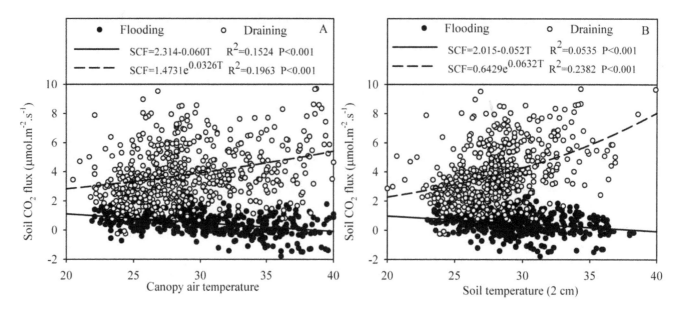

Figure 5. Relationship between soil CO_2 fluxes and temperature under both flooded and drained conditions. The solid lines represent the regression functions under flooded conditions, and the dashed lines represent the regression functions under drained conditions. (SCF: soil CO_2 fluxes; T: temperature).

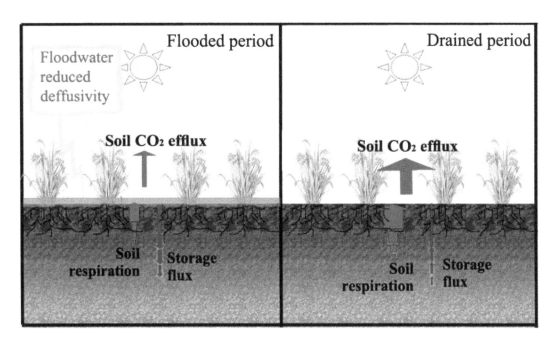

Figure 6. Schematic comparison of soil CO₂ flux processes under the flooded and the drained conditons in rice paddies.

active radiation are associated with conditions of high temperature in daytime (Fig. 3). Under drainage conditions, when soil aeration is assumed to be almost constant, soil temperature is considered to be a major control of soil CO_2 flux. Also the positive exponential relationship between soil CO_2 flux and temperature were observed during drainage period (Fig. 3). The results under drained conditions are similar to those of previous studies of CO_2 flux. For example, Chang et al. [38] found strong relationships between CO_2 flux and soil temperature and indicated that the rates of CO_2 emission increased exponentially with increases in soil temperature. Liu et al. [7], on the other hand, reported a significantly ($P<0.01$) linear relationship between soil CO_2 flux and soil temperature at a depth of 5 cm.

Conclusions

From the comparison of soil CO_2 fluxes under draining and flooding conditions we conclude that: (1) the net effluxes of CO_2 from the paddy soil were lower when the paddy was flooding than when it was draining, (2) the enhanced fluxes of CO_2 from the draining soil were due to removal of the barrier to gas transport from the soil surface to the air caused by the floodwater, and (3) there was a negative relation between soil CO_2 flux and temperature under flooding condition, whereas a positive relation under draining condition. The present study also showed how flooding and draining cycles affect the exchanges of CO_2 during the rice cultivation period. We need more measurements for multiple years to assess the long-term effect of an intermittent flooding and draining practice on the exchanges of CO_2 in rice paddy fields.

Author Contributions

Conceived and designed the experiments: FC YL. Performed the experiments: KYW YT SLL. Analyzed the data: ZGL. Contributed reagents/materials/analysis tools: GSZ. Wrote the paper: YL.

References

1. Robertson GP, Paul EA, Harwood RR (2000) Greenhouse gases in intensive agriculture: Contributions of individual gases to the radiative forcing of the atmosphere. Science 289: 1922–1925.
2. Li CF, Zhou DN, Kou ZK, Zhang ZS, Wang JP, et al. (2012) Effects of Tillage and Nitrogen Fertilizers on CH4 and CO_2 Emissions and Soil Organic Carbon in Paddy Fields of Central China. Plos One 7: e34642.
3. Guo JP, Zhou CD (2007) Greenhouse gas emissions and mitigation measures in Chinese agroecosystems. Agricultural and Forest Meteorology 142: 270–277.
4. Rodhe H (1990) A Comparison of the Contribution of Various Gases to the Greenhouse-Effect. Science 248: 1217–1219.
5. Xu M, Qi Y (2001) Soil-surface CO_2 efflux and its spatial and temporal variations in a young ponderosa pine plantation in northern California. Global Change Biology 7: 667–677.
6. Saito M, Miyata A, Nagai H, Yamada T (2005) Seasonal variation of carbon dioxide exchange in rice paddy field in Japan. Agricultural and Forest Meteorology 135: 93–109.
7. Liu Y, Li SQ, Yang SJ, Hu W, Chen XP (2010) Diurnal and seasonal soil CO_2 flux patterns in spring maize fields on the Loess Plateau, China. Acta Agriculturae Scandinavica Section B-Soil and Plant Science 60: 245–255.
8. Maier M, Schack-Kirchner H, Hildebrand EE, Schindler D (2011) Soil CO_2 efflux vs. soil respiration: Implications for flux models. Agricultural and Forest Meteorology 151: 1723–1730.
9. Yu XX, Zha TS, Pang Z, Wu B, Wang XP, et al. (2011) Response of Soil Respiration to Soil Temperature and Moisture in a 50-Year-Old Oriental Arborvitae Plantation in China. Plos One 6: e28397.
10. Li CF, Kou ZK, Yang JH, Cai ML, Wang JP, et al. (2010) Soil CO_2 fluxes from direct seeding rice fields under two tillage practices in central China. Atmospheric Environment 44: 2696–2704.
11. FAO (2010) FAOSTATS. Food and Agriculture Organization, Rome, Italy. http://faostat.fao.org/default.aspx (Accessed July, 2010).
12. Solomon S, Qin D, Manning M, Chen Z, Marquis M, et al. (eds.) (2007) Contribution of Working Group I to the Fourth Assessment Report of the Intergovernmental Panel on Climate Change. Cambridge, UK: Cambridge University Press.
13. Lee CH, Do Park K, Jung KY, Ali MA, Lee D, et al. (2010) Effect of Chinese milk vetch (Astragalus sinicus L.) as a green manure on rice productivity and methane emission in paddy soil. Agriculture Ecosystems & Environment 138: 343–347.
14. Ruddiman WF, Thomson JS (2001) The case for human causes of increased atmospheric CH₄. Quaternary Science Reviews 20: 1769–1777.
15. Fuller DQ, van Etten J, Manning K, Castillo C, Kingwell-Banham E, et al. (2011) The contribution of rice agriculture and livestock pastoralism to prehistoric methane levels: An archaeological assessment. Holocene 21: 743–759.

16. Miyata A, Leuning R, Denmead OT, Kim J, Harazono Y (2000) Carbon dioxide and methane fluxes from an intermittently flooded paddy field. Agricultural and Forest Meteorology 102: 287–303.

17. Iqbal J, Hu RG, Lin S, Hatano R, Feng ML, et al. (2009) CO_2 emission in a subtropical red paddy soil (Ultisol) as affected by straw and N-fertilizer applications: A case study in Southern China. Agriculture Ecosystems & Environment 131: 292–302.

18. Wassmann R, Papen H, Rennenberg H (1993) Methane Emission from Rice Paddies and Possible Mitigation Strategies. Chemosphere 26: 201–217.

19. Tyagi L, Kumari B, Singh SN (2010) Water management - A tool for methane mitigation from irrigated paddy fields. Science of the Total Environment 408: 1085–1090.

20. Li CL, Zhang JD, Hou ZQ (eds.) (1993) Techniques for high yield cultivations of several main crops. China Scientific and Technological Publishing House Beijing. (In Chinese).

21. Cheng YQ, Yang LZ, Cao ZH, Ci E, Yin SX (2009) Chronosequential changes of selected pedogenic properties in paddy soils as compared with non-paddy soils. Geoderma 151: 31–41.

22. Kogel-Knabner I, Amelung W, Cao ZH, Fiedler S, Frenzel P, et al. (2010) Biogeochemistry of paddy soils. Geoderma 157: 1–14.

23. Campbell CS, Heilman JL, McInnes KJ, Wilson LT, Medley JC, et al. (2001) Diel and seasonal variation in CO2 flux of irrigated rice. Agricultural and Forest Meteorology 108: 15–27.

24. Maier M, Schack-Kirchner H, Hildebrand EE, Holst J (2010) Pore-space CO_2 dynamics in a deep, well-aerated soil. European Journal of Soil Science 61: 877–887.

25. Itoh M, Sudo S, Mori S, Saito H, Yoshida T, et al. (2011) Mitigation of methane emissions from paddy fields by prolonging midseason drainage. Agriculture Ecosystems & Environment 141: 359–372.

26. Cai ZC, Tsuruta H, Rong XM, Xu H, Yuan ZP (2001) CH4 emissions from rice paddies managed according to farmer's practice in Hunan, China. Biogeochemistry 56: 75–91.

27. Minamikawa K, Sakai N (2006) The practical use of water management based on soil redox potential for decreasing methane emission from a paddy field in Japan. Agriculture Ecosystems & Environment 116: 181–188.

28. Bronson KF, Neue HU, Singh U, Abao EB (1997) Automated chamber measurements of methane and nitrous oxide flux in a flooded rice soil.1. Residue, nitrogen, and water management. Soil Science Society of America Journal 61: 981–987.

29. Cai ZC, Tsuruta H, Gao M, Xu H, Wei CF (2003) Options for mitigating methane emission from a permanently flooded rice field. Global Change Biology 9: 37–45.

30. Cao GM, Tang YH, Mo WH, Wang YA, Li YN, et al. (2004) Grazing intensity alters soil respiration in an alpine meadow on the Tibetan plateau. Soil Biology & Biochemistry 36: 237–243.

31. Han GX, Zhou GS, Xu ZZ, Yang Y, Liu JL, et al. (2007) Biotic and abiotic factors controlling the spatial and temporal variation of soil respiration in an agricultural ecosystem. Soil Biology & Biochemistry 39: 418–425.

32. Lee X, Wu HJ, Sigler J, Oishi C, Siccama T (2004) Rapid and transient response of soil respiration to rain. Global Change Biology 10: 1017–1026.

33. Chen D, Molina JAE, Clapp CE, Venterea RT, Palazzo AJ (2005) Corn root influence on automated measurement of soil carbon dioxide concentrations. Soil Science 170: 779–787.

34. Birch HF (1964) Mineralisation of plant nitrogen following alternate wet and dry conditions. Plant Soil 20: 43–49.

35. Unger S, Maguas C, Pereira JS, David TS, Werner C (2010) The influence of precipitation pulses on soil respiration - Assessing the "Birch effect" by stable carbon isotopes. Soil Biology & Biochemistry 42: 1800–1810.

36. Inglima I, Alberti G, Bertolini T, Vaccari FP, Gioli B, et al. (2009) Precipitation pulses enhance respiration of Mediterranean ecosystems: the balance between organic and inorganic components of increased soil CO2 efflux. Global Change Biology 15: 1289–1301.

37. Denef K, Six J, Bossuyt H, Frey SD, Elliott ET, et al. (2001) Influence of dry-wet cycles on the interrelationship between aggregate, particulate organic matter, and microbial community dynamics. Soil Biology & Biochemistry 33: 1599–1611.

38. Chang SC, Tseng KH, Hsia YJ, Wang CP, Wu JT (2008) Soil respiration in a subtropical montane cloud forest in Taiwan. Agricultural and Forest Meteorology 148: 788–798.

9

CO$_2$ Efflux from Shrimp Ponds in Indonesia

Frida Sidik*, Catherine E. Lovelock

The School of Biological Sciences, The University of Queensland, St Lucia, Queensland, Australia

Abstract

The conversion of mangrove forest to aquaculture ponds has been increasing in recent decades. One of major concerns of this habitat loss is the release of stored 'blue' carbon from mangrove soils to the atmosphere. In this study, we assessed carbon dioxide (CO$_2$) efflux from soil in intensive shrimp ponds in Bali, Indonesia. We measured CO$_2$ efflux from the floors and walls of shrimp ponds. Rates of CO$_2$ efflux within shrimp ponds were 4.37 kg CO$_2$ m^{-2} y^{-1} from the walls and 1.60 kg CO$_2$ m^{-2} y^{-1} from the floors. Combining our findings with published data of aquaculture land use in Indonesia, we estimated that shrimp ponds in this region result in CO$_2$ emissions to the atmosphere between 5.76 and 13.95 Tg y^{-1}. The results indicate that conversion of mangrove forests to aquaculture ponds contributes to greenhouse gas emissions that are comparable to peat forest conversion to other land uses in Indonesia. Higher magnitudes of CO$_2$ emission may be released to atmosphere where ponds are constructed in newly cleared mangrove forests. This study indicates the need for incentives that can meet the target of aquaculture industry without expanding the converted mangrove areas, which will lead to increased CO$_2$ released to atmosphere.

Editor: Vishal Shah, Dowling College, United States of America

Funding: The funding of the travel of the first author to Indonesia was supported by Graduate School International Travel Award (GSITA), University of Queensland, Australia. The funders had no role in study design, data collection and analysis, decision to publish, or preparation of the manuscript.

Competing Interests: The authors have declared that no competing interests exist.

* E-mail: f.sidik@uq.edu.au

Introduction

Soil is one of major sources of CO$_2$ emissions to the atmosphere [1,2,3]. Soil respiration, determined by measuring the CO$_2$ efflux from soil surface, is primarily from the respiration of soil organisms and roots [1,3]. On a global scale, rates of soil respiration in vegetated biomes have a positive relationship with plant productivity, which contributes to soil metabolic activity [1]. In the absence of vegetation, e.g. when land is converted to aquaculture ponds, the microbial community plays a major role in soil respiration and can release large amounts of CO$_2$ to the atmosphere [1,4]. During pond construction and operation sediment carbon is increasingly exposed to air, microbial activity accelerates which may result in increases in CO$_2$ efflux from the soil [5,6].

Mangroves are known to be habitats that sequester and store significant amounts of carbon, referred as 'blue' carbon [7–10]. The carbon stored in mangroves is mostly found below ground, comprised of highly organic soils and roots [7,11]. The removal of the mangrove forest (aboveground biomass) leads to reduction of carbon sequestration and the release of soil carbon stocks in the form of CO$_2$ to the atmosphere [5,6,12]. Recent studies have provided global estimates of the CO$_2$ efflux contribution to global greenhouse gas (GHG) emissions due to mangrove loss in order to assess the potential implications of continuing mangrove wetland conversion and the potential for GHG mitigation schemes [5,6,12,13], yet there are few empirical studies of GHG emissions from aquaculture ponds that occur in converted mangrove areas.

The conversion of mangroves to aquaculture ponds has been a critical issue in Indonesia. With a cover of 3,112,989 ha of mangrove forests, which is the largest portion of remaining global mangrove cover [13,14], Indonesia's coasts also comprise exten- sive areas of aquaculture ponds [15,16,17]. Increasing shrimp production in the 1990s led to the expansion of mangrove forest conversion to aquaculture ponds at a rate of 3.67% per year [15]. Rapid conversion of mangrove forests to aquaculture ponds mainly occurred in regions in Sumatra and Kalimantan, however many of those areas have been abandoned in the past decades [15]. Recently, the government has begun to manage the abandoned areas and has expressed interest in increasing shrimp production through revitalisation of existing shrimp ponds [17]. There has also been a substantial effort to restore ponds to mangrove forests by both government and non-government organizations [15,18]. But there is little information of the carbon emissions that are avoided through restoration, which could provide further incentives for restoration of non-productive ponds.

This study examined CO$_2$ efflux from soil in an intensive shrimp farm in Bali, Indonesia. Shrimp aquaculture is one of major aquaculture activities in Bali. In this study we measured the CO$_2$ efflux from the soil of the floors and walls of the shrimp ponds that were established 20 years ago. Furthermore, we estimated from these measurements a potential annual CO$_2$ efflux from shrimp ponds using a dataset of aquaculture area in Indonesia. The results of this study have implications for initiatives aimed at preparedness for use of the clean development mechanism (CDM) and other emissions trading in countries with extensive aquaculture.

Results

Measures of CO$_2$ efflux within shrimp ponds showed that rates of CO$_2$ efflux from the walls were 3.15 μmol m^{-2} s^{-1} which exceeded emissions from the floors of the pond which were 1.15 μmol m^{-2} s^{-1} (Figure 1, $F_{1,28} = 25.66$, P<0.0001). Extrapolation of CO_2 efflux rates to annual CO_2 loss from shrimp ponds to atmosphere gave values of 4.37 kg CO_2 m^{-2} y^{-1} (walls) and 1.60 kg CO_2 m^{-2} y^{-1} (floors). Soil temperature varied significantly between floors and walls of the pond ($F_{1,28} = 21.81$, $P<0.0001$), ranging from 31.9°C to 37.0°C in the floors (mean of 34.5°C) and from 30.9°C to 45.1°C in the walls (mean of 40°C).

Using the above values we developed estimates of annual CO_2 efflux from shrimp ponds in Indonesia. There are two published values within Indonesian government reports of the area of shrimp ponds in Indonesia [16,17]. These gave a "low" estimate of CO_2 emissions of 5.76 Tg y^{-1} and a "high" estimate of 13.95 Tg y^{-1}.

Discussion

Our measurements of CO_2 efflux from the floors of shrimp ponds in Bali found lower rates of soil CO_2 efflux than have been measured in mangroves that were cleared in Belize [5]. The lower CO_2 efflux rates may be due to the lower carbon density in the Bali soils, which are mineral soils (0.019 g C cm^{-3}), compared to the average mangrove soil carbon density in Indonesia and Southeast Asia and the highly organic peat soils in Belize [5,19]. But we found higher rates of CO_2 efflux from walls than pond floors, which increased the overall CO_2 efflux per area from the ponds. High respiration in pond walls was coincident with warmer temperature in pond walls than pond floors. The structure of ponds, where soil was pushed up to form walls, allows the soil in the walls to receive high levels of solar radiation and high levels of aeration probably leading to increased rates of C oxidation [5]. Warmer temperatures stimulate microbial activity resulting in greater CO_2 efflux from decomposition thereby increasing the release of CO_2 to the atmosphere [1,2,5,6,20].

Our measured rates of CO_2 efflux from shrimp ponds were within the range reported in previous work by Burford and Longmore (2001) who measured CO_2 efflux from pond water surfaces rather than from the soil surface [4,21]. CO_2 efflux measured by Burford and Longmore (2001) [4] was highly variable which likely reflects variation in pond management, temporal factors or other biogeochemical factors [2,4,6,13,19].

Emissions profiles of land conversion may be variable, dependent on climatic and substrate factors, year and type of land uses [2,4,6,13]. Studies of soil CO_2 emissions from several forms of established land uses in converted Asian peat forest (Table 1) indicate comparable levels of CO_2 emissions to shrimp aquaculture [22–25].

The values of CO_2 efflux from pond floors were similar to other land uses, e.g. paddy fields, oil palm and sago palm plantation. CO_2 emissions from recently constructed ponds in converted mangrove forests may be higher than presented here. Therefore the expansion of shrimp ponds and mangrove forest conversion could result in higher magnitudes of annual CO_2 emissions from aquaculture compared to the findings from this study. In the economic analysis of the benefits of protecting mangroves for their carbon values, Siikamäki et al (2012) [19] assumed a loss of 27% (low end) and 90% (high end) of soil carbon once mangroves were converted and a mean soil carbon of 0.0418 g C cm^{-3} for Indonesia. Over 20 years loss of 27% carbon gives a CO_2 emission rate of 2.2 kg CO_2 m^{-2} y^{-1}. Our data, from ponds in mineral soils, are consistent with this estimate.

The conversion of mangrove forest to aquaculture causes significant increases in CO_2 efflux to the atmosphere, and thus strong incentives to preserve coastal wetlands are needed to avoid increasing CO_2 released to atmosphere. Restoration and conservation of coastal wetlands are the primarily mechanisms proposed to reduce CO_2 emission driven from mangrove loss [13]. However, application of these mechanisms is difficult in regions targeted for enhanced aquaculture production, as is the case in Indonesia [15,17]. Revitalisation of existing shrimp ponds is an additional option to meet the goals of increased aquaculture productivity without expanding the converted area that leads to increasing CO_2 released to atmosphere.

Our estimates of CO_2 efflux from shrimp pond aquaculture may be improved by increasing the spatial and temporal sampling of CO_2 emissions from aquaculture land use. Soil respiration in forested ecosystems varies with metabolic activity of tree roots, variation in temperature and water content of soils [2]. In aquaculture ponds live roots are eliminated and soils are saturated which may reduce variation in CO_2 efflux. However the temperature of pond soils may vary widely which could be incorporated into more complex modelling of CO_2 emissions from aquaculture ponds. Additionally, greater confidence in our scaling up requires improved estimates of the aerial extent of existing shrimp ponds in Indonesia [26]. This study used the area of existing aquaculture published in 2001 (438,010 ha) to calculate the "high" estimate of country-wide CO_2 emissions from shrimp ponds [16]. We assumed that all aquaculture land was utilised for shrimp pond production as the data indicated that shrimp (*Paneous monodon*) culture dominated in aquaculture areas in Indonesia, second only to culture of milkfish (*Chanos chanos*). Our "high" estimate scenario however is likely an underestimate because conversion of mangrove forest to ponds was greater than were reported [15]. Additionally, our "low" estimate scenario, which used more recent data (2011) of existing working shrimp ponds in 22 districts in Indonesia of 180,844 ha [17], did not include previously converted ponds that are not currently in production. Improved documentation of areas of ponds and management of ponds would enhance our confidence of estimates of CO_2 emissions contributed by aquaculture in Indonesia.

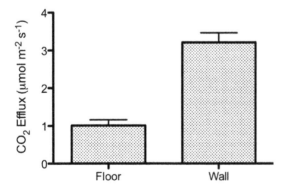

Figure 1. CO_2 efflux from the floors and walls of shrimp ponds in Bali, Indonesia. The rates of CO_2 efflux from the walls were significantly higher than pond floors ($F_{1,28} = 25.66$, $P<0.0001$).

Conclusion

The conversion of mangrove forests to shrimp ponds resulted in CO_2 losses to the atmosphere for 4.37 kg CO_2 m^{-2} y^{-1} from the walls and 1.60 kg CO_2 m^{-2} y^{-1} from the floors of ponds. Our estimate of annual CO_2 emission from shrimp ponds in Indonesia region was between 5.76 and 13.95 Tg y^{-1}. These values are comparable to CO_2 emissions from other land uses of converted

Table 1. Comparison of CO_2 emissions from land uses linked to tropical forest loss.

Type of land conversion	Location	Carbon emission	Source
Shrimp ponds	Bali, Indonesia	1.60 kg CO_2 m^{-2} y^{-1} (floors)	This study
		4.37 kg CO_2 m^{-2} y^{-1} (walls)	
Mangrove clearing	Belize	2.9 – 10.6 kg CO_2 m^{-2} y^{-1}	Lovelock et al (2011)
Paddy field	Kalimantan, Indonesia	1.4 kg CO_2 – C m^{-2} y^{-1}	Hadi et al (2005)
Abandoned paddy field	South Kalimantan, Indonesia	~1.2 – 1.5 kg CO_2 – C m^{-2} y^{-1}	Inubushi et al (2003)
Oil palm plantation	South Asia	~0.75 – 1.1 kg CO_2 m^{-2} y^{-1}	Reijnders and Huijbregts (2008)
	Sarawak, Malaysia	1.5 kg CO_2 – C m^{-2} y^{-1}	Melling et al (2005)
Sago palm plantation	Sarawak, Malaysia	1.1 kg CO_2 – C m^{-2} y^{-1}	Melling et al (2005)
Rice-soybean rotation field	Kalimantan, Indonesia	2 kg CO_2 – C m^{-2} y^{-1}	Hadi et al (2005)

lowland forests. The CO_2 emission released to atmosphere might be higher than we report here if ponds are constructed in newly cleared mangrove forests. Knowledge of the amounts of CO_2 released from shrimp ponds may contribute to preparedness for use of the clean development mechanism (CDM) and other emissions trading schemes in countries, particularly Indonesia, which have made a commitment to protect mangrove forests concurrently with commitments to meet targets of high production in the aquaculture industry.

Materials and Methods

This study was conducted in Perancak estuary, Bali, Indonesia (8° 23' 40" S, 114° 37' 39" E). The area is a coastal plain associated with the Perancak River that is comprised of a mix of paddy fields, mangrove forests and aquaculture ponds. The estuary is characterised by sedimentary limestones and alluvial platforms. The soils are related to the volcanic stratigraphy derived from the Batur volcano [27,28]. Soil organic carbon contents in this area, measured by Sidik et al (unpublished), are 0.019 g C cm^{-3} in the mangrove forests. In the 1990s, huge areas of mangrove forests were converted to intensive shrimp ponds, however, there is no literature that provides accurate information of the extent of mangroves cleared for shrimp ponds in the area. Since early 2000, these aquaculture activities have been diminished due to the global economic crisis and diseases, which have resulted in numerous shrimp ponds ceasing to be in production.

The measurements of CO_2 efflux were made on the 24 and 25 November 2012. We made measurements between 10.30 am and 2 pm local time sampling the floors and walls of each pond twice during the measurement campaign. We selected three working shrimp ponds, which were about 20 years old. All the shrimp ponds were located in the same farm with similar soils and pond management. The area of each pond was about 2000 m^2 with a depth of 1.5 m. A day before the measurements of CO_2 efflux the ponds were drained. Ponds in this farm are usually stocked with *Panaeus vanamei* and aerated with paddlewheels. We measured the CO_2 efflux from the soil in floors and walls of these ponds. CO_2 efflux from soils was measured using a LiCor 6400 portable photosynthesis system configured with the LiCor soil CO_2 flux chamber (LiCor Corp, Lincoln, NE, USA) inserted 0.5 cm into the soil. Soil temperature was measured at 2 cm depth simultaneously with CO_2 efflux. We conducted five measurements in the floor and walls of each pond. The locations of the measurements within each pond were randomly distributed over the floors and walls and were at least 1 m apart from each other. The surface of the floor of the pond was scraped to remove the microalgal film prior to each measurement after which the chamber was placed on the surface of the sediment. Differences in CO_2 efflux rates between floors and walls were assessed using ANOVA.

We extrapolated to CO_2 efflux for shrimp ponds in Indonesia by multiplying our measurements scaled up to an annual rate by the reported area of shrimp ponds. The areas of shrimp ponds were derived from recent information published by Indonesian government [16,17]. As uncertainties existed in the available data, we used a conservative approach to estimate the CO_2 efflux from shrimp ponds [26]. We multiplied the mean of CO_2 efflux from ponds by two different estimates of pond areas to generate "high" and "low" estimates of CO_2 emission from ponds. The lower area came from published information from the Ministry of Marine Affairs and Fisheries [17] and the high estimate from literature [16] derived from the Ministry of Agriculture. Our surveys of the area indicated that pond floors occupied approximately 90% of the shrimp pond footprint and walls about 10%. However, walls were three dimensional, typically 1.5 m high and 1 m wide at the top (4 linear meters). We therefore estimated total CO_2 efflux from floors and walls as (0.9 x floor efflux) + (0.1 x wall efflux x 4) to incorporate the three dimensional nature of the walls.

Ethics Statement

The field measurement was undertaken in a private shrimp farm in Perancak, Bali, with the permission from the owner.

Acknowledgments

We thank our field personnel from the Institute for Marine Research and Observation, Ministry of Marine Affairs and Fisheries: Hanggar Prasetio, Nuryani Widagti, Nyoman Surana, and Ajis for their support during the field measurements. We thank Titik Herlina and Didu Supriatna for providing access to their shrimp farm in Perancak, Bali for this study.

Author Contributions

Conceived and designed the experiments: FS CEL. Performed the experiments: FS CEL. Analyzed the data: FS. Contributed reagents/materials/analysis tools: FS CEL. Wrote the paper: FS. Edited the manuscript: CEL.

References

1. Raich JW, Schlesinger WH (1992) The global carbon dioxide flux in soil respiration and its relationship to vegetation and climate. Tellus B 44: 81–99.
2. Raich JW, Tufekcioglu A (2000) Vegetation and soil respiration: correlations and controls. Biogeochemistry 48: 71–90.
3. Raich JW, Potter CS (1995) Global patterns of carbon dioxide emissions from soils. Global Biogeochem Cy 9: 23–36.
4. Burford MA, Longmore AR (2001) High ammonium production from sediments in hypereutrophic shrimp ponds. Mar Ecol Prog Ser 224: 187–195.
5. Lovelock CE, Ruess RW, Feller IC (2011) CO_2 efflux from cleared mangrove peat. PLoS ONE 6: e21279. doi:10.1371/journal.pone.0021279.
6. Pendleton L, Donato DC, Murray BC, Crooks S, Jenkins WA, et al. (2012) Estimating global "blue carbon" emissions from conversion and degradation of vegetated coastal ecosystems. PLoS ONE 7: e43542. doi:10.1371/journal.pone.0043542.
7. Donato DC, Kauffman JB, Murdiyarso D, Kurnianto S, Stidham M, et al. (2011) Mangroves among the most carbon-rich forests in the tropics. Nature Geosci 4: 293–297.
8. Alongi DM (2002) Present state and future of the world's mangrove forests. Environ Conserv 29: 331–349.
9. Duarte CM, Middelburg JJ, Caraco N (2005) Major role of marine vegetation on the oceanic carbon cycle. Biogeosciences 2: 1–8.
10. Chmura GL, Anisfeld SC, Cahoon DR, Lynch JC (2003) Global carbon sequestration in tidal, saline wetland soils. Global Biogeochem Cy 17: 1111–1120.
11. Lovelock CE (2008) Soil respiration and belowground carbon allocation in mangrove forests. Ecosystems 11: 342–254.
12. Donato DC, Kauffman JB, Mackenzie RA, Ainsworth A, Pfleeger AZ (2012) Whole-island carbon stocks in the tropical Pacific: Implications for mangrove conservation and upland restoration. J Environ Manage 97: 89–96.
13. Crooks S, Herr D, Tamelander J, Laffoley D, Vandever J (2011) Mitigating climate changethrough restoration and management of coastal wetlands and near-shore marine ecosystems: challenges and opportunities. Environment Department Paper 121. World Bank. Washington DC.
14. Giri C, Ochieng E, Tieszen LL, Zhu Z, Singh A, et al. (2011) Status and distribution of mangrove forests of the world using earth observation satellite data. Global Ecol Biogeogr 20: 154–159.
15. Ministry of Forestry (2002) Lingkungan: udang di balik mangrove. Available at: http://www.dephut.go.id/Halaman/STANDARDISASI_&_ LINGKUNGAN_KEHUTANAN/INFO_VI02/VII_VI02.htm. Accessed 2012 December 24.
16. Puspita L, Ratnawati E, Suryadiputre INN, Meutia AA (2005) Lahan basah buatan di Indonesia. Wetlands International - Indonesia Programme. Bogor.
17. Ministry of Marine Affairs and Fisheries (2012) Revitalisasi tambak, KKP pacu produksi udang. Available at: http://www.kkp.go.id/index.php/arsip/c/7800/REVITALISASI-TAMBAK-KKP-PACU-PRODUKSI-UDANG/. Accessed 2012 December 24.
18. Wetlands International (2012) Replanting mangroves in the abandoned shrimp ponds. Available at: http://www.wetlands.org/Whatwedo/Ouractions/IndonesiaMangroverestorationinJava/OurwokinJava/tabid/2294/Default.aspx. Accessed 2013 February 11.
19. Siikamäki J, Sanchirico JN, Jardine SL (2012) Global economic potential for reducing carbon dioxide emissions from mangrove loss. Proc Natl Acad Sci USA 109:14369–14374.
20. Chimner RA (2004) Soil respiration rates of tropical peatlands in Miconesia and Hawaii. Wetlands 24: 51–56.
21. Burford M, Costanzo SD, Dennison WC, Jackson CJ, Jones AB, et al. (2003) A synthesis of dominant ecological processes in intensive shrimp ponds and adjacent coastal environments in NE Australia. Mar Poll Bull 46: 1456–1469.
22. Hadi A, Inubushi K, Furukawa Y, Purnomo E, Rasmadi M, Tsuruta H (2005) Greenhouse gas emissions from tropical peatlands of Kalimantan, Indonesia. Nutr Cycl Agroecosys 71: 73–80.
23. Inubushi K, Furukawa Y, Hadi A, Purnomo E, Tsuruta H (2003) Seasonal changes of CO_2, CH_4 and N_2O fluxes in relation to land-use change in tropical peatlands located in coastal area of South Kalimantan. Chemosphere 52: 603–608.
24. Reijnders L, Huijbregts MAJ (2008) Palm oil and the emission of carbon based greenhouse gases. J Clean Prod16: 477–482.
25. Melling L, Hatano R, Goh KJ (2005) Soil CO2 flux from three ecosystems in tropical peatland of Sarawak, Malaysia. Tellus B 57: 1–11.
26. Friess DA, Webb EL (2011) Bad data equals bad policy: how to trust estimates of ecosystem loss when there is so much uncertainty? Environ Conserv 38: 1–5.
27. Tanaka T, Sunarta N (1994) Relationship between regional changes of soil physical properties and volcanic stratigraphy on the southern slope of Batur volcano in the island of Bali, Indonesia. Environ Geol 23: 182–191.
28. McTaggart WD (1989) Hydrologic management in Bali. Singapore J Trop Geo 9: 96–111.

10

Role of Megafauna and Frozen Soil in the Atmospheric CH$_4$ Dynamics

Sergey Zimov*, Nikita Zimov

Northeast Science Station, Pacific Institute for Geography, Russian Academy of Sciences, Cherskii, Russia

Abstract

Modern wetlands are the world's strongest methane source. But what was the role of this source in the past? An analysis of global ^{14}C data for basal peat combined with modelling of wetland succession allowed us to reconstruct the dynamics of global wetland methane emission through time. These data show that the rise of atmospheric methane concentrations during the Pleistocene-Holocene transition was not connected with wetland expansion, but rather started substantially later, only 9 thousand years ago. Additionally, wetland expansion took place against the background of a decline in atmospheric methane concentration. The isotopic composition of methane varies according to source. Owing to ice sheet drilling programs past dynamics of atmospheric methane isotopic composition is now known. For example over the course of Pleistocene-Holocene transition atmospheric methane became depleted in the deuterium isotope, which indicated that the rise in methane concentrations was not connected with activation of the deuterium-rich gas clathrates. Modelling of the budget of the atmospheric methane and its isotopic composition allowed us to reconstruct the dynamics of all main methane sources. For the late Pleistocene, the largest methane source was megaherbivores, whose total biomass is estimated to have exceeded that of present-day humans and domestic animals. This corresponds with our independent estimates of herbivore density on the pastures of the late Pleistocene based on herbivore skeleton density in the permafrost. During deglaciation, the largest methane emissions originated from degrading frozen soils of the mammoth steppe biome. Methane from this source is unique, as it is depleted of all isotopes. We estimated that over the entire course of deglaciation (15,000 to 6,000 year before present), soils of the mammoth steppe released 300–550 Pg (10^{15} g) of methane. From current study we conclude that the Late Quaternary Extinction significantly affected the global methane cycle.

Editor: Ben Bond-Lamberty, DOE Pacific Northwest National Laboratory, United States of America

Funding: The authors have no support or funding to report.

Competing Interests: The authors have declared that no competing interests exist.

* E-mail: sazimov55@mail.ru

Introduction

Ice core analyses indicate that during the Pleistocene-Holocene transition (18,000 to 11,000 year before present (BP)), coincident with a rise in Greenland's temperatures, atmospheric methane content increased from ~1000 Tg (1 Tg = 10^{12} g) to ~2000 Tg (Fig. 1A) [1–5]. The modern atmospheric lifetime of methane is approximately 10 years [6] (atmospheric methane lifetime is the ratio between atmospheric methane content and global annual methane flux). If the hypothesis that oxidation rates of methane in the atmosphere did not vary substantially in the past is accepted, and the lifetime of methane in the atmosphere stayed stable, then global CH$_4$ emissions during Pleistocene-Holocene transition (18–11 ka BP) increased from ~100 Tg/yr to ~200 Tg/yr. In the Holocene, relative to the late Pleistocene, the climate was stable while global methane emission was not (Fig. 1B). From these data, it can be assumed that during 15–6 ka BP strong methane sources existed. By 6 ka BP, these sources vanished or substantially decreased in strength. This, in turn, caused a decrease in atmospheric concentrations of methane. Later, in the second part of the Holocene, other sources appeared (or were activated) and global emissions of methane rose by approximately 50 Tg/yr.

A comparison of ice core data from Greenland and Antarctica indicated that in the Last Glacial Maximum (LGM), atmospheric methane concentrations were roughly equal in the southern and northern hemispheres (Fig. 1B) [5]. Which indicate approximately equal methane production in both northern and southern hemisphere. However, simultaneous with a CH$_4$ concentration rise, an interhemispheric CH$_4$ gradient appeared, indicating the arrival of a strong northern source of methane. Based on this gradient, a northern methane source of 40–50 Tg/yr for the Bølling-Allerod (~14.5–13 ka BP), 40 Tg/yr for the Younger Dryas (~13–11.8 ka BP), and 60–70 Tg/yr for the Preboreal (~11.8–9 ka BP) was determined [2–5,8]. In contrast to the atmospheric methane concentration, the interhemispheric gradient stayed relatively stable over the course of the Holocene.

The origins of the dynamics noted above are actively debated. The rise in atmospheric methane during Pleistocene-Holocene transition can be explained by the expansion of boreal or tropical wetlands [1], the destabilization of gas clathrates [7], or the production of CH$_4$ following the thawing of organic rich permafrost under anaerobic conditions [8].

Between 5 and 1 ka BP in a stable climate, methane emissions increased by at least 50 Tg/yr (Fig. 1B). The reason for this phenomenon is unclear to us. We are not convinced by the hypothesis that humans early farming have caused this rise [9]. Modern rice paddies and livestock produce 70 and 90 Tg CH$_4$/yr

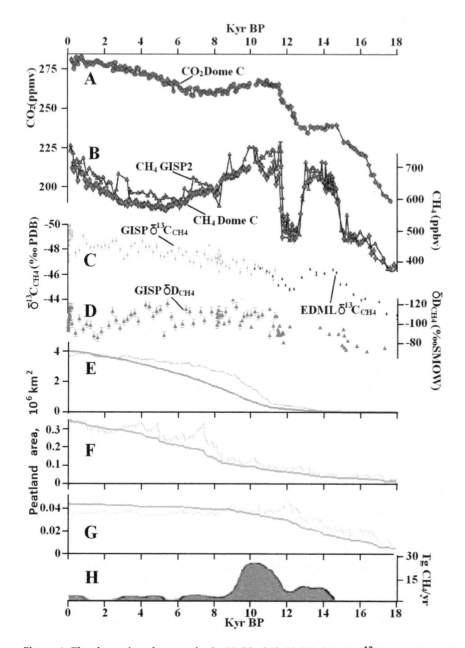

Figure 1. The dynamics of atmospheric: A) CO₂ [17]; B) CH₄ [17]; C) δ¹³CH₄ [12,13]; and D) δDCH₄ [11,13]. GISP1&2–Greenland Ice Sheet Projects. EDML - EPICA Dronning Maud Land Ice Core. The dynamics of peatland areas: E) boreal [17,18]; F) tropical [18]; and G) southern [18]. Thin lines represent emissions from these wetlands based on the model of wetland succession (arbitrary unit). H) Methane emissions from northern Siberian permafrost [8].

respectively [6].During 5 to 1 ka BP, the human population was two orders of magnitude lower than at present [10], therefore we think that increase in human population is not a sufficient explanation for the rise. Additionally, the first rice paddies were cultivated in areas of natural wetlands, and domestic animals replaced wild animals in pastures.

The isotopic composition of atmospheric methane obtained from ice core analyses allows us to understand the dynamics of primary methane sources [6,11,12]. Methane from different sources is different in isotopic composition (Fig. 2). The dynamic of isotopic methane signature from the LGM to the present is now known (Fig. 1C and 1D) [11–13]. This allows evaluation of the input to global methane emissions from different sources.

An attempt to estimate primary CH₄ sources during Pleistocene-Holocene transition was made by *Fischer et al.* [12]. The authors employed a model that describes the dynamics of atmospheric methane, its interhemispheric gradient, and the ¹³C and deuterium budgets. However Monte Carlo approaches allowed the dynamics for only two sources to be calculated and according to their results, biomass burning emissions in the LGM for the 3.7 year atmospheric CH₄ lifetime were 65 Tg/yr, and for the 5.6 year lifetime equalled 41 Tg/yr (the same as the biomass burning emissions of today [6]). Boreal wetland emissions during the LGM were 4 Tg/yr and later increased to 54 Tg/yr in the Preboreal, regardless of lifetime duration [12]. The boreal wetland dynamic was predictable, since in the model the presence of an interhemispheric gradient was fully dependent on the presence of a

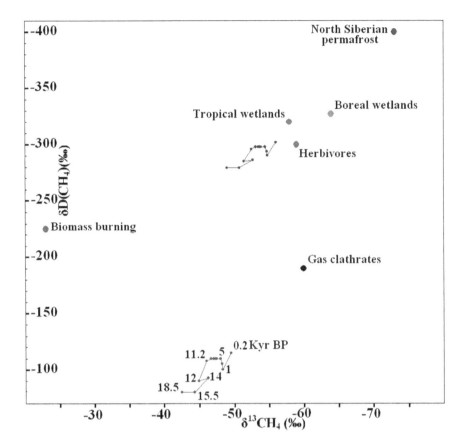

Figure 2. The isotopic signatures (¹³C and deuterium) of primary natural sources [12]. The blue line represents the dynamics of the isotopic content of methane in the atmosphere above Greenland from the LGM to the pre-industrial [11–13]; and purple line represents the dynamic of the isotopic content of the weighted average source calculated for conditions as in the scenario in Fig. 3A.

single northern source – boreal wetlands (permafrost was not considered). The Fischer model assumed that interhemispheric gradient in the LGM was small and, therefore the model showed that boreal wetland emissions were small, as well. Over the course of Pleistocene-Holocene transition interhemispheric gradient substantially increased and therefore, the model was supposed to show and did show a strong rise in boreal wetland emissions regardless of the dynamic of the isotopic content of atmospheric methane and its lifetime.

Below, we present our isotopic model. The general idea and equations of the model are similar to the Fischer model and most input parameters are the same. However, the following differences between our and Fischer's models allowed us to constrain or estimate all main methane sources from the LGM to the present: 1) we obtained measurements not only for the Pleistocene-Holocene transition but also for the Holocene; 2) we added into the model an additional northern source – permafrost; 3) we did not add restrictions connected with the interhemispheric gradient; interhemispheric methane gradient dynamics data was utilized only to test the results of our modelling: we checked by modelling whether the strengths of the northern sources corresponded with the values of the interhemispheric gradient; 4) we set emissions dynamics for boreal and tropical wetlands using a model based on a large global database of basal peat initiation dates. These allowed us to reduce the number of unknowns in the isotopic model and, in contrast to Fischer's results, allowed us to calculate our system of equations, not with the Monte Carlo approach, but analytically.

Background

Before analysing the isotopic model, we want to discuss current knowledge about the glacial–interglacial dynamic of the main methane sources. Below we compare the results obtained from the modelling with these initial estimates.

Gas Clathrates

The storage of gas clathrates in the bottom sediments of seas and oceans is approximately 10000–20000 Pg [14]. But we doubt that this source seriously influenced the methane concentration in the atmosphere. Our observations of methane bubbling in different bottom sediments [15], have indicated that if CH_4 production in bottom sediments is low, bubbling occurs only at low pressure. In lakes, bubbling occurs at record low atmospheric pressure. In rivers, channels, and on the shelf of the East-Siberian Sea it occurs at low water levels. Record low pressure in the sediments occurs rarely and during that short time period releases the methane which had accumulated in the sediments for several months or even years. As soon as the pressure (water level) increases, bubbling immediately stops [15]. We assume that the same processes are correct for gas clathrates on the continental slope and the ocean floor. During deep glacial ocean regression the pressure in the bottom sediments declined much more strongly than in lakes and rivers. Gas clathrates emissions during the regression should have increased, with the maximum emissions being at the deepest regression, and after the LGM peak, the emissions should have immediately stopped as soon as ocean levels began to rise. However, for atmospheric methane dynamics we did not determine such an outcome. Gas clathrate methane has a very high concentration of deuterium isotope [11,12] and, therefore, even minor changes in the emissions of this source into the atmosphere, should substantially change the deuterium content in the atmospheric methane; however, before and after the LGM, the deuterium content in the atmosphere remained stable (Fig. 1C) [11].

Gas clathrate destabilization during ocean regression must have taken place [14]. But this methane did not penetrate to the atmosphere. Field experiments indicate that methane bubbles of a few centimetres in diameter lose ~20% of their methane content per 10 m of updrift (methane within the bubble is replaced by nitrogen and oxygen) [15]. Therefore, methane from the sea sediments (from hundreds of metres in depth and more) was dissolved in the ocean and oxidized to CO_2 [16]. During Pleistocene-Holocene transition the deuterium content of atmospheric CH_4 decreased (Fig. 1C). Amongst natural sources, the δDCH_4 of gas clathrates is the heaviest (Fig. 2). Therefore, the destabilization of gas clathrates could not be the cause of an atmospheric CH_4 rise [11], since a source with low deuterium content is required [11].

Wetlands

Wetland CH_4 emission dynamics can be reconstructed using the dynamics of peat accumulation [17,18]. The conditions required for methanogenesis and peat accumulation are similar – plant remains should appear under anaerobic conditions (i.e., isolated by a layer of water). Some wetlands become episodically covered with water. In such places peat does not accumulate, but methane is still occasionally emitted. The conditions for the existence of peatlands and episodic wetlands are similar – the more humid the climate and the less the drainage, the wetter the soils. If, in a specific region due to climate change, the area of peatlands increases, then the area of episodic wetlands will increase as well. Therefore, the dynamics of wetland areas as well as their CH_4 emissions should be correlated with the dynamics of peatland areas.

The dynamics of peatland areas for three climatic zones are shown in Figures 1E, 1F and 1G (thick lines). These graphs were built based on 1,700 ^{14}C basal peat initiation dates [17,18]. This is the world's biggest global database of peat initiation dates and it can allow reconstruction of the world wetland dynamics. From these graphs it can be concluded that during Pleistocene-Holocene transition, tropical and boreal peatlands were rare and also the rates of peat accumulation were an order of magnitude slower than during the stages of active peatland expansion [18]. Basal peat dates have not been determined for the LGM, but there are peatlands older than the LGM [17,18]. This means that in the LGM there were few peatlands.

There is no other evidences of abundancy of wetlands during the LGM [19]. An analysis of paleo-vegetation maps indicates that, during the LGM, forested areas were ten times smaller than during the Holocene [19,20]. Sphagnum spores (a reliable proxy for moist conditions) are abundant during the Holocene, but they are absent in pollen spectra during the LGM [20,21]. For northern territories with underlying permafrost, another reliable indicator of anaerobic conditions is soil methane [22]. In the frozen soils of the Holocene and other interglacial periods, methane is frequently present [22]. In soils that experience thawing, methane is always abundant, while in all of the frozen soils that accumulated during the LGM it is absent [22]. One of the additional reasons for glacial landscape dryness is a lowering of the erosion basis. At ocean regression in the Pleistocene, river mouths lowered, inclinations of the rivers increased and river beds deepened. Following the erosion basis, drainage of the territory increased and the area of

floodplains reduced. Most of the current vast moist plains in the glaciation were well drained uplands, and only when deglaciation finished and the water level (ground water level) reached the modern level, did landscape saturation occur. The Holocene climate was relatively stable but because of slow processes like land grading, silting of river beds and depressions, the area of peatlands increased over the course of Holocene. Accounting for all the above, we accept that wetlands during the LGM were rare. Now, knowing the dynamic of the wetlands area, we will try to reconstruct the methane emission dynamic from the wetlands.

The work of *McDonald et al.* [17] noted that, during the initial stage of peatland formation (high productive grass-herb formation), CH_4 emissions were higher than during later low productive stages. Therefore, for reconstructing peatland methane emissions, we utilized a model in order to take peatland succession into consideration. In the model we accept that maximum methane emission occurs at the time of peatland initiation, and declines afterwards. We assumed various scenarios of successions and in all cases the peatland succession model indicates that methane emission graph differ slightly from the dynamics graph of the peatland area. Only by accepting a very long stage of succession are the differences visible. In Figures 1E, 1F, and 1G the thin lines were built by assuming that, at the moment of initiation of every peatland, methane production was at a maximum, and very slowly declined. In 600 years, emissions were halved; in 1,600 years they were a quarter of their initial level, and six times lower during late succession. Even in the cases of long and deep succession, differences in graph shapes were not that pronounced.

Peat is a very good proxy for anaerobic conditions [17,18]. It is likely that the graphs in Figures 1D–F are the most precise data reflecting the dynamics of methane from wetlands. These graphs reflect the dynamic of methane emission in arbitrary units, but if we know the modern (pre-industrial) emission of boreal and tropical wetlands then, based on these graphs, we can estimate wetland methane emissions for any other time slice (up to the LGM). Utilizing these data we can determine whether wetlands could be the reason for the methane rise during Pleistocene-Holocene transition.

Area of southern wetlands (mostly Patagonian) smoothly grew from 18 to 12 ka BP (Fig. 1G), but their area was ~1% of the modern boreal wetland area. Therefore, it is unlikely they had an effect on the global dynamic of atmospheric methane. One can clearly see that the strong expansion of boreal wetlands only began around 11 ka BP, three thousand years after the methane atmospheric concentration reached Holocene levels and a strong interhemispheric gradient appeared (Fig. 1B). Tropical wetland expansion took place at an even later date. From these dynamics we can conclude that both tropical and boreal wetlands had little influence on the methane concentration rise during Pleistocene-Holocene transition. Between 11 and 7 ka BP numerous young wetlands appeared worldwide (Figs. 1D–F), so maximum wetland emissions should be expected during this period of time. However, atmospheric methane concentrations did not rise and instead declined (Fig. 1B), indicating that Pleistocene-Holocene transition other stronger sources existed that sharply declined in strength from 11 to 6 ka BP.

Permafrost

As shown during Pleistocene-Holocene transition few wetlands existed and the main source of methane could have been permafrost [8].

In northern Siberia during glaciation, mammoth steppe ecosystem existed [8,23,24]. In this ecosystem, water was often a limiting resource. Grass roots looking for water pierced the entire soil layer (down to the permafrost) where the temperature was never substantially higher than 0°C. Due to low decomposition, organic carbon accumulated not only on the surface but also in the deep active layer [24]. During glaciation on the planes of Siberia, massive loess strata (yedoma) accumulated. As dust accumulated on the soil surface, the lower soil horizons, together with organic remains, became incorporated into the permafrost. The carbon content in yedoma, despite low humus, is tens of kg/m^3 and sometimes reaches 1 ton/m^2 [25,26].

During deglaciation the depth of the summer thaw increased and ice wedges began to melt. In such places, water filled depressions, and migrating thermokarst lakes appeared. During the course of deglaciation these lakes eroded half of the yedoma. At the present time, yedoma is only preserved on slopes or in the form of hills where new lakes cannot form. On lake bottoms, all yedoma became thawed and turned into many metres of silt. Part of the organic material was then transformed by methanogens into CH$_4$. The diffusion of CH$_4$ from sediment is/was very slow, and the main source is/was bubbling [27,28].

Permafrost degradation is a very strong methane source (up to hundreds grams of CH$_4$ per square meter per year) [27,28]. On the one hand, the CH$_4$ emitted from permafrost can be considered as a fossil since it is strongly depleted in ^{14}C. (Note that the ^{14}C content of atmospheric methane decreased during Pleistocene-Holocene transition [29].) On the other hand, the ^{13}C and the deuterium content of methane from the Siberian permafrost is substantially lower than for all other sources (δ^{13}CH$_4$ = −73‰ (−58 to −99‰), δDCH$_4$ = −400‰ (−338 to −479‰) [22,27,28].

Permafrost methane emission is the only source that fulfils all of the following demands: it is the northern source; it is depleted in both deuterium and ^{14}C; and its dynamics (Fig. 1H) are correlated with methane atmospheric dynamics in Pleistocene-Holocene transition (Fig. 1B).

During the LGM, the mammoth steppe resting on permafrost was the largest biome. It expanded from France to Canada and from the Arctic islands to China [8,20]. In many regions of this biome, sediment deposition took place. As these sediments were accumulating they became incorporated into the permafrost. Such vast territories (besides the yedoma territory in the north of Siberia and Alaska) could be found in Europe, America, Siberia and China. Deposition of frozen sediments usually occurred with the accumulation of loess (area of 3*10^6 km^2, see Fig. 1 in ref. [8]) but also at the foot of slopes and in river valleys. On all of these territories thick strata, similar to yedoma, accumulated; similar in carbon content, but with fewer ice wedges [8,24,25]. In places where there was no notable sedimentation, soils were thin. At the beginning of the glaciation, when permafrost appeared, it was ~1.6 m below the surface [24]. During glaciation, active layers decreased to ~0.6–0.8. Therefore, the bottom portion of the soils became incorporated into the permafrost, even without sedimentation. Since these thin soils had developed for many thousands of years, they accumulated humus and the C concentration within them was two times higher than in yedoma [24]. During deglaciation, frozen soils of the mammoth steppe and "southern yedoma" thawed, not only underneath lakes (as in the north of Siberia), but ubiquitously; they lost their ice and passed the stage of anaerobic decomposition, which was followed by the drainage and aerobic decomposition stages – resulting in the loss of Pleistocene carbon and the transition to Holocene soils and loess [24,25].

Methane emissions from northern Siberian permafrost have a very light isotopic signature. The light isotopic content of this CH$_4$ is due to the following factors: (i) methane is released quickly and does not become fractionated by methanotrophs; (ii) methane production occurs very slowly at very low temperatures [22]; and

(iii) methane production occurs in deep sediments through the decomposition of Pleistocene organic matter in Pleistocene water [30,31]. The content of heavy isotopes declines in precipitation water as clouds move from south to north and from oceans to the interior of continents. Therefore, in northern Siberia the isotopic content of Pleistocene precipitation was very low, as indicated in the yedoma ice [30,31]. Since the hydrogen used in photosynthesis and methanogenesis originates from precipitating water, the deuterium content in CH$_4$ obtained from yedoma was also low. Our experimental methane isotope data was obtained for the coldest permafrost and for the most remote from the warm oceans sites in northern Siberia. The isotopic signature of methane from other "warmer" regions should be substantially heavier with isotopes. During the beginning of deglaciation, southern permafrost was largely degrading. For example, in Europe and China the signature of this source must have been substantially closer to the signature of boreal wetlands. With deglaciation, as permafrost degraded further on the north, the isotopic signature of the permafrost source must have turned lighter. The possible dynamics of this signature are shown in Table 1.

Unlike wetlands where annual emissions of carbon into the atmosphere are roughly equilibrated with carbon consumption with photosynthesis, degrading permafrost is only a source of carbon. The permafrost carbon reservoir was filled for tens of thousands of years within glacial steppes. Decomposition of this organic matter occurred in different conditions. The type of surface (thermokars lake, glacial lake, sea, wetland, or dry land) is not important for methanogenesis to occur during permafrost thawing. Organic materials decompose year round in deep sediments.

Local permafrost thawing, only in northern Siberia, could have released up to 26 Tg CH$_4$/yr (Fig. 1H) [8]. We may expect that at least the same amounts were released from the total thawing of American, European, west and south Siberian, and Chinese permafrost, as well as permafrost from other regions, including Patagonian permafrost. However, in the southern hemisphere permafrost was rare, having mostly a northern source. Consequently, the permafrost source is enough to maintain the interhemispheric methane gradient [8]. The dynamics of permafrost emissions should reflect not the dynamics of temperatures in the northern hemisphere, but the dynamics of the above- and belowground deglaciation (lots of frozen soils were hidden under glaciers [32]). The dynamic of methane emissions from thawing European permafrost should be close to that of the north Siberian permafrost (Fig. 1H). After 6 ka BP, this source should have quickly depleted.

Modern permafrost emissions, connected with permafrost degradation under lakes and shallow seas, contribute approximately 10 Tg CH$_4$/yr [8,33]. It is unlikely that such a high emission could be recorded for the entire Holocene. Most likely, modern high permafrost emissions are connected with the climate warming and the active exploration of northern Siberia in the 20th century. Before that, northern Siberia was poorly populated. In most ecosystems, the soil surface was covered with a thick layer of moss. Moss is a good heat insulator, and the depth of summer thaw under the moss in the yedoma region is only 20–40 cm. But if the moss layer is burned then the depth of the summer thaw (active layer) increases several fold. This often causes erosion and permafrost degradation [34]. This in turn leads to the creation of numerous ponds and thermokarst lakes [28]. In the 20th century active exploration, mining operations and city and town construction took place. Annual precipitation on the yedoma territories is approximately 200 mm/yr (~100 mm in summer) – it is a very dry region. According to our estimates, around half of

Table 1. The parameters values assumed in the model for A and B scenarios for all investigated time slices.

ka BP	18	15.5	14	12	11.2	10.6	9.2	8.4	7.4	5	3	1	0.2
ATM, ppb	370	470	600	500	730	730	730	660	650	580	610	675	730
ATM δC13, ‰	−42.5	−44.3	−46.25	−44.9	−46	−46.6	−47	−47.25	−47.4	−48	−48.2	−48.3	−49.5
ATM δD, ‰	−80	−80	−93	−90	−108	−110	−110	−110	−110	−110	−105	−100	−115
WT δC13, ‰	−58	−58	−58	−58	−58	−58	−58	−58	−58	−58	−58	−58	−58
WT δD, ‰	−320	−320	−320	−320	−320	−320	−320	−320	−320	−320	−320	−320	−320
WB δC13, ‰	−64	−64	−64	−64	−64	−64	−64	−64	−64	−64	−64	−64	−64
WB δD, ‰	−327	−327	−327	−327	−327	−327	−327	−327	−327	−327	−327	−327	−327
HV δC13, ‰	−59	−59 (−61.5)	−59 (−63)	−59 (−63)	−59 (−65)	−59 (−65)	−59 (−65)	−59 (−65)	−59 (−65)	−59 (−65)	−59 (−65)	−59 (−65)	−59 (−65)
HV δD, ‰	−300	−300	−300	−300	−300	−300	−300	−300	−300	−300	−300	−300	−300
BB δC13, ‰	23	−23	−23	−23	−23	−23	−23	−23	−23	−23	−23	−23	−23
BB δD, ‰	−225	−225	−225	−225	−225	−225	−225	−225	−225	−225	−225	−225	−225
GC δC13, ‰	−60	−60	−60	−60	−60	−60	−60	−60	−60	−60	−60	−60	−60
GC δD, ‰	−190	−190	−190	−190	−190	−190	−190	−190	−190	−190	−190	−190	−190
PF δC13, ‰	−65.7	−65.7	−65.7	−65.7	−69.3	−69.8	−70.6	−71.3	−72	−72	−72	−72	−72
PF δD, ‰	−340	−340	−340	−340	−370	−374	−380	−387	−393	−393	−393	−393	−393

ka BP – thousands of years before present; ATM, ppb - atmospheric methane concentration in parts per billion units; δC13 and δD isotopic signatures of atmospheric methane or methane from specific sources. WT – tropical wetlands; WB – boreal wetlands; HV – herbivores methane source; BB –biomass burning methane source; GC – geological methane source; PF – permafrost methane source. If values differed between the A and B scenarios, the value for the second scenario is presented in brackets.

all forests in the yedoma regions burned in the 20th century. This disturbance caused increased active layer depth, rise of autumn-winter CO_2 emissions from the soils and rise of seasonal amplitudes of atmospheric CO_2 [35]. We suppose that must have increased the methane flux from the permafrost region as well.

Herbivores

The LGM is a period of permafrost zone expansion and the accumulation of carbon in permafrost [25,36]. ^{14}C dating of thermokarst lake initiation dates indicated stable permafrost in the LGM (Fig. 1H). Therefore, we can infer that there was no methane emission from degrading permafrost during the LGM. As noted above, wetland and gas clathrates emissions were also minor. Therefore, some other sources must have constituted the main contribution during the LGM. From the list of well-known methane sources [12] only one big source is left – herbivores.

Lets make some investigation and rough estimates on whether could this source have played important role the late Pleistocene? Modern methane emissions from wild herbivores are approximately 2–6 Tg/yr [37], but herbivores were more abundant in the past. During the last deglaciation, tens of megafauna species became extinct, and a hypothesis even exists which states that during the Younger Dryas CH_4 decline was caused by a decline in megafauna in North America [38].

Current animal populations in grasslands are mostly limited by human regulations, while in the Pleistocene, before man invented reliable hunting tools and learned how to hunt all animal species, herbivore biomass was limited by the forage available for animals. During the winter/dry season, under high animal diversity, everything that grew during the summer/rainy season was eaten [23,34,39,40]. Everything uneaten by bulls and horses was eaten by omnivorous goats. If available food resources were left unused on the pastures in the spring, mortality related to starvation was rare and the animal population grew in the following summer, and by the following year all the available forage was eaten. For all wild ruminant animals (e.g., bulls, antelope, deer, goats, etc.), methane production is 28.3 kg per ton of forage, or ~100 kg/yr per ton of animal weight [37]. For non-ruminant animals (e.g., proboscideans, horses, pigs, etc.), CH_4 production is ~4.5 times smaller [37].

Currently, the area of pasture and cropland on Earth is 35.3×10^6 km^2. Harvests from these territories maintain a "megafauna" population of ~1.2 Pg (34 tons per km^2 of pastures), one-third of which is human biomass and two-thirds of which comprises domestic animals, mainly ruminants, that produce ~90 Tg CH_4/yr [37]. During the LGM, forested areas were ten times smaller than during the Holocene, and the area of grass–herb dominated ecosystems, reached 70×10^6 km^2 [19]. I.e. pasture area was twice of modern pastures and agricultural fields.

Vegetative periods on wild pastures are longer than on tilled fields. On pastures the entire biomass is utilized, and not only seeds and fruits. In the LGM there were no "technical" plant species, like cotton. On pasture, nutrients are not removed from the fields with the harvest but quickly and uniformly returned to the bio cycle. Taking all this into account, we can suppose that herbivore methane emissions in the LGM could be no less than the modern 90 Tg/yr.

We obtained more precise estimates of animal densities and their methane emissions for northern Siberia, where, because of high rates of yedoma accumulation, high numbers of late Pleistocene animal skeletons and bones are preserved [34,40]. We did estimates with different methods and for different regions. As a result, it was ascertained that in the late Pleistocene (over the 40 ka period) on the plains of the Siberian north, on each square kilometre approximately 1,000 of mammoths, 20,000 bison, 30,000 horses and 80,000 reindeer had died. From that it was calculated that, on average, for every moment of time on each square kilometre one mammoth, five bison, 7.5 horses and 15 reindeer roamed. Accounting for other more rare animals, the biomass of herbivores equalled 10.5 ton/km^2 [34,40]. Half of that biomass belonged to ruminant bison and reindeer with the rest being non-ruminant mammoths and horses [34,40].To maintain this biomass, approximately 100 tons of dry biomass per annum was supposed to be consumed from each square kilometre.

Even today such a high animal density in high- and mid-lattitudes can be observed. The modern density of semi-wild Yakutian horses on the highly productive meadows of northern Siberia –30 horses/km^2 [34,40], which roughly corresponds with our calculations. The same density can be observed in the pastures of the" Pleistocene Park" in the Kolyma river lowland [34,40]. In the "Oostvaardersplassen" park in the northern Netherlands, animal density is not regulated by man, but determined only by pasture productivity. There animal density is stable around 100 hoofed animals per square kilometre [41].

In the LGM, animal density in the north was substantially lower than on average for the period of the late Pleistocene [34,40]. Therefore, we can accept that on each square kilometre of plane land (yedoma) pasture in LGM only 50–70 tons of forage were consumed. North of Siberia is the coldest and driest part of the mammoth steppe biome and this biome in turn comprised the most severe steppes in the world. In warmer and wetter climates, productivity (and consequently animal density) was probably higher. If it is assumed that, on average, on the planet herbivores consumed 100–140 tons/km^2, the area of pastures was $70 * 10^6$ km^2 and ruminant animals constituted half of all the big herbivores, then global methane herbivore emissions could potentially reach 120–170 Tg/yr.

One of the features of the herbivore source is a quick and strong response to changes in precipitation. If in arid climates precipitation increased, then the productivity of grasses and herbs would increase in the same year, and in 1–2 years the number of ungulate and herbivore methane emission would increase respectively. This most likely occurred in the beginning of the Bølling warming (similarly, increase in herbivore density and herbivore methane productions should have been observed during all events of increased precipitation over the course of the Pleistocene). Activation of other sources, such as wetlands, or permafrost thawing appeared later. Eventually forest expansion caused pasture areas to decline and the herbivore methane flux have probably decreases. Additionally, the warming favoured the expansion of man to the north and to America. Due to increased hunting pressure, a number of herbivores declined at that time [34,40].

Biomass Burning

A direct proxy of fire intensity dynamics is the amount of charcoal in stratigraphic columns, but such data are rather sparse. Relatively reliable data are published only for the territory of the USA [42]. They showed a strong rise in fire activity in the Bølling-Allerod and Preboreal [42]. For north-eastern USA, data are obtained simultaneously with data on the abundance of dung fungus spores, which characterize the amounts of dung in the sediments and consequently the animal density. These data indicated that during the LGM, in north-eastern USA fires were rare, herbivores consumed everything, and there was nothing to burn [43]. After herbivore disappearance, fires became "persistent" [43]. Similar dynamics have been recorded for various regions ([39] and references therein).

Methods. A Methane Isotope Model

During methane oxidation, isotopic fractionation causes the isotopic content of atmospheric CH_4 to differ from the isotopic content of the weight averaged CH_4 source through the values of KIE (kinetic isotopic effects).

The KIE coefficients of $\delta^{13}CH_4$ and δDCH_4 are only moderately dependent on temperature and methane lifetime, and can be considered to be constants, allowing one to reconstruct the dynamics of the weighted average CH_4 source's isotopic content by knowing the dynamics of the isotopic content of atmospheric methane [6,11,12].

To reconstruct the dynamics of the main sources of CH_4, for every time slice we investigated we used the following set of equations [6,11,12]:

$$WB_i + WT_i + BB_i + GC_i + PF_i + HV_i = 1 \qquad (1)$$

$$WB_i \cdot \delta_{WB} + WT_i \delta_{WT} + BB_i \cdot \delta_{BB} + GC_i \cdot \delta_{GC} + PF_i \cdot \delta_{PF} + HV_i \cdot \delta_{HV} = \delta_{ATMi} \cdot (KIE_{13C}/1000 + 1) + KIE_{13C} \qquad (2)$$

$$WB_i \cdot D_{WB} + WT_i \cdot D_{WT} + BB_i \cdot D_{BB} + GC_i \cdot D_{GC} + PF_i \cdot D_{PF} + HV_i \cdot D_{HV} = D_{ATMi} \cdot (KIE_D/1000 + 1) + KIE_D \qquad (3)$$

where WB, WT, BB, GC, PF, HV are CH_4 sources from respectively boreal wetlands, tropical wetlands, biomass burning, gas clathrates, permafrost and herbivores, units are the ratio of total global emission;D and δ are the isotopic signatures of all of the corresponding sources, in ‰ (presented in Table 1 and fig. 2 [12]);KIE_{13C} and KIE_D are the coefficients of atmospheric isotopic fractionation, in ‰.i, number of time slice (see column labels in the Table 1 for details on investigated time slices).

These formulas show the relationship between the isotopic signatures of methane sources and the global atmosphere. As input parameters we have the isotopic record from the Greenland ice cores [11,12,13] (blue line in Fig. 2 and Table 1)(i.e., we know the isotopic signatures of the atmosphere above Greenland), but for our calculation globally averaged values are necessary. Therefore, as a first step we need to find a connection between the global atmosphere and the atmosphere above Greenland. In the southern hemisphere, non-tropical sources are minor. Therefore, the situation where the entire methane atmosphere becomes isotopically lighter, while becoming heavier above Greenland during the same period of time, is not possible. The dynamics of Greenland methane should be roughly parallel with the global methane dynamics. The $\delta^{13}CH_4$ from Greenland is different, by less than 1‰, from the $\delta^{13}CH_4$ of Antarctica and the averaged atmosphere, and remained the same for the period of time spanning from deglaciation to 1 ka BP [13]. In Figure 1, one can see that the values of $\delta^{13}CH_4$, obtained for the Younger Dryas through the Preboreal periods in Greenland (GISP 2) and Antarctica (EDML), are similar. We assumed that during the LGM the difference also remained small.

The δDCH_4 of the atmosphere above Greenland 1 ka BP, differed from Antarctic values by ~12‰, and from averaged atmospheric values by ~8‰ [13]. LGM emissions of isotopically light methane from boreal wetlands and permafrost were near zero and the interhemispheric methane gradient was low. Herbivores and biomass burning doesn't influence inter-hemispheric methane gradient. Therefore, the LGM interhemispheric gradient of methane isotopes should also be low. As a first approximation, we can assume that interhemishperic gradients of atmospheric methane concentrations and the isotopic contents are proportional to each other. We assumed that the δDCH_4 from Greenland differed by 8‰ from the averaged atmospheric values for the Holocene, by 4‰ for the Younger Dryas (12 ka BP), by 6‰ for the Bølling-Allerod period (14 ka BP), and by 2‰ for the LGM and 15.5 ka BP.

As a result, we ascertained the dynamic of isotopic signatures of global atmospheric methane. During atmospheric methane oxidation isotopic fractionation takes place, therefore methane in the atmosphere is much heavier than the source methane. This fractionation is described by parameters of KIE for both deuterium and ^{13}C. The *Fischer et al.* [12] model utilized KIE values of −6.8‰ for $\delta^{13}CH_4$ and −218‰ for deuterium. We utilized this pair of KIEs in our first scenario (Fig. 3A).

Inclusion of both isotopic dynamic of atmospheric CH_4 and KIEs allows the calculation of the right hand sides of Equations 2 and 3. It is the dynamic of the isotopic content of the weighted average global source (purple line in Fig. 2).

We have three equations and six sources (strengths of six sources of methane) for each time slice. But for the LGM time slice we know that the emissions from permafrost, tropical and boreal wetlands are close to zero. Therefore, for the LGM we can solve the equations directly: herbivore emissions constituted 69% of global emissions, biomass burning 39% and gas clathrates emission 2%. Without knowing the lifetime of methane in the atmosphere in the LGM it is impossible to reconstruct the size of the total global methane emission. But if we were to accept that total herbivore methane emissions in the LGM were 100 Tg/yr, then global emissions were 145 Tg/yr, and accounting for the methane content in the atmosphere in the LGM (~1000 Tg CH4), the life of methane was ~6.9 years.

The third step is to set values of tropical and boreal wetland methane emissions for all time slices. The wetland succession model we created based on peatland initiation dates (thin lines in Figs. 1D and 1E) allow the detection of the relative dynamic of these sources. Recent estimates for modern tropical and boreal wetland emissions are 110 and 55 Tg CH_4/yr [44]. Accepting these values for the pre-industrial time slice allows the reconstruction of the wetland methane source from the LGM to the present for all time slices. In Equation 1 all source units are the ratios of total global methane flux. To estimate the strength of wetlands sources not in Tg/yr units but in the ratios of total, for every time slice the methane lifetime should be considered.

Lifetime is an unknown and poorly constrained parameter in our model [6]. But solving the system of equations for numerous time slices, from the LGM to the present, and having reasonable (or any) solutions is possible only in narrow range of input parameters, including lifetime.

Accepting tropical and boreal wetland emissions of 110 and 55 Tg CH_4/yr [44] for the pre-industrial time slice, we automatically set the upper limit of methane lifetime values. If we were to accept a lifetime of more than 8.6 years then late Holocene solutions of Equations 1–3 would be lost. In turn, if the lifetime is shorter than 7.6 years, herbivore emissions during the late Holocene would exceed reasonable values (25 Tg/yr). At that time, wild herbivores were already rare and domestic animals were still few. Therefore, for the first scenario, we assumed 8 years for all time slices. It is likely that this value is correct for the entire

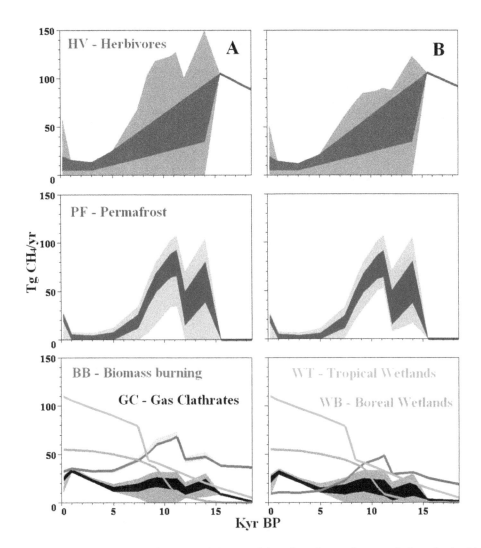

Figure 3. Model estimated ranges of the possible values for methane emissions for herbivore, permafrost, biomass burning and gas clathrates for time slices from the LGM to the pre-industrial. Dark shaded areas represent calculated ranges obtained following the additional restriction placed on herbivore and permafrost emissions. Wetland dynamics are provided as generalized graphs in Figs. 1F and 1G. A – scenario calculated for KIEs −6.8‰ and −218‰ [12]; B- scenario calculated for KIEs −10.8‰ and −227‰ [6].

Holocene. For Pleistocene-Holocene transition and the LGM it could be different. However wetland emissions were minor at that time; therefore changes in lifetime have little effect on the relative strength of other sources. If required, we can change the lifetime of methane during Pleistocene-Holocene transition and in the LGM, which would consequently change the strength of all sources in Figure 3 at that time. If we change the lifetime of methane in the atmosphere, it would change the strength of all sources in Tg/yr units, but emissions in units of the ratio from global emissions would stay intact.

After accepting boreal and tropical wetland emissions values for every time slice as ratios of global methane emission we obtain a system with three equations (Eq. 1–3) and four unknowns (biomass burning, gas clathrates, permafrost and herbivores). This does not allow us to calculate the dynamic of the source, but it allows us to estimate possible ranges of strengths of these sources for every time slice (Fig. 3). True estimates could only be made for the LGM and 15.5 ka BP time slices, where permafrost source was taken as 0.

The model's solution can be presented visually. For each time slice, in Figure 2, global emissions should be divided between individual sources, in the way "centre of gravity" would coincide with the position of the weighted average source (knots in the purple line). Model reactions to different changes can also be evaluated visually. "Remote" sources – biomass burning, gas clathrates and permafrost – have a "long lever", therefore the shift in strength by a few Tg/yr of these sources, noticeably shifts the "centre of gravity". On the other hand, an error in the isotopic signature of these sources does not influence solutions very much. For example, a few per mil shift in the biomass burning source signature in Figure 2 would still leave this source far to the left from the "centre of gravity". Tropical wetlands and herbivores have a "short lever" therefore the change in strength by a few Tg/yr of these sources is not noticeable in the solutions, while an error in the isotopic signatures (position of these sources in Fig. 2), can substantially bias our solutions.

Modelling results are the most sensitive to changes in KIEs. These two parameters define the position of the purple line on Figure 2. If this line were slightly shifted towards the biomass burning, then the ratio of this source in global emissions would increase (and the ratio of other sources would consequently decrease). If the purple line were shifted to the bottom right, then the ratio of gas clathrates would increase. But if the purple line were further shifted towards the gas clathrates, then for fixed wetland emissions and lifetimes, the solutions in our model would be lost – the dynamic of methane isotopes in the atmosphere (Figs. 1B and 1C) does not allow substantial gas clathrate sources, regardless of the assumed KIE parameters.

In the available literature, KIE coefficients are broadly varied [6]. For $\delta^{13}CH_4$ an estimate of $-6.8‰$ exists, but a more recent value of $-10.8‰$ also exists [6]. The KIE for deuterium is $-218\pm50‰$ [5]. The *Fischer et al.* [12] model utilized KIE values of $-6.8‰$ and $-218‰$. We utilized this pair of KIEs in our first scenario as well (Fig. 3A). In this scenario, we had very high biomass burning methane emissions in the LGM (Fig. 3A). But this contradicts the high density of animals at that time – actively grazed pastures do not supply much organic material for fires [43].

For the second scenario (Fig. 3B) we used KIEs of $-10.8‰$ and $-227‰$, shifting the purple line in Figure 2 to the upper right. For these KIEs, solutions existed for all time slices. Moreover, we obtained solutions for the modern methane budget as well, but in order to do so had to make "permafrost corrections" in the modern budget. The ratio of fossil sources (coal and gas) in the modern budget is calculated through the $^{14}CH_4$ budget [6]. The ^{14}C age of methane in the atmosphere allows the calculation of the ratio of all sources that are fossil sources. Permafrost is also a fossil source (strongly depleted in ^{14}C) but it was not considered in reconstructions of the modern methane budget. Unlike gas and coal heavy with isotopes [6], permafrost is very light. For the modern budget it was necessary to add permafrost, and correspondingly reduce the ratio of other fossil sources. The current permafrost source is relatively small (~ 10 Tg/yr) but it has very distinct deuterium and ^{13}C signals compared to coal and gas; therefore taking even a small weight from one long "lever" and placing it with the opposite long "lever" (Fig. 2) substantially changes the isotopic content of global emissions. In the modern budget several additional anthropogenic sources are considered. By adding 10 Tg/yr emissions from permafrost to it, for KIEs of $-10.8‰$ and $-227‰$ we have managed to obtain values of all modern sources in the frames of their known ranges. Therefore, we saw no reason not to use KIE values of $-10.8‰$ and $-227‰$ for paleo reconstructions.

Since herbivores are a strong source with an isotopic signature close to the weighted average source signature, the parameter with the second strongest impact on our model results was the isotopic signature of this source. In the first scenario we used $\delta^{13}C$ for herbivores, as in the Fischer model, for all time slices. *Fischer et al.* used a $\delta^{13}C$ for HV of $-59‰$, as for modern animals, with a high proportion of C_4 plants (e.g. corn etc.) in the diet [6,45]. This value is likely to be true for the LGM also, since, during low atmospheric CO_2, the ratio of C_4 plants increased and the $\delta^{13}C$ in C_3 plants decreased by 2‰ [46]. However, for other time slices, CO_2 concentrations were higher. For Pleistocene-Holocene transition we used the herbivore signature in compliance with atmospheric CO_2 concentrations [46]. So, in the second scenario for the LGM, we assumed $-59‰$ (as in Fischer model); for the time slice of 15.5 ka BP we used $-61.5‰$, for the Bølling-Allerod and Younger Dryas we used $-63‰$, and for the Preboreal and Holocene we used 65‰. In calculating the modern budget we have utilized the same value as Fischer, $-59‰$. Detailed data of all assumed isotopic signatures of all sources can be found in Table 1.

For the second scenario, using the same restrictions as for the first, the lifetime was determined to be 9 years.

Results

Modelling results for both scenarios are presented in Fig. 3, we see that for the herbivore and permafrost ranges in our model an additional correction was required. The model returned clearly overestimated maximum values for these parameters. During deglaciation, permafrost emissions could not be substantially greater than the northern flux obtained by the interhemispheric gradient. During the Holocene, forested areas increased and pasture areas decreased. Additionally, numerous herbivore species became extinct. If current herbivore emissions are 90 Tg/yr, then 200 years ago, when human and domestic animal populations were much lower, emissions could not be as high as 50 Tg/yr. Any corrections for any ranges can easily be made using Figure 3. Since the equations in our model are linear, the values for all of the calculated sources are interconnected. The herbivore source is located inside the triangle of permafrost, gas clathrates and biomass burning (Fig. 2). Therefore, the herbivore source maximum corresponds to the minimum for permafrost, gas clathrates and biomass burning, and vice versa. If, for any time slice in Figure 3, we, for example, restrict the range of permafrost by one-third from the top, then the gas clathrates and biomass burning ranges would be restricted by one-third from the top as well, and the herbivore source range would also be restricted by one-third, but from the bottom. If we accept a middle value from the range of one specific source then the values of the other three sources would also be located in the middle of their ranges.

For our work, we only made the most obvious and least controversial correction (see paragraph above) – we restricted the ranges of herbivores and permafrost from the top. However, reducing the range of herbivores from the top, restricted the permafrost, gas clathrates and biomass burning source ranges from the bottom; while restricting the range of permafrost sources from the top led to restricted range of biomass burning and gas clathrates sources from the top and restricted herbivore source ranges from the bottom. As a result we obtained relatively narrow ranges for all sources (the dark shaded sector in Fig. 3).

Testing our model we changed all of the initial parameters over a wide range, and implemented additional sources. In all cases, if solutions existed, the dynamics of the sources were similar to the results in Figure 3. We are not confident that 110 and 55 Tg $CH4$/yr are correct estimates, and tested our model for other values of modern wetlands emissions. This required accepting different lifetimes of methane. However, the main results of our work stayed intact.

All modelling was done on Maple v.10.0 software.

Biomass burning emissions in all scenarios had a very narrow range (Fig. 3), and responded strongly to variations in initial parameters but in all cases strongly increased from 15.5 to 12 ka BP. As was noted earlier, similar dynamics for fires were reconstructed using charcoal records [42].

Gas clathrate emissions, as we predicted, were low for all of the scenarios. For the end of the Holocene, for all scenarios, we determined a clear peak of gas clathrates (as suggested by Sowers, [13]).

Earlier we mentioned that methane from the bottoms of deep seas could not penetrate into the atmosphere therefore GC emissions during the Holocene (Fig. 3) could be connected with the land [47] or with shallow Siberian seas [33]. During the LGM, their bottoms were land and were armoured with a gas-proof thick frozen layer. Only when these layers thawed could trapped gas escape. Soils of the mammoth steppes are relatively thin (10–60 m) and thaw quickly. Gas clathrates are located deep, and, in order to thaw hundreds of metres of permafrost, thousands of years are required. Therefore, as hundreds of metres of permafrost degraded in the second part of the Holocene, gas clathrates emissions could grew (Fig. 3).

In all scenarios global **permafrost** emissions closely followed atmospheric methane dynamics and permafrost degradation in the Siberian north (Fig. 1H). If emissions from boreal wetlands and permafrost were combined then the dynamic of this united northern source would be compiled with the interhemispheric CH_4 gradient. By integrating the permafrost emission data (Fig. 3), we found that the source released ~300–550 Pg CH_4 (225–412 Pg C) to the atmosphere from 15 to 6 ka BP. During the Holocene, permafrost sources quickly diminished, while boreal wetlands emissions grew. This explains why the interhemispheric gradient was relatively stable in the Holocene.

In pre-industrial times (200–400 years ago), a sharp decline in the deuterium content in atmospheric methane took place (Fig. 1C). The model was sensitive to this decline and showed that this decline was caused by the activation of permafrost emissions (Fig. 3). Increased degradation of permafrost at that time made sense; at that time, active colonization (exploration) of Siberia and Alaska was underway and in the north many settlements and towns, connected with a transport network, appeared. Sable trapping became an activity not only for colonizers, but also for indigenous people for the first time [48]. With colonization, fire intensification must have taken place and permafrost degradation must have been activated.

Herbivore emissions during the Pleistocene, for a methane lifetime of 8–9 years, were 90–100 Tg/yr. If we were to accept a shorter lifetime, then herbivore emissions could be even higher. These estimates are close to our estimates obtained via global forage productivity (see Herbivore section in Background). Herbivore emissions until 15.5 ka BP were close to the capability of world pastures, decreased during the Holocene, and reached a minimum during the late Holocene. Wild animals were already rare, and domestic animals were still found in small numbers.

Conclusions

Today wetlands are main natural source of methane and it is natural and easiest to suppose that it was same in the past. However we haven't found any proofs of this hypothesis. The main conclusion of our work is that permafrost and herbivores in the past were not only important but also the main methane sources. The main fact from which these two conclusions follow is that in the LGM and during Pleistocene-Holocene transition there is little evidence of peat or anaerobic conditions in the soils. From that it follows that, at that time, wetland methane emissions were at least several times weaker then today. Therefore, some other sources were major at that time. Since, besides boreal wetlands, we know of only one northern methane source (permafrost), we can conclude that permafrost was responsible for the interhemispheric gradient dynamic in deglaciation. In the LGM, permafrost was stable and the main methane source (in the absence of wetlands) could only be herbivores.

These results were obtained independently from isotopic model. The model in turn also depends on data about the lack of wetlands in the LGM and during Pleistocene-Holocene transition, but it is independent of data about the interhemispheric gradient.

The isotopic model we used is simple and broadly accepted [6,11,12]. We have only added the unknown permafrost source, and utilized the most obvious restrictions – herbivore methane emissions in the Holocene could not be higher than in the LGM and in the end of the Holocene could not exceed 25 Tg/yr;

permafrost emissions could not be higher than indicated by the interhemispheric gradient.

The model has allowed the dynamic of all sources to be reconstructed in more detail. Additionally, noting that only the dynamic was reliably reconstructed, absolute values of the strength of each source have been estimated very approximately. Our estimates rely on modern estimates of wetland emissions. If true values differ from the assumed, then to solve the model equations different methane lifetimes should be taken. Our model estimates are rough, we did not consider some of the sources; for example, in the late Holocene some of the wetlands were transformed into rice paddies [9].

We know the values of the isotopic signatures of all sources and the KIE, possibly with substantial errors. These values can be corrected in the future. But if the position of each source in Figure 2 relative to each other will not change, then with new corrected values either solution of the model will not exist (wrong incoming values) or the model will show that herbivores were the main source in the LGM and permafrost during deglaciation.

Discussion

The cold and well aerated soils of mammoth steppe are unlikely to have produced methane in the Pleistocene, and are likely to have accumulated carbon [24,34,40]. This would certainly have affected concentrations of greenhouse gases in the atmosphere. Even modern permafrost soils are rich in carbon and are estimated to contain 1672 Pg C [49] globally. In the LGM, the mammoth steppe was the biggest biome and permafrost covered a much larger area. Estimating carbon storage in permafrost during the LGM has been a significant challenge [24], however, we can now make rough estimates using methane emissions from permafrost sources estimated in this study. Integrating the curve in figure 3, we estimate that permafrost emitted 400 Pg CH4 (300 Pg C) into the atmosphere during the deglaciation. What was the total carbon loss from the northern soils over that time period? Even in fully anaerobic conditions only part of the stored organic (28±12%) can be transformed into methane [50]. When thick (tens of meters) frozen ice-rich soils (yedoma and its southern analogues) thaw they are initially anaerobic, however, as permafrost degradation continues, underlying gravel and sand thaws and water begins to drain from shallower soil layers. These soils become aerobic and carbon is transformed into the $CO2$. However on most territories occupied by the mammoth steppe, soils (both including active layer and soils incorporated into the permafrost) were shallow (less than 2–3 meters deep), and carbon stored in these soils was mostly decomposed under aerobic conditions when the climate warmed. Therefore, if we accept that 15% of permafrost carbon loss was transformed into methane, we can estimate that permafrost soils lost 2000 Pg C during deglaciation (15–6 ka BP).

As current climate warming proceeds, permafrost temperatures are increasing and widespread permafrost degradation is projected. Currently, permafrost soils contain 1672 Pg of carbon [49]. What portion of this carbon will be transformed into methane as these soils begin to thaw? We assumed 15% during deglaciation. We could not take smaller value, since that would have returned an unrealistically big estimate of the permafrost carbon reservoir in the Pleistocene [24]. In modern permafrost, a higher proportion of carbon has already been thawed in the Holocene and then refrozen. It is unlikely that at future thawing such soils would produce lots of methane because much of the easily-decomposed forms of carbon have already been processed. We estimate that no more than 10% of organic carbon in modern permafrost is likely to be transformed into methane as the climate continues to warm. Thus, we estimate that modern permafrost pool has a potential to release 167 Pg C in the form of methane. If future permafrost degradation (in contrast to the Pleistocene-Holocene transition) is rapid and a quarter of this methane is released within 200 years, average methane emissions will be ~250 Tg/yr. This value will make degrading permafrost world biggest methane source.

If our calculations of animal densities in the Pleistocene are correct (or at least close), it appears that wild nature solely on recycled resources managed to sustainably maintain biomass higher than the entire biomass of modern civilization (humans and all domestic animals) which is supported only by burning enormous amounts of non-refundable resources. This results draws attention to the ineffectiveness of current human management of the precious, finite resources available on our planet.

Author Contributions

Conceived and designed the experiments: ZS ZN. Performed the experiments: ZN. Analyzed the data: ZS. Contributed reagents/materials/analysis tools: ZN. Wrote the paper: ZS ZN.

References

1. Chappellaz J, Blunier T, Raynaud D, Barnola JM, Schwander J, et al. (1993) Synchronous changes in atmospheric CH4 and Greenland climate between 40 and 8 kyr BP. Nature 366, 443.

2. Chappellaz J, Blunier T, Kints S, Dallenbach A, Barnola JM, et al. (1997) Changes in the atmospheric CH_4 gradient between Greenland and Antarctica during the Holocene. J.Geophys. Res. 102, 15987.

3. Severinghaus JP, Sowers T, Brook EJ, Alley RB, Bender ML (1998) Timing of abrupt climate change at the end of the Younger Dryas interval from thermally fractionated gases in polar ice. Nature 391, 141.

4. Brook EJ, Harder S, Severinghaus JP, Steig EJ, Sucher CM (2000) On the Origin and Timing of Rapid Changes in Atmospheric Methane During the Last Glacial Period. Global Biogeochem. Cylces 14, 559.

5. Dällenbach A, Blunier T, Flückiger J, Stauffer B, Chappellaz J, et al. (2000) Changes in the atmospheric CH_4 gradient between Greenland and Antarctica during the Last Glacial and the transition to the Holocene. Geophys. Res. Let. 27, 7, 1005, doi:10.1029/1999GL010873.

6. Quay P, Stutsman J, Wilbur D, Snover A, Dlugokencky E, et al. (1999) The isotopic composition of atmospheric methane. Global Biogeochem Cycles 13, 445.

7. Kennet JP, Cannariato KG, Hendy IL, Behl RJ (2003) Methane hydrates in Quaternary climate change AGU, Washington, D.C.

8. Walter KM, Edwards ME, Grosse G, Zimov SA, Chapin III FS (2007) Thermokarst lakes as a source of atmospheric CH4 during the last deglaciation. Science 318, 633.

9. Ruddiman WF, Guo Z, Zhou X, Wud H, Yu Y (2008) Early rice farming and methane emissions. Quaternary Science Reviews 27, 1291–1295 doi:10.1016/j.quascirev.2008.03.007.

10. Ferretti DF, Miller JB, White JWC, Etheridge DM, Lassey KR, et al. (2005) Unexpected Changes to the Global Methane Budget over the Past 2000 Years. Science 309, 1714–1717.

11. Sowers T (2006) Late quaternary atmospheric CH4 isotope record suggests marine clathrates are stable. Science 311, 838.

12. Fischer H, Behrens M, Bock M, Richter U, Schmitt J, et al. (2008) Changing boreal methane sources and constant biomass burning during the last termianation. Nature 452, 864.

13. Sowers T (2010) Atmospheric methane isotope records covering the Holocene period. Quat. Sci. Rev. 29, 213.

14. Kennett JP, Cannariato KG, Hendy IL, Behl RJ (2000) Carbon isotopic evidence for methane hydrate instability during quaternary interstadials. Science 288, 128.

15. Zimov SA, Voropaev YV, Davydov SP, Zimova GM, Davydova AI, et al. (2001). In Paepe R, Melnikov V editors. Permafrost Response on Economic Development, Environmental Security and Natural Resources, Kluwer Academic Publishers, 511–524.

16. Kessler JD, Valentine DL, Redmond MC, Du M, Chan EW, et al. (2011) A Persistent Oxygen Anomaly Reveals the Fate of Spilled Methane in the Deep Gulf of Mexico. Science 331, 312–315.

17. MacDonald GM, Beilman DW, Krementski KV, Sheng Y, Smith LC, et al. (2006) Rapid early development of circumarctic peatlands and atmospheric CH4 and CO2 variations. Science 314, 285.

18. Yu Z, Loisel J, Brosseau DP, Beilman DW, Hunt SJ (2010) Global peatlands dynamic since the Last Glacial Maximum. Geophys. Res. Lett 37, L13402.

19. Adams JM, Faure H (1998) A new estimate of changing carbon storage on land since the last glacial maximum, based on global land ecosystem reconstruction. Global and Plan. Change 16–17, 3.

20. Adams JM, Faure H, Faure-Denard L, McGlade JM, Woodward FI (1990) Increases in terrestrial carbon storage from the Last Glacial Maximum to the present. Nature 348, 711–714.

21. Gajewski K, Viau A, Sawada M, Atkinson D, Wilson S (2001) Sphagnum peatland distribution in North America and Eurasia during the past 21,000 years. Global Biogeochemical Cycles 15: 297–310.

22. Rivkina EM, Kraev GN, Krivushin KV, Laurinavichus KS, Fyodorov-Davydov DG, et al. (2006) Methane in permafrost of northern arctic. Earth Cryosphere 10, 23–41.

23. Zimov SA (2005) Pleistocene Park: return of mammoth's ecosystem. Science 308, 796–798.

24. Zimov NS, Zimov SA, Zimova AE, Zimova GM, Chuprynin VI, et al. (2009) Carbon storage in permafrost and soils of the mammoth tundra-steppe biome: role in the global carbon budget. Geophys. Res. Lett. 36, L02502, doi:10.1029/2008GL036332.

25. Zimov SA, Schuur EAG, Chapin, III FS (2006) Permafrost and the global carbon budget. Science 312, 1612.

26. Schirrmeister L, Siegert C, Kuznetsova T, Kuzmina S, Andreev A, et al. (2002) Paleoenvironmental and paleoclimatic records from permafrost deposits in the Arctic region of Northern Siberia. Quat. Int. 89, 97–118.

27. Zimov SA, Voropaev YV, Semiletov IP, Davidov SP, Prosiannikov SF, et al. (1997) North Siberian lakes: a methane source fueled by Pleistocene carbon. Science 277. 800–802.

28. Walter KM, Zimov SA, Chantom JP, Verbyla D, Chapin, III FS (2006) Methane bubbling from Siberian thaw lakes as a positive feedback to climate warming. Nature 443, 71.

29. Petrenko VV, Smith AM, Brook EJ, Lowe D, Riedel K, et al. (2009) $^{14}CH_4$ measurements in Greenland ice: investigating Last Glacial Termination CH_4 sources. Science 324, 506.

30. Vasil'chuk YuK, Kotlyakov VM (2000) Principles of Isotope Geocryology and Glaciology. Moscow University Press. 616 p.

31. Brosius LS, Walter Anthony KM, Grosse G, Chanton JP, Farquharson LM, et al. (2012) Using the deuterium isotope composition of permafrost meltwater to constrain thermokarst lake contributions to atmospheric CH4 during the last deglaciation. J. Geoph. Res. 117, G01022, doi:10.1029/2011JG001810.

32. Zeng N (2003) Glacial-interglacial atmospheric CO2 change – the glacial burial hypothesis. Advances in Atmospheric Sciences 20, 677–693.

33. Shakhova N, Semiletov I, Salyuk A, Yusupov V, Kosmach D, et al.(2010) Extensive Methane Venting to the Atmosphere from Sediments of the East Siberian Arctic Shelf. Science 327, 1246–1250.

34. Zimov SA, Zimov NS, Chapin III FS (2012), The Past and Future of the Mammoth Steppe Ecosystem, In: Louys J editor. Paleontology in Ecology and Conservation. Springer-Verlag Berlin Heidelberg 2012. 193–224.

35. Zimov SA, Daviodov SP, Zimova GM, Davidova AI, Chapin III FS, et al. (1999) "Contribution of Disturbance to Increasing Seasonal Amplitude of Atmospheric CO2. Science 284, 5422.

36. Zech R, Huang Y, Zech M, Tarozo R, Zech W (2010) A permafrost glacial hypothesis to explain atmospheric CO2 and the ice ages during the Pleistocene. Clim. Past Discuss. 6, 2199–2221. 201.

37. Crutzen P, Aselmann I, Seiler W (1986) Methane Production By Domestic Animals, Wild Ruminants, Other Herbivorous Fauna, and Humans. Tellus 38B (3–4), 271–284.

38. Smith FA, Elliott SM, Lyons SK (2010) Methane emissions from extinct megafauna. Nature Geoscience 3, 374.

39. Johnson CN (2008) Ecological consequences of Late Quaternary extinctions of megafauna. Proc. R.Soc. B doi:10.1098/rsb.2008.1921.

40. Zimov SA, Zimov NS, Tikhonov AN, Chapin III FS (2012) Mammoth steppe: a high-productivity phenomenon. Quat. Sci. Rev. 57, 26–45.

41. Vera FVM (2009) Large-scale nature development – the Oostvaardersplassen. British Wildlife, June 2009.

42. Marlon JR, Bartlein PJ, Walsh MK, Harrison SP, Brown KJ, et al. (2009) Wildfire responses to abrupt climate change in North America. PNAS 106, 2519.

43. Gill JL, Williams JW, Jackson ST, Lininger KB, Robinson GS (2009) Pleistocene megafaunal collapse, novel plant communities, and enhanced fire regimes in North America. Science 326, 1100.

44. Bousquet P, Ringeval B, Pison I, Dlugokencky EJ, Brunke E-G, et al. (2011) Source attribution of the changes in atmospheric methane for 2006–2008 Atmos. Chem. Phys., 11, 3689.

45. Cicerone RJ, Oremland RS (1988) Biogeochemical aspects of atmospheric methane. Global Biogeochem. Cycles 2, 299.

46. Van de Water PK, Leavitt SW, Betancourt JL (1994) Trends in stomatal density and $^{13}C/^{12}C$ ratios of Pinus flexilis needles during last glacial-interglacial cycles. Science 264, 239.

47. Walter-Anthony KM, P Anthony, Grosse G, Chanton J (2012) Geologic methane seeps along boundaries of Arctic permafrost thaw and melting glaciers. Nature Geoscience, doi:10.1038/ngeo1480.

48. Syroechkovskii VE (1986) Severnii Olen'. Agropromizdat. Moscow (in Russian).

49. Tarnocai C, Canadell JG, Schuur EAG, Kuhry P, Mazhitova G, et al. (2009) Soil organic carbon pools in the northern circumpolar permafrost region. Global Biogeochem Cycles 23: 2023.

50. Walter Anthony KM, Zimov SA, Grosse G, Jones MC, Anthony P, et al. (2014) Permafrost thaw by deep lakes: from a methane source to a Holocene carbon sink. Nature. Submitted.

11

The Dimethylsulfide Cycle in the Eutrophied Southern North Sea: A Model Study Integrating Phytoplankton and Bacterial Processes

Nathalie Gypens[1]*, **Alberto V. Borges**[2], **Gaelle Speeckaert**[1], **Christiane Lancelot**[1]

1 Ecologie des Systèmes Aquatiques, Université Libre de Bruxelles, Brussels, Belgium, **2** Unité d'Océanographie Chimique, Université de Liège, Liège, Belgium

Abstract

We developed a module describing the dimethylsulfoniopropionate (DMSP) and dimethylsulfide (DMS) dynamics, including biological transformations by phytoplankton and bacteria, and physico-chemical processes (including DMS air-sea exchange). This module was integrated in the MIRO ecological model and applied in a 0D frame in the Southern North Sea (SNS). The DMS(P) module is built on parameterizations derived from available knowledge on DMS(P) sources, transformations and sinks, and provides an explicit representation of bacterial activity in contrast to most of existing models that only include phytoplankton process (and abiotic transformations). The model is tested in a highly productive coastal ecosystem (the Belgian coastal zone, BCZ) dominated by diatoms and the Haptophyceae *Phaeocystis*, respectively low and high DMSP producers. On an annual basis, the particulate DMSP (DMSPp) production simulated in 1989 is mainly related to *Phaeocystis* colonies (78%) rather than diatoms (13%) and nanoflagellates (9%). Accordingly, sensitivity analysis shows that the model responds more to changes in the sulfur:carbon (S:C) quota and lyase yield of *Phaeocystis*. DMS originates equally from phytoplankton and bacterial DMSP-lyase activity and only 3% of the DMS is emitted to the atmosphere. Model analysis demonstrates the sensitivity of DMS emission towards the atmosphere to the description and parameterization of biological processes emphasizing the need of adequately representing in models both phytoplankton and bacterial processes affecting DMS(P) dynamics. This is particularly important in eutrophied coastal environments such as the SNS dominated by high non-diatom blooms and where empirical models developed from data-sets biased towards open ocean conditions do not satisfactorily predict the timing and amplitude of the DMS seasonal cycle. In order to predict future feedbacks of DMS emissions on climate, it is needed to account for hotspots of DMS emissions from coastal environments that, if eutrophied, are dominated not only by diatoms.

Editor: Douglas Andrew Campbell, Mount Allison University, Canada

Funding: The present work is a contribution to the Fonds de la Recherche Fondamentale Collective DMS-SNS project (1882638) funded by the Fonds de la Recherche Scientifique. The funders had no role in study design, data collection and analysis, decision to publish, or preparation of the manuscript.

Competing Interests: The authors have declared that no competing interests exist.

* E-mail: ngypens@ulb.ac.be

Introduction

Dimethylsulfide (DMS) is a volatile sulfur (S) compound that plays an important role in the global S cycle and may control climate by influencing cloud albedo through the emission of atmospheric aerosols [1]. However, the significance of this feedback remains uncertain [2], as the present knowledge of mechanisms controlling DMS production is insufficient to allow a realistic description of DMS(P) production in Earth System models [3], and predict with confidence the impact of future climate change on surface ocean DMS [4], [5], [6].

In marine ecosystems, phytoplankton are the primary producers of dimethylsulfoniopropionate (DMSP), the precursor of the DMS (e.g. [7]). However, the amount of DMSP synthesized by cells varies among phytoplankton classes and species [8], [9], as well as with the physiological status [10], [11]. Overall, Bacillariophyceae (diatoms) synthesize less DMSP than Dinophyceae and Hapto-phyceae [8]. The metabolical role of DMSP in marine organisms is still unclear [7]. DMSP has been suggested to play a role as an osmoprotectant [12], [13], as a cryoprotectant [14], [15], [16], and as a nitrogen salvage mechanism during growth limitation

[11], [17]. The DMS and/or acrylic acid derived from DMSP cleavage might also act for phytoplankton as an antioxidant [18], [19], [20], as a deterrent for zooplankton [21], [22], [23], or as an anti-viral [24]. The conversion of DMSP to DMS and acrylic acid is catalysed by phytoplankton DMSP-lyases [25]. The intracellular DMSP is also released in the water column as dissolved DMSP (DMSPd) through various phytoplankton mortality processes, including cell lysis [26], [27], grazing pressure [22], [28], and viral infection [29]. Once in the water column, DMSPd is available for assimilation and degradation by bacterioplankton and part of the DMSPd is cleaved into DMS through bacterial metabolism [30], [31], [32]. Although largely variable, phytoplankton and bacterial lyases might contribute almost equally to the DMS production in marine ecosystems [7], [33], [34]. Yet, the main part of DMSPd is degraded by bacteria through the demethylation/demethiolation pathways for fulfilling their S and/or carbon (C) needs [35]. Once produced, DMS can also be consumed by bacteria to satisfy S and mainly C needs [36], photooxided [37], [38], or emitted to the atmosphere across the air-sea interface [39], [40]. The relative importance of these processes is variable and depends on physical forcing factors, but observational evidence suggests that microbial

consumption and photooxidation are the main DMS fates [38], [41], [42]. Because DMS production results from the balance of several complex processes, the link between DMSP production and atmospheric DMS emission is not direct and statistical relationships between DMS concentrations and other environmental variables (such as chlorophyll a (Chl a), nutrients, irradiance or mixed layer depth) are uncertain and generally regional in scope [39], [40].

Several mechanistic models of different biological complexity (reviewed by Le Clainche et al. [43]) have been therefore developed to better assess and understand DMS production and controlling factors in marine ecosystem [44], [45], [46], [47], [48], [49], [50], [51], [52], [53], [54], [55], [56], [57], [58]. All these models couple a biogenic S module composed of two or three state variables (DMS, particulate DMSP (DMSPp) and/or DMSPd) to a C- or nitrogen- (N) based ecological model of the plankton community [43], [59]. Most of them subdivide phytoplankton into several functional groups characterized by a specific DMSP cell quota (S:C) in agreement with observations [7]. S:C quota is generally considered as a constant with the exception of models of Le Clainche et al. [52] and Polimene et al. [58] that include variation of S:C with light intensity. The representation of heterotrophic compartments is generally less complex [43] and only some recent modelling studies include an explicit representation of the bacteria (e.g. [50], [56], [57], [58]). To the best of our knowledge the DMSP/DMS model of Archer et al. [50] is the only attempt to link the DMSP/DMS fate to bacterial degradation of organic matter, distinguishing between C and DMS- and DMSP-consuming bacteria types. These authors conclude that a tight coupling between the ecological processes and the DMS cycle is required to properly model DMS emissions to the atmosphere due to both the species dependence of DMSP production and the complexity of microbial metabolic pathways leading to the production of DMS.

Accordingly, we integrated a module describing the DMS(P) cycle into the existing ecological MIRO model [60] that describes C and nutrients cycles in the Southern North Sea (SNS) with an explicit description of the phytoplankton and bacteria dynamics to study the microbial controls of DMS(P) production and fate including DMS emission to the atmosphere. The MIRO model is a conceptual model of the biogeochemical functioning of marine ecosystem that includes an explicit description of growth and fate of *Phaeocystis* (Haptophyceae) that is one of the most intense DMSP producers [8], [61], [62]. The model was applied to the English Channel and the SNS with a focus to the Belgian coastal waters characterized by massive spring blooms of *Phaeocystis globosa* that develops between the spring and summer diatom blooms (e.g. [63], [64], [65]) in response to excess NO_3^- river inputs [66]. This is an adequate case study of *Phaeocystis*-dominated coastal area where the model can be applied to study the link between DMSP production/cleavage by phytoplankton, DMS(P) bacterial transformation, and DMS emissions as field observations also report important DMS concentration [33], [67], [68]. The NE Atlantic Shelves (including the SNS) were indeed pointed as "hot-spot" areas for DMS concentrations (with the Atlantic Subarctic region) in the Atlantic Ocean [40].

In this paper, we first describe the concepts behind the DMS(P) mathematical model and its coupling with the ecological MIRO model (MIRO-DMS). The model is then applied in the SNS to describe the seasonal evolution of DMS(P) and the associated DMS emission to the atmosphere, and provide an annual budget of DMS(P) fluxes. Sensitivity tests on parameters are conducted to identify key microbial controls of DMS(P) production and how these change the emission of DMS to the atmosphere. Finally, we test the applicability of several published empirical relationships that predict DMS from other variables such as Chl a.

Materials and Methods

Model description

The MIRO-DMS model results from the coupling between a module describing the DMS(P) dynamics and the existing ecological MIRO model developed to represent the dynamics of the ecosystem of the North Sea dominated by *Phaeocystis* colonies [60], [69].

The ecological MIRO model, describing C, N, phosphorus (P) and silica (Si) cycles, assembles four modules describing the dynamics of three phytoplankton Functional Types (FT; diatoms, nanoflagellates and *Phaeocystis* colonies), two zooplankton FT (meso- and microzooplankton) and one bacteria FT involved in the degradation of dissolved and particulate organic matter (each with two classes of biodegradability) and the regeneration of inorganic nutrients (NO_3^-, NH_4^+, PO_4^{3-} and $Si(OH)_4$) in the water column and the sediment. Equations and parameters were formulated based on current knowledge of the kinetics and the factors controlling the main auto- and heterotrophic processes involved in the functioning of the coastal marine ecosystem (fully documented by Lancelot et al. [60] and in http://www.int-res.com/journals/suppl/appendix_lancelot.pdf).

The description of the DMS cycle requires the addition of three state variables: DMSPp associated to phytoplankton cells, DMSPd and DMS. Processes and parameters describing the DMS(P) cycle (Fig. 1) and its link with carbon rates in MIRO are described below by equations 1 to 12.

DMSPp synthesis and fate. The DMSPp is a constitutive compatible solute produced by phytoplankton cell [11]. In the MIRO-DMS model, the DMSPp cellular production and fate are similar to those of other phytoplankton functional molecules, with DMSPp production linked to phytoplankton growth, and DMSPp loss mainly resulting from cell lysis, micro/mesozooplankton grazing and sedimentation (Eq. 1). These processes are described for each phytoplankton FT (diatoms (DA), nanoflagellates (NF) and *Phaeocystis* colonies (OP) expressed in mgC m^{-3}) as in the MIRO model and a specific DMSP:C quota (SC) is attributed to the three phytoplankton types. The DMSPp (in mmolS m^{-3}) state equation is:

$$\frac{dDMSPp}{dt} = [\mu_n - lysis_n - grazing - sed_n] * SC_n \qquad (1)$$

$$for \ n = DA, \ NF \ and \ OP$$

where μ_n represents the growth of different phytoplankton types (in mgC m^{-3} h^{-1}), $lysis_n$ is the phytoplankton lysis (in mgC m^{-3} h^{-1}) (flux$_{1+2}$, Fig. 1) and SC_n is the intracellular phytoplankton S:C quotas (molS:mgC) derived from the literature (Table 1; [7]). sed_n correspond to the loss of DMSPp due to diatoms and *Phaeocystis* colonies sedimentation (in mgC m^{-3} h^{-1}) (flux$_4$, Fig. 1). In the model, the sedimentation of nanoflagellates is considered as null. $grazing$ is the predation pressure of micro and mesozooplankton on respectively on nanoflagellates (NF) and diatoms (DA) (in mgC m^{-3} h^{-1}) (flux$_3$, Fig. 1). *Phaeocystis* colonies (OP) are not subject to grazing [70].

DMSPd release and fate. The DMSPd simulated in the water column results from the DMSPp released after phytoplankton lysis and zooplankton grazing. When released, DMSPp remains partly as DMSPd in the water column but is also partly

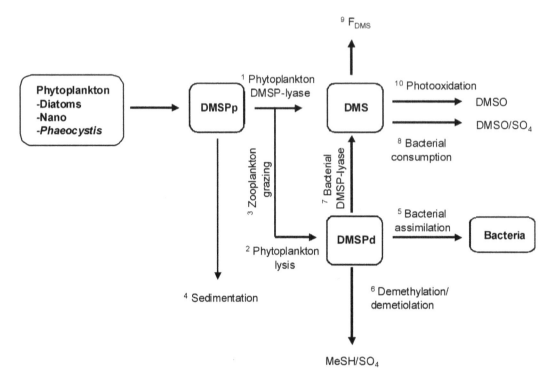

Figure 1. Diagram representing the state variables and processes of the DMS cycle incorporated into the ecological MIRO model.

directly cleaved in DMS by phytoplankton DMSP-lyases [22], [25], [61], [71], [72], [73]. The DMSPd originated from micro- and meso-zooplankton grazing is either directly released by "sloppy-feeding", excretion or egestion [21] and can represent up to 70% of the ingested DMSPp [74]. Wolfe and Steinke [22] also suggested that part of the DMSPp is directly converted to DMS. In the model, we assume that all the DMSPp ingested by micro- and meso-zooplankton is transformed into DMSPd (Eq. 1, 2). The fate of DMSPd is controlled by bacteria either through enzymatic cleavage into DMS and/or by demethylation/de-methiolation, i.e. the cleavage of DMSPd to methanethiol (MeSH) [75] and acrylate or propionate [76] for fulfilling the C and S needs of bacteria [77], [78]. In the model, the state equation of

DMSPd (in mmolS m^{-3}) is:

$$\frac{dDMSPd}{dt} = \left(1 - y_{DMS}^n\right) * lysis_n * SC_n + grazing * SC_n - \\ DMSPd_{uptake} \qquad for \; n = DA, \; NF \; and \; OP \tag{2}$$

where y_{DMS}^n corresponds to the fraction of DMSPp directly cleaved in DMS by phytoplankton DMSP-lyases. In the reference simulation, this fraction was set to 10% of the DMSPp released for each phytoplankton group [34]. $lysis_n$ (flux$_2$, Fig. 1) and $grazing$ (flux$_3$, Fig. 1) respectively are the phytoplankton cellular lysis and

Table 1. DMS(P) model parameters.

Parameter	Description	Units	Value	Reference
SC_{DA}	Diatoms S:C quota	molS:mgC (molS:molC)	0.000072 (0.00086)	Stefels et al. [7]
SC_{NF}	Nanoflagellates S:C quota	molS:mgC (molS:molC)	0.00092 (0.011)	Stefels et al. [7]
SC_{OP}	*Phaeocystis* colonies S:C quota	molS:mgC (molS:molC)	0.00092 (0.011)	Stefels et al. [7]
SC_{BC}	Bacteria S:C quota	molS:molC	0.01	Fagerbakke et al. [103]
y_{DMS}^{DA}	Part of diatoms DMSPp hydrolysed in DMS by phytoplankton lyase	-	0.1	Niki et al. [34]
y_{DMS}^{NF}	Part of nanoflagellates DMSPp hydrolysed in DMS by phytoplankton lyase	-	0.1	Niki et al. [34]
y_{DMS}^{OP}	Part of *Phaeocystis* colonies DMSPp hydrolysed in DMS by phytoplankton lyase	-	0.1	Niki et al. [34]
$K0$	Sea surface photooxydation rate	h^{-1}	0.09	Brugger et al. [91]
$Ratio_S^{BC}$	Bacteria ratio using DMS(P) as substrate for sustain their S need	-	1	

grazing (in mgC m^{-3} h^{-1}) and $DMSPd_{uptake}$ is the bacterial uptake of DMSPd (in mmolS m^{-3} h^{-1}) (flux$_{5+6+7}$, Fig. 1).

The description of $DMSPd_{uptake}$ is based on the bacterial C uptake described in MIRO, adjusted with the DMSPd stoichiometry of C substrates available to bacteria and taking consideration of the proportion of the bacterial community using the DMSPd (and DMS if necessary) for their C and S needs:

$$DMSPd_{uptake} = Ratio_{BC}^{S} * BC * bmx * \frac{SBC}{SBC + k_{SBC}} * \frac{DMSPd}{SBC} \quad (3)$$

where $Ratio_{BC}^{S}$ is the proportion of the bacterial community using the DMSPd for their S and C need. As a first approximation, we consider in the model that the whole bacterial community is able to degrade DMSP ($Ratio_{BC}^{S} = 1$). BC is the bacterial biomass, bmx is the bacterial growth, SBC are monomeric C substrates available for bacteria and k_{sbc} is the half-saturation constant for the bacterial consumption of SBC (in mgC m^{-3}).

Bacteria do not assimilate all of the DMSPd they consume, but take only the C and S they need to sustain their growth. It is known that 75 to 90% of DMSPd consumed by bacteria is degraded via demethylation and, although only 5 to 30% of metabolized DMSPd is assimilated into bacterial proteins, and this incorporation could satisfy the total S demands and between 1% and 15% of the C demands of the bacterioplankton [35], [75], [79], [80], [81]. In the model, the bacterial S need ($Sneed$, flux$_5$, Fig. 1) is estimated according to their growth, $Ratio_{BC}^{S}$, and the bacterial S:C ratio, according to:

$$S_{need} = Ratio_{BC}^{S} * BC * y_{BC} * bmx * \frac{SBC}{SBC + k_{SBC}} * SC_{BC} \quad (4)$$

where y_{BC} is the bacterial growth efficiency and SC_{BC} is the bacterial S:C ratio (Table 1).

The DMSPd not assimilated is demethylated (1- $lyase_{Bact}$, flux$_6$, Fig. 1) to produce SO$_4^{2-}$ or MeSH or cleaved by bacterial DMSP-lyase ($lyase_{Bact}$, flux$_7$, Fig. 1) as DMS and acrylate and used for the C requirements of the bacteria [35] according to:

$$bacterial\_lyase = lyase_{Bact} * (DMSPd_{uptake} - S_{need}) \quad (5)$$

where $lyase_{Bact}$ is the fraction of DMSPd consumed by bacteria which is cleaved in DMS and fixed to 10% for the reference simulation based on Niki et al. [34]. If DMSPd concentration is not sufficient to support bacterial S needs, DMS can be used as S source (Eq. 7) and bacterial DMSP-lyase activity is null.

Beside bacteria, several studies [82], [83], [84] have shown the capacity of some low DMSP-producer phytoplankton taxa to take up DMSPd. Hence, in parallel to their role of DMSP-producer, phytoplankton could also be a sink for DMS(P) cycle and therefore modify atmospheric DMS emission. However, knowledge on the DMSP-uptake phytoplankton taxa, its ecological role and governing factors and the phytoplankton competitive ability for DMSP regarding bacteria uptake is today insufficient for a proper inclusion in the model.

DMS production and fate. DMS is produced from enzymatic cleavage of DMSP by phytoplankton [11] and bacteria [85]. The major loss pathways of DMS are the bacterial consumption via the DMS monooxygenase and methyltransferase and oxidation via the DMS dehydrogenase [7], [68], [86], [87], [88]. DMS is also released to the atmosphere [39], [40] or photooxidized into dimethylsulfoxide (DMSO) [37], [38]. The DMS (in mmolS m^{-3}) state equation is:

$$\frac{dDMS}{dt} = y_{DMS}^{n} * lysis_n * SC_n + bacterial\_lyase - DMS_{uptake} - photooxydation - F_{DMS} \qquad for\ n = DA,\ NF\ and\ OP \quad (6)$$

where y_{DMS}^{n} corresponds to the fraction of DMSPp directly cleaved in DMS by phytoplankton DMSP-lyases (flux$_1$, Fig. 1), $lysis_n$ is the phytoplankton cellular lysis (in mgC m^{-3} h^{-1}), $bacterial\_lyase$ is the enzymatic cleavage of DMSPd in DMS by bacteria (flux$_7$, Fig. 1), DMS_{uptake} is the bacterial consumption of DMS (flux$_8$, Fig. 1), $photooxidation$ term is the photochemical oxidation of DMS into DMSO (flux$_{10}$, Fig. 1) and F_{DMS} is the emission of DMS to the atmosphere through the air-sea water interface (in mmolS m^{-3} h^{-1}) (flux$_9$, Fig. 1).

Although bacterial degradation of DMS is important [36], [41], [81], [88] less than 10% of S of DMS consumed is incorporated into bacterial biomass [81], [89] and satisfies 1% to 3% of the bacterial S demand. This suggests that DMS is a minor source of S for bacterioplankton, and is probably taken up by bacteria only as a supplementary substrate [81]. Bacteria predominantly metabolized DMS into non-volatile sulfur products, DMSO and SO$_4^{3-}$ [36], [81], [87], [90].

Based on that, we assume that bacterial uptake of DMS (DMS_{uptake}) will cover the bacteria S needs if $DMSPd_{uptake}$ (Eq. 3) is not sufficient. In the model, DMS_{uptake} is described from the consumption of carbon by bacteria and the DMS content of bacterial C substrates, according to:

$$DMS_{uptake} = Ratio_{BC}^{S} * bmx * BC * \frac{SBC}{SBC + k_{SBC}} * \frac{DMS}{SBC} \quad (7)$$

The photooxidation of DMS into DMSO is described considering a photooxidation constant ($K0$, [91]) modulated by the light extinction coefficient in water, according to:

$$photooxidation_z = DMS_z * K0 * \exp^{-kD.z} \quad (8)$$

where z is the water depth (m), $K0$ is the photooxidation rate in the surface (Table 1; [91]) and kD is the light extinction coefficient, and (DMS)$_Z$ is DMS at depth z. As a first approximation, the ultraviolet A (UVA) penetration in the water column is considered equal to that of photosynthetic active radiation (PAR), as PAR attenuation in the studied coastal area is mainly governed by detrital particulate and colored dissolved organic matter. This assumption corresponds to a maximum water penetration of UVA and tends to overestimate the DMS loss by photooxidation.

The DMS air-sea flux (F$_{DMS}$) is determined based on the surface DMS concentration and the gas transfer velocity (k) of DMS at in-situ temperature (k_{DMS}):

$$F_{DMS} = k_{DMS}[DMS] \quad (9)$$

with

$$k_{DMS} = k_{600}(600/Sc_{DMS})^{0.5} \quad (10)$$

where k$_{600}$ is k normalized to a Schmidt number (Sc) of 600 and

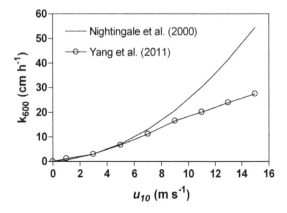

Figure 2. Gas transfer velocity (k_{600}) as a function of wind speed (u_{10}) given by the Nightingale et al. [107] parameterization, and the binned measurements of Yang et al. [93] to which was fitted a polynomial relationship (Eq. 13). The k data of Yang et al. [93] were originally reported normalized to a Schmidt number of 660 (k_{660}) and were converted to k_{600}.

Sc_{DMS} is the Sc of DMS computed according to Saltzman et al. [92]:

$$Sc_{DMS} = 2674 - 147.12T + 3.726T^2 - 0.038T^3 \quad (11)$$

where T is sea surface temperature (°C).

k_{600} (cm h^{-1}) was computed from a parameterization (Fig. 2) as a function of wind speed referenced at 10 m height (u_{10} in m s^{-1}) that we derived from the binned data reported by Yang et al. [93] in their Table 2 (data without bubble normalization):

$$k_{600} = 0.5093u_{10} + 0.2179u_{10}^2 - 0.0087u_{10}^3 \ (r^2 = 0.999, n = 8) \quad (12)$$

u_{10} data were extracted from the National Centers for Environmental Prediction (NCEP) Reanalysis Daily Averages Surface Flux (http://www.cdc.noaa.gov/) for one station in the North Sea (3.75°E 52.38°N).

Model implementation

For this application, the MIRO-DMS model was implemented in the SNS using a multi-box 0D frame delineated on the basis of the hydrological regime and river inputs (Fig. 3) [60]. In order to

Table 2. F_{DMS} computed for sensitivity tests on DMS(P) model parameters.

	Parameters	Units	Values	Annual mean [DMS] (μmolS m^{-3})	F_{DMS} (mmolS m^{-2} y^{-1})
REFERENCE				**0.9**	**0.19**
Sensitivity to phytoplankton parameters					
Test 1	SC_{NF}, SC_{OP}	mol S:molC	0.018	1.5	0.32
Test 2	SC_{NF}, SC_{OP}	mol S:molC	0.004	0.3	0.07
Test 3	SC_{DA}	mol S:molC	0.00212	0.9	0.21
Test 4	SC_{DA}	mol S:molC	0	0.8	0.18
Test 5	SC_{DA}	mol S:molC	0.0034	1.0	0.23
Test 6	y^{DA}_{DMS}, y^{NF}_{DMS}, y^{OP}_{DMS}	-	0	0.5	0.11
Test 7	y^{DA}_{DMS}, y^{NF}_{DMS}, y^{OP}_{DMS}	-	0.25	1.4	0.32
Test 8	y^{DA}_{DMS}, y^{NF}_{DMS}, y^{OP}_{DMS}	-	0.5	2.4	0.53
Test 9	y^{NF}_{DMS}, y^{OP}_{DMS}	-	0.5	2.3	0.51
Test 10	y^{DA}_{DMS}	-	0.5	0.9	0.21
Sensitivity to bacteria parameters					
Test 11	SC_{BC}	mol S:molC	1:37	0.8	0.17
Test 12	SC_{BC}	mol S:molC	1:196	0.9	0.2
Test 13	$Ratio_{BC}$	-	0.75	1.1	0.24
Test 14	$Ratio_{BC}$	-	0.5	1.4	0.32
Test 15	$Ratio_{BC}$ for DMSPd	-	0.5	0.9	0.2
Test 16	$Ratio_{BC}$ for DMS	-	0.5	1.4	0.32
Test 17	khydrolysis	-	0.25	1.6	0.35
Sensitivity to wind speed and k parameterization					
Test 18	wind forcing	m s^{-1}	3.9	0.9	0.24
Test 19	wind forcing	m s^{-1}	−25%	0.9	0.13
Test 20	wind forcing	m s^{-1}	+25%	0.9	0.26
Test 21	k parameterization	cm h^{-1}	Nightingale et al., 2002	0.9	0.19

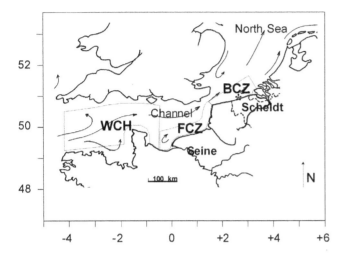

Figure 3. Map of the study area with the MIRO-DMS multi-box frame delimitation with WCH = Western Channel; FCZ = French Coastal Zone; BCZ = Belgian Coastal Zone (adapted from Gypens et al. [94]). Model results analysis will focus on the BCZ where simulated results were daily-averaged for year 1989, when DMS(P) field data are available for comparison.

take account for the cumulated nutrient enrichment of Atlantic waters by the Seine and Scheldt rivers, the model was run successively in the Western Channel (WCH) area considered as a quasi-oceanic closed system, the French coastal zone (FCZ) influenced by the Seine and Atlantic waters from the WCH, and, finally, in the Belgian coastal zone (BCZ) influenced by the direct Scheldt loads and the inflowing FCZ waters. Model simulations were performed using meteorological and river forcing for the year 1989 when DMS(P) data are available for comparison [95]. The seasonal variation of the state variables was calculated by solving the different equations expressing mass conservation according to the Euler procedure. A time step of 15 min was adopted for the computation of the numerical integration. The analysis of daily-averaged model results will be performed in the BCZ where field DMS(P) are available [95]. DMS(P) data for the year 1989 were retrieved from the Global Surface Seawater Dimethylsulfide (DMS) Database (available at http://saga.pmel.noaa.gov/dms/) and correspond to data available in the SNS between 51.0°N–52.5°N and 1.5°E–4.5°E [95].

Results

DMS(P) seasonal cycle in the Southern North Sea

Validation of the MIRO ecological model is given by Lancelot et al. [60] and Gypens et al. [69], and is not repeated here. The performance of the MIRO-DMS model is evaluated through its ability to reproduce the seasonal variations of available field data of DMSPp, DMSPd and DMS in the BCZ for the year 1989 [95]. However, due to the limited data set, a statistical validation was not attempted and we only compared qualitatively field data and model output. For this comparison, daily simulated results are compared to data of DMS(P) acquired by Turner et al. [67] during short 2–3 day cruises at monthly intervals. Data for each month ranged between 2 and 15 samples, for the purpose of the validation, they were averaged, and standard deviations are given in plots as error bars.

Changes in DMS(P) concentrations are analyzed in parallel to the evolution of the planktonic compartments (phytoplankton and bacteria) (Fig. 4). The phytoplankton evolution simulated in the

area is characterized by a succession of spring diatoms, *Phaeocystis* colonies, and summer diatoms (Fig. 4a). Spring diatoms initiate the phytoplankton bloom in early March and are followed by *Phaeocystis* colonies which reach Chl *a* concentration of 25 mgChl*a* m^{-3} (Fig. 4b) in April. Summer diatoms bloom after the *Phaeocystis* decline and remain until fall. On an annual scale, diatom and *Phaeocystis* biomass are similar, the latter being however concentrated during a short period of time, of 1 month (Fig. 4a). In association with the decline of the different phytoplankton blooms, three bacterial maxima are simulated (Fig. 4c).

In agreement with available data, the simulated DMS(P) concentrations show low values except during the spring *Phaeocystis* bloom (Fig. 4d). Simulated DMS(P) values are lower than observed DMS(P) concentrations in early April (the spring diatom bloom). As observed by Turner et al. [67], Kwint and Kramer [68] and van Duyl et al. [96] in North Sea coastal waters, DMSP and DMS concentrations increase in spring and decrease in autumn to low winter values. The maxima in DMS(P) concentrations are limited to a period of about 6 weeks (April, May) and concurred with the *Phaeocystis* bloom as also observed by Stefels et al. [33] in the same area. The model correctly reproduces the observed DMSP seasonal pattern, in particular the timing of the seasonal peak. However, the model fails to reproduce amplitude of the seasonal cycle, with simulated maximal DMSPp concentration (580 μmolS m^{-3}; Fig. 4d) three times higher than measured concentration. On the other hand, the modeled DMSPd is much lower than the field observations. This could be due to an experimental bias in older data-sets due to cell breakage leading to an over-estimation of DMSPd and an underestimation of DMSPp [97]. Indeed, the maximum simulated total DMSP (DMSPt = DMSPp + DMSPd) of 670 μmolS m^{-3} is close to the maximum observed DMSPt of 730 μmolS m^{-3}. This discrepancy could also be due to the low temporal resolution of observations (1 month, [95]), i.e. insufficient to fully capture the dynamics of the system. Indeed, data obtained with a higher sampling frequency (2 samples per week) in the Wadden Sea (Marsdiep) in 1995, show DMSPp concentrations of about 1700 μmolS m^{-3} during a *Phaeocystis* bloom that reached a maximum of 80 10^6 cell L^{-1} [96]. In agreement with these observations, the simulated maximum of DMSPp (Fig. 4d) coincides with the *Phaeocystis* colonies bloom (Fig. 4a) and reach a value of about 580 μmolS m^{-3} for a *Phaeocystis* biomass of 1600 mgC m^{-3} (Fig. 4a) corresponding to 58 10^6 cell L^{-1}. Hence, the modeled DMSP seasonal peak is bracketed by the lower values of Turner et al. [95] in the more open water of the SNS and the higher values of van Duyl et al. [96] in the near-shore coastal waters of the SNS.

The time lag of about 10 days between the simulated DMSPp and DMSPd (210 μmolS m^{-3}; Fig. 4e) peaks is due to the fact that DMSPd results from the phytoplankton lysis and grazing by zooplankton that increase at the end of the bloom. As for DMSPp, simulated DMSPd is also underestimated in comparison with the observed concentration in March during the spring diatom bloom.

The simulated DMS peak reaches a value of 28 μmolS m^{-3} (Fig. 4f) and appears in between DMSPp and DMSPd maxima. The accumulation of DMS simulated during the decay of *Phaeocystis* (Fig. 4a) is consistent with the work of Stefels and van Boekel [61] showing that phytoplankton lyases are active during the stationary phase of the bloom. Simulated and observed DMS show similar seasonal patterns but simulated concentration of DMS is lower than the maxima observed in May (50 μmolS m^{-3}, Fig. 4f). However, when spatially averaged over the SNS to take into account for the non-regular distribution of sampling stations, observed DMS concentrations show a maximal value of 25 μmolS m^{-3} (Fig. 5 in Turner et al. [95]).

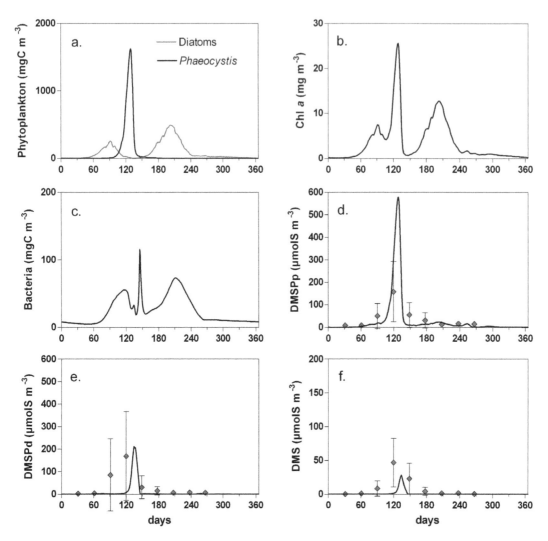

Figure 4. Seasonal evolution of diatoms and *Phaeocystis* colonies biomass (a), total chl a (b), bacteria biomass (c) and DMSPp (d), DMSPd (e) and DMS (f) concentration simulated for year 1989 in the Belgian Coastal Zone by the MIRO-DMS model and compared to monthly DMS(P) averaged data (◇) from Turner et al. [95]. The error bars represent the standard deviation of the mean.

Figure 5a shows the seasonal evolution of atmospheric DMS emissions simulated by the model in the BCZ. As expected the DMS flux to the atmosphere follows closely the temporal pattern of the simulated DMS concentrations (Fig. 4f), ranging from low values in winter to a maximal value of 37 μmolS m^{-2} d^{-1} in spring. The important daily variability simulated during F_{DMS} peak (Fig. 5a) results from wind speed variability (Fig. 5b).

Annual DMS budget

The relative importance of each processes involved in the DMS cycle was estimated based on the annual S budget (Fig. 6) obtained by integrating the daily S rates simulated by the model in the BCZ (Fig. 2) and integrated on the average depth of the study area (17 m).

MIRO-DMS estimates the total annual phytoplankton production of DMSPp at 50 mmolS m^{-2} y^{-1}, of which 13% are produced by diatoms, 9% by nanoflagellates and 78% by the *Phaeocystis* colonies. From this, 3.2 mmolS m^{-2} y^{-1} of DMSPp are directly converted in DMS by phytoplankton DMSP-lyase (mainly that of *Phaeocystis*) representing a DMS flux similar to bacterial DMSP-lyase activity. The importance of phytoplankton

DMSP-lyase was previously reported in the area by Stefels and Dijkhuizen [25] and Wolfe and Steinke [22]. The production of DMS by phytoplankton DMSP-lyase simulated in the model is three times higher than the DMS loss due to flux to the atmosphere and photochemical oxidation, as observed (between 1.5 to 4.5 times) in the Dutch coast during a *Phaeocystis* bloom [33].

DMSPd results from phytoplankton cell lysis (68%) and zooplankton grazing (32%). The dominant process is the cell lysis of *Phaeocystis*, which in itself releases almost 50% of DMSPp throughout the year. The sedimentation of DMSPp amounts to 4.3 mmolS m^{-2} y^{-1}. Bacterial uptake accounts for the majority the removal of both DMSPd and DMS inducing a rapid decrease of their concentrations in the water column. The consumption of DMSPd is sufficient to sustain the total bacteria S need (10.8 mmolS m^{-2} y^{-1}), and provides up to 16% of bacteria C requirements. In agreement with previous findings [35], [75], [80], [98], the major fate for simulated DMSPd is the demethylation/ demethiolation pathways that consumes 28.5 mmolS m^{-2} y^{-1} and results in S products other than DMS (mainly SO_4^{2-} and MeSH). Although only 8% (3.2 mmolS m^{-2} y^{-1}) of the DMSPd consumed by bacteria is cleaved to DMS, this flux represents 50% of annual

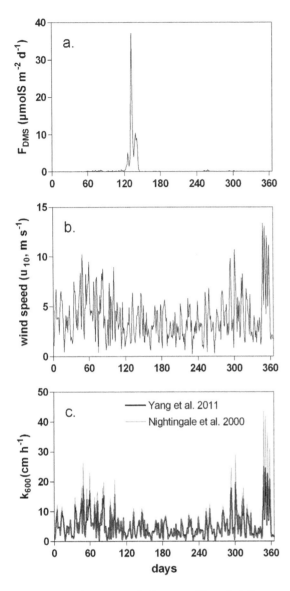

Figure 5. Daily DMS emission (μmolS m^{-2} d^{-1}) computed by the MIRO-DMS model in the Belgian Coastal Zone for year 1989 (a), wind speed (u$_{10}$) (b) and the gas transfer velocity (k_{600}) computed using the Yang et al. [93] and the Nightingale et al. [107] relationships (c).

DMS input and is similar to phytoplankton DMSP-lyase activity (Fig. 6).

Bacteria also consume directly DMS and about 83% (5.3 mmolS m^{-2} y^{-1}) of the DMS pool is consumed by bacteria and transformed in SO$_4$$^{2-}$ or DMSO. Kiene and Bates [41] found that microbial DMS consumption was generally 10 times faster than the flux of DMS to the atmosphere. This ratio is about 17 times in our model results with about 14% of the DMS converted into DMSO by photooxidation and finally only 3% of the DMS emitted to the atmosphere. Annual F$_{DMS}$ represents <1% of the DMSPp production in the water column in agreement with Archer et al. [49].

Discussion

Annual S budget simulated in the BCZ points both phytoplankton and bacteria as key controlling factors of the DMS

production. However the relative importance of these processes will results from their description and parameterization in the model. Sensitivity analyses were then carried out to estimate the impact on the atmospheric emission of DMS of the description of several biological processes compared to physical processes (wind speed and k_{600} parameterization). In particular, the impact of phytoplankton S:C quota determining the maximal DMSP production of the ecosystem, the importance of phytoplankton DMSP-lyase that represents the direct transformation pathway of DMSP into DMS and the DMS(P) bacterial uptake and lyase activity were tested.

Sensitivity to biological processes

Sensitivity to phytoplankton parameters. In our model, phytoplankton S:C quotas were fixed, corresponding to the mean values of measurements for Haptophyceae and diatoms reported by Stefels et al. [7]. Sensitivity tests were performed by varying the S:C quotas within the range of extreme values reported for each phytoplankton type by Stefels et al. [7] (Table 2). Increasing (decreasing) by 70% the *Phaeocystis* S:C value in the model (Test 1 and 2, Table 2) increases (decreases) simulated DMSP and DMS concentrations (Fig. 7; 8a) and annual F$_{DMS}$ by a similar factor (Table 2) without changing the seasonal pattern. Due to the low value of the tested diatom S:C (Tests 3 and 4; Table 2), any modification has little effect on the simulated DMS(P) (Fig. 7; 8a) and F$_{DMS}$ (Table 2). However, some diatom species are characterized by higher S:C quota [8] as *Skeletonema costatum* that is characteristic of the spring diatoms in the SNS [65]. An additional simulation was performed using S:C quota measured for this species (Test 5; Table 2). Increasing the diatom S:C quota increases annual F$_{DMS}$ (Table 2) but also results in an overestimation of simulated DMSPp in summer (Fig. 7). This suggests that in the SNS, dominant diatoms in spring and summer are characterized by different S:C quotas, and that it is essential to take into account for their specific phytoplankton DMSP content to correctly reproduce seasonal evolution of DMS(P) concentration for different FTs (diatoms versus *Phaeocystis*), but also within a FTs (spring versus summer diatoms).

One of the indirect consequences of the choice of the phytoplankton S:C is the possibility for bacteria to fulfil their S need from the consumption of DMS(P). For low phytoplankton S:C ratio (Tests 2 and 4, Table 2) only 60 to 90% of the bacterial S needs in summer and fall can be sustained by DMS(P). As a consequence, the associated bacterial DMSP-lyase activity is decreased.

In the reference simulation, 10% of the DMSP released after phytoplankton lysis is directly cleaved into DMS leading to a DMS flux (3.2 mmolS m^{-2} y^{-1}) similar to the DMS flux that comes from bacterial enzymatic cleavage (Fig. 6). However, the relative importance of both processes varies during the seasonal cycle with maximal phytoplankton DMSP-lyase activity simulated at the maximum of the *Phaeocystis* bloom and bacterial DMSP-lyase activity dominating at the decline of the bloom. As deduced by Stefels et al. [7] from the observations of van Duyl et al. [96] in the North Sea, algal DMSP-lyase activity is more important than bacterial enzymatic cleavage at high concentration of DMSPd and explains the occurrence of maximum DMS concentration before the DMSPd peak in our results (Fig. 4e,f). After the decay of the *Phaeocystis* bloom, bacteria and associated DMSP cleavage largely increase.

Most, but not all [34], DMSP-producing species of phytoplankton have DMSP-lyase activity. However, the importance of this activity is not especially correlated with intracellular DMSP concentration [72], [99]. The importance of the direct

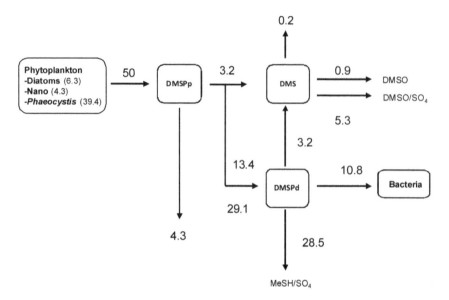

Figure 6. Annual sulfur budget in the Belgian Coastal Zone computed by the MIRO-DMS model for the year 1989 (mmolS m^{-2} y^{-1}).

transformation of DMSP into DMS, on the DMS emission was tested by varying the cleavage yield (y^n_{DMS}, Eq. 3) between 0% and 50% (Table 2). The absence of phytoplankton DMSP-lyase activity (Test 6, Table 2), delays the DMS peak by a few days, and decreases both the simulated DMS (Fig. 8b) and F_{DMS} by about 40% (Table 2). This is higher than the 25% computed by van den Berg et al. [46] based on a modeling study in the SNS. When 25% or 50% of DMSPd released from phytoplankton lysis is converted into DMS (Tests 7 to 9; Table 2), the DMS concentration and F_{DMS} largely increase compared to the reference simulation (from 1.5 to 2.5 times, Table 2). Although simulated DMSPd decreases, this effect is limited as the DMSPd pool is also provided by zooplankton grazing, and its fate controlled by bacterial activity. Increasing only diatom DMSP-lyase yield has little effect on F_{DMS} (Test 10; Table 2), indicating the dominance of *Phaeocystis* in phytoplankton DMSP-lyase activity.

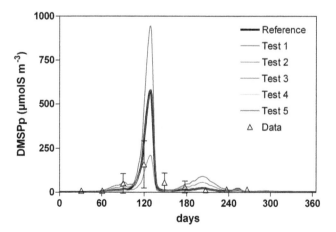

Figure 7. Seasonal evolution of DMSPp concentration simulated by the MIRO-DMS model for year 1989 for different phytoplankton *S:C* ratio (see Table 2 for the description of the sensitivity tests).

Altogether these sensitivity tests show that phytoplankton DMSP-lyase is a key process controlling both DMS concentration and F_{DMS} and even more important when associated to a high DMSP-producer such as *Phaeocystis*. It is therefore important to determine this enzymatic activity in high DMSP-producing species or among species that co-occur with high DMSP-producing species. An explicit description of DMSP-lyase activity in models could also be important if this activity varies as a function of environmental conditions.

Sensitivity to bacteria parameters. As observed by several authors (e.g. [75]), bacterial uptake is the major fate of DMSPd in the model, but only 8% of this DMSPd is cleaved into DMS by bacteria. This agrees with recent observations concluding that bacteria are not key players in DMSPd cleavage into DMS [100], [101] but play a major role in regulating the flux of DMS indirectly by the consumption and demethylation of DMSPd with production of S product other than DMS.

However, the proportion of DMSPd consumed by bacteria and transformed into DMS is function of the DMSPd concentration [96] and the bacterial S demand [35]. Indeed, previous studies suggested that the fraction of DMSPd converted into DMS increases with DMSPd concentration [75]. Lower DMSPd concentrations are completely assimilated, whereas higher concentrations result in increasing amounts of DMS produced [102]. Moreover, a strong demand for S decreases bacterial cleavage of DMSPd [35]. The sensitivity of model results to DMSPd concentration and/or bacterial S needs was estimated either by modifying the release of DMSPd by phytoplankton, the bacterial S:C quota or the proportion of the bacterial community that use DMSP as S source.

In the model, DMSPd is released in the water column by phytoplankton lysis and grazing processes (Eq. 2). The modification of the phytoplankton DMSP-lyase activity affects the F_{DMS} but also the relative contribution of phytoplankton and bacterial processes to DMS production. Increasing the direct transformation of DMSPp in DMS by phytoplankton DMSP-lyase will decrease the DMSPd bacterial uptake and the bacterial production of DMS. Increasing the cleavage yield (y^n_{DMS}, Eq. 3) up to 50% (Test 8, Table 2) will decrease bacterial DMS production by 40% but increases both DMS production by phytoplankton and

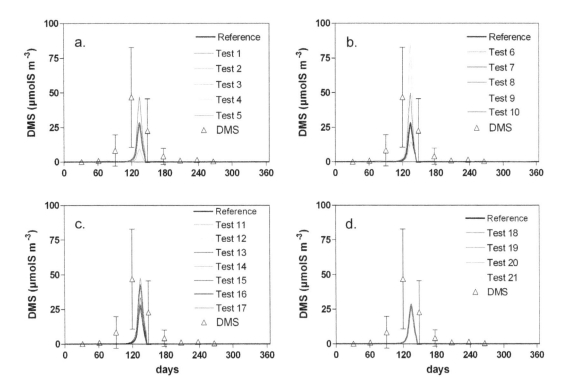

Figure 8. Seasonal evolution of DMS concentration simulated by the MIRO-DMS model for year 1989 by modifying phytoplankton S:C ratio (a), phytoplankton lyase (b), bacteria S:C content and bacterial processes (c) and wind speed (d). See Table 2 for the description of the sensitivity tests.

F_{DMS}. However, when compared to the data available in the literature [7], [34], these results overestimate the contribution of phytoplankton compared to bacteria to the DMS production (with phytoplankton contribution up to 90% of the DMS production). One possible source of overestimation of DMSPd concentrations in the model can however result from the assumption that all the DMSPp ingested by micro- and meso-zooplankton is transformed into DMSPd (Eq. 1, 2). Indeed, Wolfe and Steinke [22] also suggested that part of the DMSPp can be directly converted to DMS since digestion promotes the activity of DMSP-lyase present in the membrane of the prey. To test the impact of the direct conversion of DMS by zooplankton, 30% of DMSPp (based on Archer et al. [74]) ingested by grazing was directly transformed in DMS and added in Eq. 6. This results in an increase of DMS concentration in the water column and of F_{DMS} (0.29 mmolS $m^{-2}y^{-1}$) with little impact on DMSPd concentration.

Sensitivity tests were then conducted by varying the bacterial S:C quota between extreme values reported in the literature i.e. 1:37 and 1:196 molS:molC ([103]; Tests 11 and 12; Table 2). In our model, decreasing the bacterial S:C ratio will decrease the proportion of consumed DMSPd that will be assimilated by bacteria and increase the cleavage of DMSPd into DMS. This can enhance the F_{DMS}. However, as shown in Table 2, this parameter is not very sensitive in our application as the DMSPd produced is largely enough to fulfil the S needs of the whole bacterial community.

In a third series of tests, we modified the percentage of bacteria able to use DMSPd and/or DMS as S source. The hypothesis of 100% used in the reference simulation was based on the observation that most marine bacteria have the genetic capability to demethylate DMSP [36], [104], [105] and that DMSPd/DMS concentrations can support almost all bacteria S needs [79], [80],

[87]. However, all bacteria are not able to metabolize DMSP and/ or DMS. We therefore explore the sensitivity of the model to the bacteria diversity by decreasing this proportion to 75% or 50% (Tests 13 and 14; Table 2). As expected, the turnover rate of DMSPd and DMS decreases and the F_{DMS} increases (Table 2). The maximum concentrations of DMSPd and DMS (Fig. 7c) simulated are 270 µmolS m^{-3} and 33 µmolS m^{-3} when considering that 75% of the bacterial community is able to degrade DMS(P) and 375 µmolS m^{-3} and 40 µmolS m^{-3} for a fraction of 50%. The simulated DMS emissions to the atmosphere also increase with an annual F_{DMS} of about 0.24 and 0.32 mmolS $m^{-2} y^{-1}$, respectively, compared to 0.19 mmolS $m^{-2} y^{-1}$ in the reference simulation (Table 2). This increase of DMS emission results from the combination of bacterial DMSP cleavage and the decrease of bacterial DMS uptake. In these simulations, bacterial DMSP-lyase activity shows a small increase (up to 3.4 and 3.6 mmolS $m^{-2} y^{-1}$ compared to 3.2 mmolS $m^{-2} y^{-1}$ in the reference simulation), and the increase of F_{DMS} mainly results from the decrease of bacterial DMS uptake and the accumulation of DMS in the water column. This is confirmed by results obtained by modifying only DMSPd (Test 15) or DMS bacterial uptake (Test 16). These results are consistent with the observations that suggest that bacterial DMS uptake may be a quantitatively important sink for DMS from the surface ocean [36], [81], [87], [90].

In the model, bacterial cleavage of DMSP in DMS represents 10% of the uptake of DMSPd not assimilated by bacteria. To test the importance of bacterial DMSP-lyase activity, this fraction was set to 25% inducing an increase of almost two fold of both the concentration of DMS and F_{DMS}.

Due to their importance on both DMSPd and DMS transformation, bacterial processes need to be accurately described and/or parameterized in ecosystem models. Note that in the present version of the model we only considered one bacterial community, and we did not individually represent the DMS- or DMSP-consumers although this simplification also induces possible uncertainties and underestimation of F_{DMS} resulting from the maximal hypothesis of bacterial uptake ($Ratio^{BC}_S = 1$). This is particularly important for the direct bacterial uptake of DMS. Similarly, the bacteria state variable lumps both bacteria and Achaea that might also be important for the demethylation/demethiolation processes [106].

Sensitivity to physical processes: Wind speed and k_{600} parameterization

Besides biological processes, F_{DMS} is also function of the k_{600} that depends on the intensity of wind speed and how it is translated into turbulence (depending on the parameterization). Additional tests were performed to estimate the sensitivity of the simulated atmospheric emission of DMS to wind speed and k_{600} parameterization (Table 2). Changing wind speed will mainly affect the F_{DMS} that change up to 37% (Test 20, Table 2) with little change for DMS concentrations (Fig. 8d). Due to very low values of wind speed (Fig. 5b) during the *Phaeocystis* bloom and the peak production of DMS (Fig. 4a,d), the use of a constant annual mean wind speed will increase annual F_{DMS} (Test 18; Table 2). Indeed, to accurately compute F_{DMS} it is required to use high temporal resolution u_{10} data [49]. However, considering the low effect of F_{DMS} compared to bacterial DMS consumption, this has little impact on the dissolved DMS concentration (Fig. 8d).

In the reference simulation we used a parameterization of k_{600} based on the data reported by Yang et al. [93]. Several other parameterizations of k_{600} exist and for the purpose of a sensitivity analysis, we chose the one of Nightingale et al. [107] that has been used in the recent F_{DMS} climatology of Lana et al. [40]. Nightingale et al. [107] parameterize k_{600} as a function of u_{10}, according to:

$$k_{600} = 0.33u_{10} + 0.22u_{10}^2 \qquad (13)$$

The k values used in the Nightingale et al. [107] parameterization were determined from two dual tracer (^{3}He and SF_6) release experiments in the SNS, and this parameterization has been shown to be also applicable in open ocean conditions [108]. The k values of Yang et al. [93] were obtained from measurements of [DMS] and direct measurements of F_{DMS} by eddy-covariance during 2 experiments in the Pacific Ocean and 3 experiments in the Atlantic Ocean. The k_{600} values of Nightingale et al. [107] and Yang et al. [93] strongly diverge at $u_{10} > 8$ m s^{-1} (Fig. 2). This has been attributed to reduced bubble-mediated transfer at high wind speeds of highly soluble DMS compared to enhanced bubble-mediated transfer of sparingly soluble gases such as ^{3}He and SF_6.

The net annual F_{DMS} computed with the Yang et al. [93] derived parameterization (Eq. 12) and the Nightingale et al. [107] parameterization (Eq. 13) are not different in the area during the simulation period. This is due to the fact that during the period of high DMS concentrations (during the *Phaeocystis* bloom) wind speed is low (average 3.3 ± 1.7 m s^{-1}, Fig. 5b), and the k_{600} values computed from the two relationships are very close (Fig. 5c). The two k_{600} relationships only significantly diverge for $u_{10} > 8$ m s^{-1} (Fig. 2), and such u_{10} values only occur during winter and fall in the SNS (Fig. 5b) when [DMS] is very low or zero (Fig. 4f). Since wind speeds > 8 m s^{-1} are rare events in the area (<6% of observations), the annual average of k_{600} computed from the Yang

et al. [93] relationship (5.20 cm h^{-1}) is only ~9% lower than the one computed using the Nightingale et al. [107] relationship (5.67 cm h^{-1}).

Comparison of DMS and F_{DMS} modelled by the mechanistic MIRO-DMS model and derived from empirical relationships (statistical models)

In order to achieve global [109], [110], [111], [112], [113], [114], [115] or regional [116] estimates of F_{DMS}, several empirical relationships have been derived from DMS field data and variables such Chl a, NO_3^-, T, primary production, solar radiation, or mixed layer depth that can be derived at higher spatial and temporal resolution from climatologies, remote sensing or models. We tested if some of these empirical relationships that are assumed universal and generic were applicable to the SNS that is representative of a temperate eutrophied coastal system. Several empirical parameterizations that allow to compute DMS concentration in marine waters (Table 3) were applied in the area using MIRO-DMS outputs (Chl a, NO_3^-) and compared to DMS concentration obtained with the MIRO-DMS, and with the available DMS observations in area (Fig. 9a).

DMS concentrations simulated with the algorithm of Simó and Dachs [111] show maximal DMS concentrations similar to those simulated by the model during the *Phaeocystis* bloom (Fig. 9a). However, they overestimate F_{DMS} along the seasonal cycle (Fig. 9b), in particular due to an overestimation of the DMS concentrations related to spring and summer diatom blooms (Fig. 9a). Neither Anderson et al. [109] nor Lana et al. [114] relationships can reproduce the amplitude of DMS seasonal cycle and DMS peak associated to *Phaeocystis* bloom (Fig. 9a). As for the Simó and Dachs [111] relationship, the Anderson et al. [109] and Lana et al. [114] relationships over-estimate the DMS concentration associated with the diatom spring and summer blooms. In the area, the mixed layer depth is constant (= total depth, since it is a permanently well-mixed shallow system) and the seasonal evolution of DMS concentrations (Fig. 9a) simulated by all these relationships is controlled by the evolution of Chl a (Fig. 9c), without any distinction in DMSP cellular content among phytoplankton groups. To take into account of this variability we also tested two additional relationships respectively developed by Aumont et al. [110] and revised by Belviso et al. [112] based on a similar data-set. The Fp ratio representing the community structure index (and corresponding to the ratio of the diatoms and dinoflagellates to the total Chl a) used in these relationship was computed based diatoms and non-diatoms (nanoflagellates and *Phaeocystis* colonies) Chl a simulated by the MIRO-DMS model. Results obtained with both relationships largely overestimated DMS concentrations in the area during *Phaeocystis* bloom (with DMS values up to 400 nM with the Belviso et al. [112] equation and unrealistic values up to 5000 nM with the Aumont et al. [110] equation). Both relationships were established from data-sets with total Chl a values <4 µg L^{-1} (and non-diatom Chl a values lower than 1 µg L^{-1}), well below the maximum values in the SNS, up to 25 µg L^{-1} (Fig. 4b). Based on these results, we conclude that these relationships are not adapted to ecosystems dominated by high biomass of non-siliceous species, typically in eutrophied coastal environments.

The F_{DMS} computed from DMS derived the various empirical parameterizations are higher than F_{DMS} computed with MIRO-DMS, about 6 times higher for Lana et al. [114] and about 10 to 15 times higher for Anderson et al. [109] and Simó and Dachs [111] relationships. These F_{DMS} values are also largely higher than the maximal F_{DMS} previously estimated in the area [33], [46], [67], [117]. Despite the fact that these relationships give lower

Table 3. Empirical relationships tested in the MIRO-DMS, and the corresponding annual mean of [DMS] and F_{DMS}. Fp is the community structure index computed as the ratio between the diatoms and non-diatoms (nanoflagellates and *Phaeocystis* colonies) Chl a simulated by the MIRO-DMS and z in the depth of the mixed layer (m) that is constant in the MIRO-DMS application (17m).

Equations	Reference	[DMS]	F_{DMS}
		(μmolS m^{-3})	(mmol S m^{-2} y^{-1})
[DMS] = 2.29 for log10(CJQ)<1.72	Anderson et al. [109]	2.2	2.23 for k_{NO3} = 0.8
[DMS] = 8.24 [log10(CJQ)−1.72]+2.29 for log10(CJQ)>1.72		2.5	2.63 for k_{NO3} = 2
where C = Chl a (mgm^{-3}), J = mean daily irradiance (Wm^{-2})			
and Q = NO$_3$/(NO$_3$+k$_{NO3}$) (mmolm^{-3})			
[DMS] = −ln (z)+5.7 for Chl a/z<0.02	Simó and Dachs [111]	3.1	3.03
[DMS] = 55.8 (Chl a/z)+0. for Chl a/z>0.02			
[DMS] = 2.356+0.614 * Chla	Lana et al. [114]	1.1	1.21
DMSPp = (20*Chla*Fp)+21 for Chla'<0.3 mg m^{-3}	Belviso et al. [112]	-	-
DMSPp = (20*Chla*Fp)+(356.4 * Chla −85.5)			
for Chla'>0.3 mg m^{-3}			
DMS:DMSP = 0.231−3.038Fp−16 Fp^2			
−38.05Fp^3+41.12Fp^4−16.32Fp^5			
DMSPp = (20*Chla*Fp)+	Aumont et al. [110]	-	-
(13.64+0.10769* (1+24.97*(1-Fp)*Chla)$^{2.5}$)			
DMS:DMSP = 0.015316+0.005294/(0.0205+Fp) for Fp<0.6			
DMS:DMSP = 0.674*Fp−0.371 for Fp>0.6			

seasonal maxima DMS concentrations (with the exception of the Simo and Dachs [111] relationship), they compute DMS concentrations through the year during both diatom and *Phaeocystis* blooms. MIRO-DMS only simulates DMS during the *Phaeocystis* bloom, when wind speed and k_{600} are low (Fig. 5b,c), while DMS is very low during the rest of the year.

Conclusions

The application in the BCZ of the newly developed biogeochemical model MIRO-DMS shows that modelled F_{DMS} is more sensitive to the description and parameterization of biological than abiotic processes. The results confirm the importance of accounting for specific phytoplankton cellular DMSP between different FTs (*Phaeocystis versus* diatoms) but also within a FT (spring *versus* summer diatoms) to describe DMSP and DMS concentrations in marine ecosystems. Due to their elevated S:C quota and their

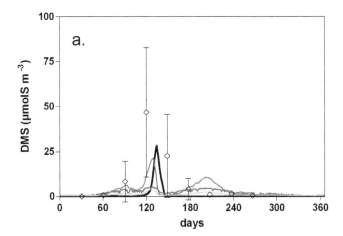
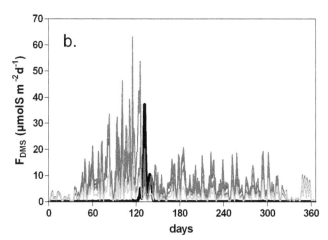

Figure 9. Seasonal evolution of DMS concentration (a) and flux (b) computed in the BCZ for year 1989 using the MIRO-DMS model (black) and the empiric relationship of Simó and Dachs [111] (grey), Anderson et al. [109] with a k$_{NO3}$ of 0.8 and 2 mmolN m^{-3} (blue) and Lana et al. [114] (green) and compared to available data (◇) from Turner et al. [95]. The error bars represent the standard deviation on the mean.

major contribution (50%) to the annual primary production, *Phaeocystis* colonies are responsible of 78% of the annual production of DMSP in the BCZ. This work is an additional modelling effort to explicitly include bacterial processes in transforming DMS(P), and shows their contribution in processing DMSP and as a sink of DMS that is much higher than DMS removal by photooxidation and F_{DMS}.

Current empirical relationships to predict DMS from Chl *a* [109], [110], [111], [112], [114] were unable to satisfactorily reproduce the seasonal cycle of DMS in timing and amplitude in the SNS in comparison with field data and MIRO-DMS simulations. In the data-sets from which these empirical relationships were established, the high Chl *a* values were related to diatoms unlike eutrophied coastal environments such as the SNS where high biomass is not associated to diatoms. Therefore, future projections of F_{DMS} and the investigation of the potential feedback on climate require to use modeling tools that accurately represent DMS(P) dynamics in coastal environments that are hotspots of DMS emissions, in particular, in eutrophied coastal environments dominated by high biomass non-diatom blooms. Further, bacterial processing of DMS(P) needs to be correctly represented in models.

The potential feedbacks of DMS emissions on climate will depend on the impact of climate change on the phytoplankton composition and biomass, as postulated by the CLAW hypothesis [1], but also of the response of the bacterial communities to global changes, and how they will modulate the sinks of DMS in seawater (emission to the atmosphere *versus* bacterial consumption/transformation).

Acknowledgments

We are grateful to Sébastien Milleville for his work during his master thesis. AVB is a senior research associate at the FNRS. We acknowledge two anonymous reviewers and the editor for constructive comments on the manuscript.

Author Contributions

Conceived and designed the experiments: NG CL AVB. Performed the experiments: NG. Analyzed the data: NG GS AVB CL. Contributed reagents/materials/analysis tools: NG GS AVB CL. Wrote the paper: NG GS AVB CL.

References

1. Charlson RJ, Lovelock JE, Andreae MO, Warren SG (1987) Oceanic phytoplankton, atmospheric sulphur, cloud albedo and climate. Nature 326:655–661.

2. Quinn PK, Bates TS (2011) The case against climate regulation via oceanic phytoplankton sulphur emissions. Nature, 480, 51–56.

3. Halloran PR, Bell TG, Totterdell IJ (2010) Can we trust empirical marine DMS parameterisations within projections of future climate? Biogeosciences 7:1645–1656. doi:10.5194/bg-7-1645-2010.

4. Cameron P, Elliott S, Malrud M, Erickson D, Wingenter O (2011) Changes in dimethyl sulfide oceanic distribution due to climate change. Geophys Res Lett 38(7), doi: 10.1029/2011GL047069.

5. Levasseur M. (2011) If Gaia could talk. Nature Geosci. 4: 351–352.

6. Six KD, Kloster S, Ilyina T, Archer SD, Zhang K, et al. (2013) Global warming amplified by reduced sulphur fluxes as a result of ocean acidification. Nature Climate Change doi:10.1038/nclimate1981.

7. Stefels J, Steinke M, Turner SM, Malin G, Belviso S (2007) Environmental constraints on the production and removal of the climatically active gas dimethylsulphide (DMS) and implications for ecosystem modelling. Biogeochemistry 83(1-3): 245–275.

8. Keller MD, Bellows WK, Guillard RRL (1989) Dimethyl sulfide production in marine phytoplankton. In: Saltzman ES, Cooper WJ (eds) Biogenic sulfur in the environment. American Chemical Society, Washington DC, pp 167–182.

9. Matrai PA, Keller MD (1994) Total organic sulfur and dimethylsulfoniopropionate in marine phytoplankton: intracellular variations. Mar Biol 119:61–68.

10. Keller MD, Korjeff-Bellows W (1996) Physiological aspects of the production of dimethylsulfoniopropionate (DMSP) by marine phytoplankton. In: Kiene RP, Visscher P, Keller M, Kirst G (eds) Biological and environmental chemistry of DMSP and related sulfonium compounds. Plenum Press, New York, p 131—142.

11. Stefels J (2000) Physiological aspects of the production and conversion of DMSP in marine algae and higher plants. J Sea Res 43:183–197.

12. Vairavamurthy A, Andreae MO, Iverson RL (1985) Biosynthesis of dimethylsulfide and dimethylpropiothetin by Hymenomonas carterae in relation to sulfur source and salinity variations. Limnol Oceanogr 30: 59–70.

13. Dickson DMJ, Kirst GO (1987) Osmotic adjustment in marine eukaryotic algae: The role of inorganic ions, quaternary ammonium, tertiary sulphonium and carbohydrate solutes. I. Diatoms and a Rhodophyte. New Phytologist 106: 645–655.

14. Karsten U, Wiencke C, Kirst GO (1992) Dimethylsulphonio-propionate (DMSP) accumulation in green macroalgae from polar to temperate regions: Interactive effects of light versus salinity and light versus temperature. Polar Biol 12: 603–607.

15. Karsten U, Kuck K, Vogt C, Kirst GO (1996) Dimethylsulfoniopropionate production in phototrophic organisms and its physiological function as a cryoprotectant. In: Kiene, R.P., Visscher, P.T., Keller, M.D. et Kirst, G.O. (Eds.), Biological and environmental chemistry of DMSP and related sulfonium compounds. Plenum Press, New York, pp. 143–153.

16. Kirst GO (1996) Osmotic adjustment in phytoplankton and macroalgae. The use of dimethylsulfoniopropionate (DMSP). In: Kiene, R.P., Visscher, P.T., Keller, M.D. et Kirst, G.O. (Eds.), Biological and environmental chemistry of DMSP and related sulfonium compounds. Plenum Press, New York, pp. 121–129.

17. Spielmeyer A, Pohnert G. (2012) Daytime, growth phase and nitrate availability dependent variations of dimethylsulfoniopropionate in batch cultures of the diatom Skeletonema marinoi. J Exp Mar Biol Ecol 413: 121–130.

18. Sunda W, Kieber DJ, Kiene RP, Huntsman S (2002) An antioxidant function for DMSP and DMS in marine algae. Nature 418: 317–320.

19. Harada H, Vila-Costa M, Cebrian J, Kiene RP (2009) Effects of UV radiation and nitrate limitation on the production of biogenic sulfur compounds by marine phytoplankton. Aquat Bot 90:37–42.

20. Archer SD, Ragni M, Webster R, Airs RL, Geider RJ (2010) Dimethyl sulfoniopropionate and dimethyl sulfide production in response to photoinhibition in Emiliania huxleyi. Limnol Oceanogr 55: 1579–1589.

21. Dacey JWH, Wakeham SG (1986) Oceanic dimethyl sulfide: production during zooplankton grazing on phytoplankton. Science 233: 1314–1316.

22. Wolfe GV, Steinke M (1996) Grazing-activated production of dimethyl sulfide (DMS) by two clones of Emiliania huxleyi. Limnol Oceanogr 41:1151–1160.

23. Fredrickson KA, Strom SL (2009) The algal osmolyte DMSP as a microzooplankton grazing deterrent in laboratory and field studies. J Plankton Res 31 (2): 135–152.

24. Evans C, Malin G, Wilson WH, Liss PS (2006) Infectious titres of Emiliania huxleyi virus EhV-86 are reduced by exposure to millimolar dimethyl sulphide and acrylic acid. Limnol. Oceanogr. 51:2468–2471.

25. Stefels J, Dijkhuizen L (1996) Characteristics of DMSP-lyase in Phaeocystis sp. (Haptophyceae). Mar Ecol Prog Ser 131: 307–313.

26. Nguyen BC, Belviso S, Mihalopoulos N, Gostan J, Nival P (1988) Dimethyl Sulfide Production During Natural Phytoplanktonic Blooms. Mar Chem 24:133–141.

27. Leck C, Larsson U, Bagander LE, Johansson S, Hajdu S (1990) Dimethylsulfide in the Baltic Sea: annual variability in relation to biological activity. J Geophys Res 95:3353–3364.

28. Belviso S, Kim S-K, Rassoulzadegan F, Krajka B, Nguyen BC, et al. (1990) Production of dimethylsulfonium propionate (DMSP) and dimethylsulfide (DMS) by a microbial food web. Limnol Oceanogr 35: 1810–1821.

29. Malin G, Wilson WH, Bratbak G, Liss PS, Mann NH (1998) Elevated production of dimethylsulfide resulting from viral infection of cultures of Phaeocystis pouchetii. Limnol Oceanogr 43: 1389–1393.

30. Todd JD, Rogers R, Li YG, Wexler M, Bond PL, Sun L. et al. (2007) Structural and regulatory genes required to make the gas dimethyl sulfide in bacteria. Science 315: 666–669.

31. Todd JD, Curson ARJ, Dupont CL, Nicholson P, Johnston AWB (2009) The dddP gene, encoding a novel enzyme that converts dimethylsulfoniopropionate into dimethyl sulfide, is widespread in ocean metagenomes and marine bacteria and also occurs in some Ascomycete fungi. Environ Microbiol 11:1376–1385.

32. Curson ARJ, Rogers R, Todd JD, Brearley CA, Johnston AWB (2008) Molecular genetic analysis of a dimethylsulfoniopropionate lyase that liberates the climate-changing gas dimethylsulfide in several marine alpha-proteobacteria and Rhodobacter sphaeroides. Environ Microbio 10: 757–767.

33. Stefels J, Dijkhuizen L, Gieskes WWC (1995) DMSP-lyase activity in a spring phytoplankton bloom off the Dutch coast, related to Phaeocystis sp. abundance. Mar Ecol Prog Ser 123:235–243.

34. Niki T, Kunugi M, Otsuki A (2000) DMSP-lyase activity in five marine phytoplankton species: its potential importance in DMS production. Marine Biol 136:759–764.

35. Kiene RP, Linn LJ, Bruton JA (2000) New and important roles for DMSP in marine microbial communities. J Sea Res 43:209–224.

36. Vila-Costa M, Del Valle D, González JM, Slezak D, Kiene R, et al. (2006) Phylogenetic identification and metabolism of DMS-consuming bacteria in seawater. Environ Microbiol 8: 2189–2200.

37. Brimblecombe P, Shooter D (1986) Photo-oxidation of dimethylsulfide in aqueous solution. Mar Chem 19:343–353.

38. Kieber DJ, Jiao J, Kiene RP, Bates TS (1996) Impact of dimethylsulfide photochemistry on methyl sulfur cycling in the equatorial Pacific Ocean. J Geophys Res 101: 3715–3722.

39. Kettle A, Andreae M, Amouroux D, Andreae T, Bates T, et al. (1999) A global database of sea surface dimethylsulfide (DMS) measurements and a procedure to predict sea surface DMS as a function of latitude, longitude, and month, Global Biogeochem Cy 13: 399–444.

40. Lana A, Bell TG, Simó R, Vallina SM, Ballabrera-Poy J, et al. (2011) An updated climatology of surface dimethlysulfide concentrations and emission fluxes in the global ocean. Global Biogeochem Cy 25(1). Art no GB1004. doi:10.1029/2010gb003850.

41. Kiene RP, Bates TS (1990) Biological removal of dimethyl sulfide from seawater. Nature 345:702–705.

42. Simó R, Pedrós-Alió C (1999) Short-term variability in the open ocean cycle of dimethylsulfide. Glob. Biogeochem. Cycles 13: 1173–1181.

43. Le Clainche Y, Vezina A, Levasseur M, Cropp RA, Gunson JR, et al. (2010) A first appraisal of prognostic ocean DMS models and prospects for their use in climate models. Global Biogeochem Cy 24, GB3021, doi:10.1029/2009GB003721.

44. Gabric A, Murray N, Stone L, Kohl M (1993) Modeling the production of dimethylsulfide during a phytoplankton bloom. J Geophys Res Oceans 98:22805–22816.

45. Gabric A, Gregg W, Najjar R, Erickson D, Matrai P (2001) Modeling the biogeochemical cycle of dimethylsulfide in the upper ocean: a review, Chemosphere - Global Change Science 3: 377–392.

46. van den Berg AJ, Turner SM, van Duyl FC, Ruardij P (1996) Model structure and analysis of dimethylsulphide (DMS) production in the southern North Sea, considering phytoplankton dimethylsulphoniopropionate- (DMSP) lyase and eutrophication effects. Mar Ecol Prog Ser 145:233–244.

47. Laroche D, Vezina AF, Levasseur M, Gosselin M, Stefels J, et al. (1999) DMSP synthesis and exudation in phytoplankton: a modeling approach. Mar Ecol Prog Ser 180: 37–49.

48. Jodwalis CM, Benner RL, Eslinger DL (2000) Modeling of dimethyl sulfide oceanmixing, biological production and sea-to-air flux for high latitudes. J Geophys Res, 105,D11, 14,387–14,399.

49. Archer SD, Gilbert FJ, Nightingale PD, Zubkov MV, Taylor AH, et al. (2002) Transformation of dimethylsulphoniopropionate to dimethyl sulphide during summer in the North Sea with an examination of key processes via a modelling approach. Deep-Sea Res. II, 49: 3067–3101.

50. Archer SD, Gilbert FJ, Allen JI, Blackford J, Nightingale PD (2004) Modelling of the seasonal patterns of dimethylsulphide production and fate during 1989 at a site in the North Sea. Can J Fish Aquat Sci 61:765–787.

51. Lefevre M, Vezina A, Levasseur M, Dacey JWH (2002) A model of dimethylsulfide dynamics for the subtropical North Atlantic. Deep-Sea Res Part I 49:2221–2239.

52. Le Clainche Y, Levasseur M, Vézina A, Dacey JWH, Saucier FJ (2004) Behaviour of the ocean DMS(P) pools in the Sargasso Sea viewed in a coupled physical-biogeochemical ocean model. Can J Fish Aquat Sci 61(5): 788–803.

53. Six KD, Maier-Reimer E (2006) What controls the oceanic dimethylsulfide (DMS) cycle? A modeling approach. Global Biogeochem Cy 20, GB4011, doi:10.1029/2005GB002674.

54. Bopp L, Aumont O, Belviso S, Blain S (2008) Modeling the effect of iron fertilization on dimethylsulfide emissions in the Southern Ocean. Deep Sea Res. Part II 55, doi:10.1016/j.dsr2.2007.12.002.

55. Steiner N, Denman K (2008) Parameter sensitivities in a 1-D model for DMS and sulphurcycling in the upper ocean, Deep Sea Res. Part I 55: 847–865, doi:10.1016/j.dsr.2008.02.010.

56. Toole DA, Siegel DA, Doney SC (2008) A light-driven, onedimensional dimethylsulfide biogeochemical cycling model for the Sargasso Sea. J. Geophys. Res 113:G02009. doi:10.1029/2007JG000426.

57. Vallina SM, Simo' R, Anderson TR, Gabric A, Cropp R, Pacheco JM (2008) A dynamic model of oceanic sulphur (DMSO) applied to the Sargasso Sea: simulating the dimethylsulfide (DMS) summer paradox. J Geophys Res 113:G01009. doi: 10.1029/2007JG000415.

58. Polimene L, Archer SD, Butenschön M, Allen JI (2012) A mechanistic explanation for the Sargasso Sea DMS "summer paradox". Biogeochemistry 110: 243–255. doi: 10.1007/s10533-011-9674-z.

59. Vézina A.F. (2004) Ecosystem modelling of the cycling of marine dimethyl-sulfide: a review of current approaches and of the potential for extrapolation to global scales. Can J Fish Aquat Sci 61: 845–856.

60. Lancelot C, Spitz Y, Gypens N, Ruddick K, Becquevort S, et al. (2005) Modelling diatom and Phaeocystis blooms and nutrient cycles in the Southern Bight of the North Sea: the MIRO model. Mar Ecol Prog Ser 289: 63–78.

61. Stefels J, van Boekel WHM (1993) Production of DMS from dissolved DMSP in axenic cultures of the marine phytoplankton species Phaeocystis sp. Mar Ecol Prog Ser 97:11–18.

62. Liss PS, Malin G, Turner SM, Holligan PM (1994) Dimethyl sulfide and Phaeocystis - A Review. J Mar Syst 5:41–53.

63. Lancelot C, Mathot S (1987) Dynamics of a Phaeocystis-dominated spring bloom in Belgian coastal waters. I. Phytoplanktonic activities and related parameters. Mar Ecol Prog Ser 37: 239–248.

64. Cadée GC, Hegeman J (2002) Phytoplankton in the Marsdiep at the end of the 20th century; 30 years monitoring biomass, primary production, and Phaeocystis blooms. J Sea Res 48: 97–110.

65. Rousseau V, Leynaert A, Daoud N, Lancelot C (2002) Diatom succession, silicification and silicic acid availability in Belgian coastal waters (Southern North Sea). Mar Ecol Prog Ser 236, 61–73.

66. Lancelot C (1995) The mucilage phenomenon in the continental coastal waters of the North Sea. Sci Total Environ 165:83–102.

67. Turner SM, Malin G, Liss PS, Harbour DS, Holligan PM (1988) The seasonal variation of dimethyl sulphide and DMSP concentrations in nearshore waters. Limnol Oceanogr 33: 364–375.

68. Kwint RLJ, Kramer KJM (1996) Annual cycle of the production and fate of DMS and DMSP in a marine coastal system. Mar Ecol Prog Ser 134: 217–224.

69. Gypens N, Lacroix G, Lancelot C (2007) Causes of variability in diatom and Phaeocystis blooms in Belgian coastal waters between 1989 and 2003: a model study. J Sea Res 57: 19–35.

70. Gasparini S, Daro MH, Antajan E, Tackx M, Rousseau V, et al. (2000) Mesozooplankton grazing during the Phaeocystis globosa bloom in the Southern Bight of the North Sea. J Sea Res 43:345–356.

71. Steinke M, Daniel C, Kirst GO (1996) DMSP lyase in marine macro- and microalgae: intraspecific differences in cleavage activity. In: Kiene RP, Visscher PT, Keller MD, Kirst GO (eds) Biological and environmental chemistry of DMSP and related sulfonium compounds. Plenum Press, New York, pp 317–324.

72. Steinke M, Wolfe GV, Kirst GO (1998) Partial characterisation of dimethylsulfoniopropionate (DMSP) lyase isozymes in 6 strains of Emiliania huxleyi. Mar Ecol Prog Ser 175:215–225.

73. Steinke M, Malin G, Archer SD, Burkill PH, Liss PS (2002) DMS production in a cocclithophorid bloom: evidence for the importance of dinoflagellate DMSP lyases. Aquat Microb Ecol 26:259–270.

74. Archer SD, Widdicombe CE, Tarran GA, Rees AP, Burkill PH (2001) Production and turnover of particulate dimethylsulphoniopropionate during a coccolithophore bloom in the northern North Sea. Aquat Microb Ecol 24: 225–241.

75. Kiene RP, Linn LJ (2000) On the fate of DMSP-sulfur in seawater: tracer studies with dissolved 35S-DMSP. Geochim Cosmochim Acta 64: 2797–2810.

76. Taylor BF, Visscher PT (1996) Metabolic pathways involved in DMSP degradation. In: Kiene, R.P., Visscher, P.T., Keller, M.D., Kirst, G.O. (Eds.). Biological and Environmental Chemistry of DMSP and Related Sulfonium Compounds, Plenum, New York, pp. 265–276.

77. Kiene RP, Linn LJ, González J, Moran MA, Bruton JA (1999) Dimethylsulfo-niopropionate and methanethiol are important precursors of methionine and protein-sulfur in marine bacterioplankton. Appl Environ Microbiol 65:4549–4558.

78. Simó R (2001) Production of atmospheric sulfur by oceanic plankton: biogeochemical, ecological and evolutionary links. Trends Ecol Evol 16(6):287–294.

79. Simó R, Archer SD, Pedrós-Alió C, Gilpin L, Stelfox-Widdicombe CE (2002) Coupled dynamics of dimethylsulfoniopropionate and dimethylsulfide cycling and the microbial food web in surface waters of the North Atlantic. Limnol Oceanogr 47: 53–61.

80. Zubkov MV, Fuchs BM, Archer SD, Kiene RP, Amann R, et al. (2001) Linking the composition of bacterioplankton to rapid turnover of dissolved dimethyl-sulfoniopropionate in an algal bloom in the North Sea. Environ Microbiol 3:304–11.

81. Zubkov MV, Fuchs BM, Archer SD, Kiene RP, Amann R, et al. (2002) Rapid turnover of dissolved DMS and DMSP by defined bacterioplankton communities in the stratified euphotic zone of the North Sea. Deep-Sea Res Part II 49:3017–3038.

82. Vila-Costa M, Simó R, Harada H, Gasol JM, Slezak D, Kiene RP (2006) Dimethylsulfoniopropionate uptake by marine phytoplankton. Science 314: 652–654.

83. Spielmeyer A, Gebser B, Pohnert G (2011) Investigations of the uptake of dimethylsulfoniopropionate by phytoplankton. ChemBioChem 12: 2276–2279.

84. Ruiz-González C, Galí M, Sintes E, Herndl GJ, Gasol JM, et al. (2012) Sunlight effects on the osmotrophic uptake of DMSP-sulfur and Leucine by polar phytoplankton. PLoS ONE 7(9): e45545. doi:10.1371/journal.pone.0045545.

85. Kiene RP (1990) Dimethylsulfide production from dimethylsulfoniopropionate in coastal seawater samples and bacterial cultures. Appl environ Microbiol 56:3292–3297.

86. Kiene R.P. (1993) Microbial sources and sinks for methylated sulfur compounds in the marine environment. In Microbial Growth on C1 Compounds, Vol. 7. Kelly, D.P., and Murrell, J.C. (eds). London, UK: Intercept, pp. 15–33 Kiene, R.P., 1996. Production of methanethiol from dimethylsulfoniopropionate in marine surface waters. Mar Chem 54: 69–83.

87. Kiene RP, Linn LJ (2000) Distribution and turnover of dissolved DMSP and its relationship with bacterial production and dimethylsulfide in the Gulf of Mexico. Limnol Oceanogr 45:848–861.

88. Wolfe GV, Levasseur M, Cantin G, Michaud S (1999) Microbial consumption and production of dimethyl sulfide (DMS) in the Labrador Sea. Aquat Microb Ecol 18:197–205.

89. Zubkov M, Linn LJ, Amann R, Kiene RP (2004) Temporal patterns of biological dimethylsulfide (DMS) consumption during laboratory-induced phytoplankton bloom cycles. Mar Ecol Prog Ser 271:77–86.

90. del Valle DA, Kieber DJ, Kiene RP (2007) Depth dependent fate of biologically-consumed dimethylsulfide in the Sargasso Sea. Mar Chem 103: 197–208.

91. Brugger A, Slezak D, Obermosterer I., Herndl GJ. (1998) Photolysis of DMS in the northern Adriatic Sea dependence on substrate concentration, irradiance and DOC concentration. Mar Chem 59: 321–331.

92. Saltzman ES, King DB, Holmen K, Leck C (1993) Experimental determination of the diffusion coefficient of dimethylsulfide in water. J Geophys Res 98: 16481–16486.

93. Yang M, Blomquist BW, Fairall CW, Archer SD, Huebert BJ (2011) Air-sea exchange of dimethylsulfide in the Southern Ocean: Measurements from SO GasEx compared to temperate and tropical regions, J Geophys Res 116, C00F05, doi:10.1029/2010JC006526.

94. Gypens N, Borges AV, Lancelot C (2009) Effect of eutrophication on air–sea CO2 fluxes in the coastal Southern North Sea: A model study of the past 50 years. Global Change Biol. 15: 1040–1056.

95. Turner SM, Malin G, Nightingale PD, Liss PS (1996) Seasonal variation of dimethyl sulphide in the North Sea and an assessment of fluxes to the atmosphere. Mar Chem 54:245–262.

96. van Duyl FC, Gieskes WWC, Kop AJ, Lewis WE (1998) Biological control of short-term variations in the concentration of DMSP and DMS during a Phaeocystis spring bloom. J Sea Res 40:221–231.

97. Kiene RP, Slezak D (2006) Low dissolved DMSP concentrations in seawater revealed by small volume gravity filtration and dialysis sampling. Limnol. Oceanogr. Methods 4: 80–95.

98. Howard EC, Henriksen JR, Buchan A, Reisch CR, Bürgmann H, et al. (2006) Bacterial taxa that limit sulfur flux from the ocean. Science 314, 649–652.

99. Yoch DC (2002) Dimethylsulfoniopropionate: Its sources, role in the marine food web, and biological degradation to dimethylsulfide. Appl Environ Microbiol 68:5804–5815.

100. Slezak D, Kiene RP, Toole DA, Simó R, Kieber DJ (2007) Effects of solar radiation on the fate of dissolved DMSP and conversion to DMS in seawater. Aquat Sci 69: 377–393.

101. Vila-Costa M, Kiene RP, Simó R (2008) Seasonal variability of the dynamics of dimethylated sulfur compounds in a coastal northwest Mediterranean site. Limnol Oceanogr 53:198–211.

102. Hatton AD, Shenoy DM, Hart MC, Mogg A, Green DH (2012). Metabolism of DMSP, DMS and DMSO by the cultivable bacterial community associated with the DMSP producing dinoagellate Scrippsiella trochoidea. Biogeochemistry 110(1–3):131–146.

103. Fagerbakke KM, Heldal M, Norland S (1996) Content of carbon, nitrogen, oxygen, sulfur and phosphorus in native aquatic and cultured bacteria. Aquat Microb Ecol 10: 15–27.

104. González JM, Kiene RP, Moran MA (1999) Transformation of sulfur compounds by an abundant lineage of marine bacteria in the alpha-subclass of the class Proteobacteria. Appl Environ Microbiol 65: 3810–3819.

105. Howard EC, Sun SL, Biers EJ, Moran MA (2008) Abundant and diverse bacteria involved in DMSP degradation in marine surface waters. Environ Microb 10: 2397–2410.

106. Offre P, Spang A, Schleper C (2013) Archaea in Biogeochemical Cycles. Annu. Rev. Microbiol. 67:437–57.

107. Nightingale PD, Malin G, Law CS, Watson AJ, Liss PS, et al. (2000) In situ evaluation of air-sea gas exchange parameterizationsusing novel conservative and volatile tracers. Global Biogeochem Cy 14(1): 373–387.

108. Ho DT, Law CS, Smith MJ, Schlosser P, Harvey M, et al. (2006) Measurements of airsea gas exchange at high wind speeds in the Southern Ocean: Implications for global parameterizations. Geophys. Res. Lett., 33, L16611, doi:10.1029/2006GL026817.

109. Anderson TR, Spall SA, Yool A, Cipollini P, Challenor PG, et al. (2001) Global fields of sea surface dimethylsulphide predicted from chlorophyll, nutrients and light. J Mar Syst, 30: 1–20.

110. Aumont O, Belviso S, Monfray P (2002) Dimethylsulfoniopropionate (DMSP) and dimethylsulfide (DMS) sea surface distributions simulated from a global three dimensional ocean carbon cycle model. J Geophys Res Oceans 107(C4):3029. doi:10.1029/1999JC000111.

111. Simó R, Dachs J (2002) Global ocean emission of dimethylsulfide predicted from biogeophysical data. Global Biogeochem Cy 16: 1078.

112. Belviso S, Bopp L, Moulin C, Orr JC, Anderson TR, et al. (2004), Comparison of global climatological maps of sea surface dimethyl sulfide, Global Biogeochem Cy 18, GB3013, doi:10.1029/2003GB002193.

113. Vallina S, Simó R (2007) Strong relationship between DMS and the solar radiation dose over the global surface ocean, Science, 315, 506–508, doi:10.1126/science.1133680.

114. Lana A, Simó R, Vallina SM, Dachs J (2012) Re-examination of global emerging patterns of ocean DMS concentration. Biogeochemistry 110:173–182. DOI 10.1007/s10533-011-9677-9

115. Miles CJ, Bell TG, Suntharalingam P (2012) Investigating the inter-relationships between water attenuated irradiance, primary production and DMS(P). Biogeochemistry 110:201–213. DOI 10.1007/s10533-011-9697-5

116. Watanabe YW, Yoshinari H, Sakamoto A, Nakano Y, Kasamatsu N, et al. (2007) Reconstruction of sea surface dimethylsulfide in the North Pacific during 1970s to 2000s. Mar Chem 103: 347–358.

117. Liss PS, Watson AJ, Liddicoat MI, Malin G, Nightingale PD, et al. (1993) Trace gases and air-sea exchange. Philos Trans R Soc London A 343: 531–541.

12

Carbon Cycling of Lake Kivu (East Africa): Net Autotrophy in the Epilimnion and Emission of CO_2 to the Atmosphere Sustained by Geogenic Inputs

Alberto V. Borges[1]*, Cédric Morana[2], Steven Bouillon[2], Pierre Servais[3], Jean-Pierre Descy[4], François Darchambeau[1]

1 Chemical Oceanography Unit, Université de Liège, Liège, Belgium, 2 Department of Earth and Environmental Sciences, KU Leuven, Leuven, Belgium, 3 Ecologie des Systèmes Aquatiques, Université Libre de Bruxelles, Bruxelles, Belgium, 4 Research Unit in Environmental and Evolutionary Biology, University of Namur, Namur, Belgium

Abstract

We report organic and inorganic carbon distributions and fluxes in a large (>2000 km^2) oligotrophic, tropical lake (Lake Kivu, East Africa), acquired during four field surveys, that captured the seasonal variations (March 2007–mid rainy season, September 2007–late dry season, June 2008–early dry season, and April 2009–late rainy season). The partial pressure of CO_2 (pCO_2) in surface waters of the main basin of Lake Kivu showed modest spatial (coefficient of variation between 3% and 6%), and seasonal variations with an amplitude of 163 ppm (between 579±23 ppm on average in March 2007 and 742±28 ppm on average in September 2007). The most prominent spatial feature of the pCO_2 distribution was the very high pCO_2 values in Kabuno Bay (a small sub-basin with little connection to the main lake) ranging between 11213 ppm and 14213 ppm (between 18 and 26 times higher than in the main basin). Surface waters of the main basin of Lake Kivu were a net source of CO_2 to the atmosphere at an average rate of 10.8 mmol m^{-2} d^{-1}, which is lower than the global average reported for freshwater, saline, and volcanic lakes. In Kabuno Bay, the CO_2 emission to the atmosphere was on average 500.7 mmol m^{-2} d^{-1} (~46 times higher than in the main basin). Based on whole-lake mass balance of dissolved inorganic carbon (DIC) bulk concentrations and of its stable carbon isotope composition, we show that the epilimnion of Lake Kivu was net autotrophic. This is due to the modest river inputs of organic carbon owing to the small ratio of catchment area to lake surface area (2.15). The carbon budget implies that the CO_2 emission to the atmosphere must be sustained by DIC inputs of geogenic origin from deep geothermal springs.

Editor: Moncho Gomez-Gesteira, University of Vigo, Spain

Funding: This work was funded by the Fonds National de la Recherche Scientifique (FNRS) under the CAKI (Cycle du carbone et des nutriments au Lac Kivu, 2.4.598.07) and the MICKI (Microbial diversity and processes in Lake Kivu, 1715859) projects, and contributes to the European Research Council Starting Grant AFRIVAL (African river basins: catchment-scale carbon fluxes and transformations, 240002) and to the Belgian Federal Science Policy Office EAGLES (East African Great Lake Ecosystem Sensitivity to changes, SD/AR/02A) projects. The funders had no role in study design, data collection and analysis, decision to publish, or preparation of the manuscript.

Competing Interests: AVB is a senior research associate at the FRS-FNRS. There are no patents, products in development or marketed products to declare.

* Email: alberto.borges@ulg.ac.be

Introduction

Freshwater ecosystems are frequently considered to be net heterotrophic, whereby the consumption of organic carbon (C) is higher than the autochthonous production of organic C, and excess organic C consumption is maintained by inputs of allochthonous organic C [1]. Net heterotrophy in freshwater ecosystems promotes the emission of carbon dioxide (CO_2) to the atmosphere [2], [3], [4], [5], [6], [7], [9], [10], with the global emission from continental waters estimated at ~0.75 PgC yr^{-1} [4] (0.11 PgC yr^{-1} from lakes, 0.28 PgC yr^{-1} from reservoirs, 0.23 PgC yr^{-1} from rivers, 0.12 PgC yr^{-1} from estuaries, and 0.01 PgC yr^{-1} from ground waters). More recent studies provided even higher CO_2 emission estimates. Tranvik et al. [7] revised the CO_2 emission from lakes to 0.53 PgC yr^{-1}, while Battin et al. [6] estimated CO_2 emission from streams at 0.32 PgC yr^{-1}.

Aufdenkampe et al. [8] estimated a total CO_2 emission of 0.64 PgC yr^{-1} for lakes and reservoirs, a total of 0.56 PgC yr^{-1} for rivers and streams, and a massive 2.08 PgC yr^{-1} for wetlands. Raymond et al. [10] estimated an emission of 1.8 PgC yr^{-1} for streams and rivers and 0.32 PgC yr^{-1} for lakes and reservoirs. Such emissions of CO_2 from continental waters exceed the net sink of C by terrestrial vegetation and soils of ~1.3 PgC yr^{-1} [4] as well as the sink of CO_2 in open oceans of ~1.4 PgC yr^{-1} [11].

However, our present understanding of the role of lakes on CO_2 emissions could be biased because most observations were obtained in temperate and boreal (humic) systems, and mostly in medium to small sized lakes, during open-water (ice-free) periods. Much less observations are available from hard-water, saline, large, or tropical lakes. Tropical freshwater environments are indeed under-sampled compared to temperate and boreal systems in terms of C dynamics in general, and specifically in terms of CO_2

dynamics. In an extensive compilation of CO_2 concentration data from 4902 lakes globally [12], there were only 148 data entries for tropical systems (~3%). Yet, about 50% of freshwater and an equivalent fraction of organic C is delivered by rivers to the oceans at tropical latitudes [13]. Tropical lakes represent about 16% of the total surface of lakes [14], and Lakes Victoria, Tanganyika, and Malawi belong to the seven largest lakes by area in the world. Current estimates assume that areal CO_2 fluxes are substantially higher in tropical systems than in temperate or boreal regions (often ascribed to higher temperatures) [8]. Thus, according to the zonal distribution given by Aufdenkampe et al. [8], tropical inland waters account for ~60% of the global emission of CO_2 from inland waters (0.45 PgC yr^{-1} for lakes and reservoirs, 0.39 PgC yr^{-1} for rivers and streams, and 1.12 PgC yr^{-1} for wetlands). It is clear that additional data are required to verify and re-evaluate more accurately the CO_2 fluxes from tropical systems.

Pelagic particulate primary production (PP) of East African great lakes, as reviewed by Darchambeau et al. [15], ranges from ~30 mmol m^{-2} d^{-1} for the most oligotrophic conditions (north basin of Lake Tanganyika) to ~525 mmol m^{-2} d^{-1} for the most eutrophic conditions (Lake Victoria). The comparatively fewer data on bacterial production (BP), available only for Lake Tanganyika, suggest that PP and BP are seasonally closely coupled [16]. However, with an average pelagic BP of ~25 mmol m^{-2} d^{-1} [16], the bacterial C demand would exceed the production of particulate organic C (POC) by phytoplankton in Lake Tanganyika. This has led to speculate about additional C supply to bacterioplankton, for instance, from dissolved organic C (DOC) exudation by phytoplankton [16].

Lake Kivu (2.50°S 1.59°S 29.37°E 28.83°E) is one of the East African great lakes (2370 km^2 surface area, 550 km^3 volume). It is a deep (maximum depth of 485 m) meromictic lake, with an oxic mixolimnion down to 70 m maximum, and a deep monolimnion rich in dissolved gases and nutrients [17], [18], [19]. Deep layers receive heat, salts, and CO_2 from deep geothermal springs [19]. Seasonality of the physical and chemical vertical structure and biological activity in surface waters of Lake Kivu is driven by the oscillation between the dry season (June-September) and the rainy season (October-May), the former characterized by a deepening of the mixolimnion [20]. This seasonal mixing favours the input of dissolved nutrients and the development of diatoms, while, during the rest of the year, the phytoplankton assemblage is dominated by cyanobacteria, chrysophytes and cryptophytes [15], [21], [22]. Surface waters of Lake Kivu are oligotrophic, and, consequently, PP is at the lower end of the range for East African great lakes (on average ~50 mmol m^{-2} d^{-1}) [15].

Extremely high amounts of CO_2 and methane (CH_4) (300 km^3 and 60 km^3, respectively, at 0°C and 1 atm) [19] are dissolved in the deep layers of Lake Kivu. This is due to a steep density gradient at 260 m depth that leads to residence times in the order of 1000 yr in the deepest part of the lake [19], [23]. Stable isotope and radiocarbon data suggest that the CO_2 is mainly geogenic [24]. While the risk of a limnic eruption is minimal [25], large scale industrial extraction of CH_4 from the deep layers of Lake Kivu is planned [26], [27] which could affect the ecology and biogeochemical cycling of C of the lake and change for instance the emission of greenhouse gases such as CH_4 and CO_2. The net emission of CH_4 to the atmosphere from Lake Kivu was quantified by Borges et al. [28], and was surprisingly low - among the lowest ever reported in lakes globally - considering the large amounts of CH_4 stored in deep waters. Here, we report a data-set obtained during four surveys covering the seasonality of CO_2 dynamics and fluxes, in conjunction with mass balances of C, and process rate measurements (PP and BP) in the epilimnion of Lake Kivu, with

the aim of quantifying the exchange of CO_2 with the atmosphere and determining the underlying drivers, in particular, the net metabolic status.

Materials and Methods

Official permission was not required for sampling in locations where measurements were made and samples acquired. The field studies did not involve endangered or protected species. The full data-set is available in Table S1.

2.1 Field sampling and chemical analysis

In order to capture the seasonal variations of the studied quantities, four cruises were carried out in Lake Kivu on 15/03-29/03/2007 (mid rainy season), 28/08-10/09/2007 (late dry season), 21/06-03/07/2008 (early dry season), 21/04-05/05/2009 (late rainy season), and 19/10/10-27/10/10 (early rainy season) for a selection of variables. Sampling was carried out at 15 stations distributed over the whole lake and in Kabuno Bay, and at 12 rivers draining into Lake Kivu and representing the outflow of the lake (Ruzizi River, Fig. 1). The core of data presented hereafter was obtained in 2007–2009, while from the 2010 cruise only vertical DOC and POC data obtained at two stations are presented.

Vertical profiles of temperature, conductivity and oxygen were obtained with a Yellow Springs Instrument (YSI) 6600 V2 probe. Calibration of sensors was carried out prior to the cruises and regularly checked during the cruises. The conductivity cell was

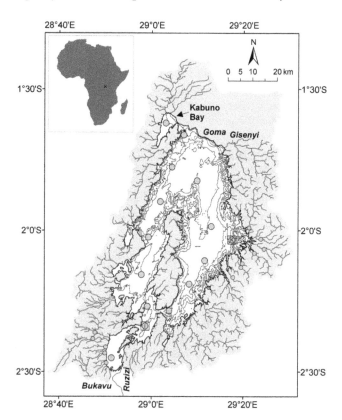

Figure 1. Map of Lake Kivu, showing bathymetry (isobaths at 100 m intervals), catchment area (shaded in grey), rivers, and sampling stations (small circles indicate the rivers). Primary production and bacterial production measurements were made at the stations identified with a square (I = Ishungu; K = Kibuye), adapted from [29].

calibrated with a 1000 µS cm^{-1} (25°C) YSI standard. The oxygen membrane probe was calibrated with humidity saturated ambient air. Salinity was computed from specific conductivity according to Schmid and Wüest [29].

Sampling for the partial pressure of CO_2 (pCO_2) was carried out at 1 m depth. Measurements of pCO_2 were carried out with a non-dispersive infra-red (NDIR) analyzer coupled to an equilibrator [30] through which water was pumped with a peristaltic pump (Masterflex E/S portable sampler). In-situ temperature and temperature at the outlet of the equilibrator were determined with Li-Cor 1000-15 probes. The NDIR analyzer (Li-Cor, Li-820) was calibrated with five gas standards: pure N_2 and four CO_2:N_2 mixtures with a CO_2 molar fraction of 363, 819, 3997 and 8170 ppm (Air Liquide Belgium).

For the determination of pH, CH_4 concentrations, $\delta^{13}C$ of dissolved inorganic C (DIC) (δ^{13}C-DIC), and total alkalinity (TA), water was sampled with a 5 L Niskin bottle (Hydro-Bios). Samples were collected every 10 m from 10 to 60–80 m depending on the cruise and station, except for CH_4 which was only sampled at 10 m. Additional samples for pH, $\delta^{13}C_{DIC}$ and TA were collected at 5 m in Kabuno Bay. Water for CH_4 analysis was collected in 50 ml glass serum bottles from the Niskin bottle with tubing, left to overflow, poisoned with 100 µl of a saturated $HgCl_2$ solution, and sealed with butyl stoppers and aluminium caps. Water samples for the analysis of $\delta^{13}C_{DIC}$ were taken from the same Niskin bottle by gently overfilling 12 ml glass Exetainer vials, poisoned with 20 µl of a saturated $HgCl_2$ solution, and gas-tight capped. A water volume of 50 ml was filtered through a 0.2 µm pore size polyethersulfone (PES) syringe filters and was stored at ambient temperature in polyethylene bottles for the determination of TA. POC and DOC samples were obtained from surface waters in June 2008 and April 2009, and along a depth profile in October 2010. DOC was filtered on 0.2 µm PES syringe filters, stored at ambient temperature in 40 mL glass vials with polytetrafluoroethylene coated septa, and poisoned with 50 µL of H_3PO_4 (85%). POC was filtered on 0.7 µm pore 25 mm diameter Whatman GF/F glass fiber filters (pre-combusted 5 h at 500°C), stored dry. Sampling of river surface waters followed the same procedures outlined above (sampling depth ~20 cm) with the addition of water sampling for total suspended matter (TSM). Samples for TSM were obtained by filtering 50–200 mL of water on pre-combusted pre-weighted 47 mm diameter GF/F glass fiber filters, stored dry.

Measurements of pH in water sampled from the Niskin bottle were carried out with a Metrohm (6.0253.100) combined electrode calibrated with US National Bureau of Standards (NBS) buffers of pH 4.002 (25°C) and pH 6.881 (25°C) prepared according to Frankignoulle and Borges [31]. Measurements of TA were carried out by open-cell titration with HCl 0.1 M according to Gran [32] on 50 ml water samples, and data were quality checked with Certified Reference Material acquired from Andrew Dickson (Scripps Institution of Oceanography, University of California, San Diego). DIC was computed from pH and TA measurements using the carbonic acid dissociation constants of Millero et al. [33]. For the analysis of δ^{13}C-DIC, a He headspace was created in 12 ml glass vials, and ~300 µl of H_3PO_4 (99%) was added to convert all DIC species to CO_2. After overnight equilibration, part of the headspace was injected into the He stream of an elemental analyser – isotope ratio mass spectrometer (EA-IRMS) (Thermo-Finnigan Flash1112 and ThermoFinnigan Delta+XL, or Thermo FlashEA/HT coupled to Thermo Delta V) for δ^{13}C measurements. The obtained δ^{13}C data were corrected for the isotopic equilibration between gaseous and dissolved CO_2 using an algorithm similar to that presented by Miyajima et al. [34], and

calibrated with LSVEC and NBS-19 certified standards or internal standards calibrated with the former. Concentrations of CH_4 were determined by gas chromatography with flame ionization detection, as described by Borges et al. [28]. DOC and δ^{13}C-DOC were measured with a customized Thermo HiperTOC coupled to a Delta+XL IRMS. POC and δ^{13}C-POC from filters were determined on a Thermo EA-IRMS (various configurations, either Flash1112, FlashHT with Delta+XL or DeltaV Advantage). Quantification and calibration of δ^{13}C data was performed with IAEA-C6 and acetanilide which was internally calibrated versus international standards.

PP and BP were measured at 2 stations: Kibuye (2.05°S 29.29°E) and Ishungu (2.34°S 28.98°E) (Fig. 1). PP was measured using the ^{14}C method [35] as described by Darchambeau et al. [15] in a pooled sample prepared from discrete samples (2 L) spaced every 5 m in the mixed layer. The mixed layer depth (MLD) was determined from vertical profiles of temperature and oxygen. The ^{14}C incubations were assumed to provide an estimate of net PP of the particulate phase (PNPP). Chlorophyll a (Chl-a) of the pooled sample was measured according to Descy et al. [36] by high-performance liquid chromatography analysis of extracts in 90% acetone from samples filtered on Macherey-Nägel GF5 (0.7 µm nominal pore size) filters (3–4 L). BP was estimated every 5 m in the mixed layer from tritiated thymidine (^{3}H-Thy) incorporation rates [37]. Samples (20 mL) were incubated in duplicate for 2 h in the dark at in-situ temperature in the presence of ^{3}H-Thy (~80 Ci mmol) at saturating concentration (~50 nmol L^{-1}). After incubation, cold trichloroacetic acid (TCA) was added (final concentration 5%) and the samples were kept cold until filtration through a 0.2 µm pore-size cellulose nitrate membrane. Filters were preserved in the dark at −20°C. The radioactivity associated with the filters was estimated by liquid scintillation using a Beckman counter LS 6000. Cell production was calculated from the ^{3}H-Thy incorporation rate using a conversion factor of 1.2×10^{18} cells produced per mole of ^{3}H-Thy incorporated into cold TCA insoluble material. This conversion factor was determined experimentally in batch experiments in which the increase of bacterial abundance and ^{3}H-Thy incorporation were followed simultaneously (data not shown) and was similar to the one used by Stenuite et al. [16] to calculate BP in Lake Tanganyika. Cellular production was multiplied by the average bacterial C content per cell (15 fgC cell^{-1}) [16] to obtain BP data. Daily BP was estimated from the experimental values considering constant activity over 24 h, and expressed per unit area (mmol m^{-2} d^{-1}), by integrating over the euphotic zone. Bacterial respiration (BR) rates were computed from BP using a bacterial growth efficiency computed from BP according to the model of Del Giorgio and Cole [38].

2.2 Bulk DIC mass balance model

TA and DIC mass balance models were constructed in order to determine the major processes controlling CO_2 dynamics in surface waters, and to evaluate the net metabolic balance of the epilimnion (net autotrophy or net heterotrophy).

The TA mass balance was constructed assuming a steady-state, according to:

$$F_{TA\_river} + F_{TA\_70m\_10m} + F_{TA\_upwelling} = F_{TA\_Ruzizi} + F_{TA*} \quad (1)$$

where F_{TA_river} is the input of TA from rivers, F_{TA_Ruzizi} is the output of TA by the Ruzizi river, $F_{TA\_70m\_10m}$ is the flux of TA from the monimolimnion to the mixolimnion by eddy diffusion, $F_{TA\_upwelling}$ is the flux of TA from the monimolimnion to the mixolimnion by upwelling, and F_{TA*} is the closing term.

$F_{\text{TA\_river}}$ was computed from discharge-weighted average TA in the 12 sampled rivers draining into Lake Kivu (TA_{river}), and total freshwater discharge from rivers (Q_{river}), according to:

$$F_{TA\_river} = TA_{river} Q_{river} \qquad (2)$$

$F_{\text{TA\_Ruzizi}}$ was computed from TA measured in the Ruzizi River (TA_{Ruzizi}), and the flow of the Ruzizi River (Q_{Ruzizi}), according to:

$$F_{TA\_Ruzizi} = TA_{Ruzizi} Q_{Ruzizi} \qquad (3)$$

$F_{\text{TA\_70m\_10m}}$ was computed from the gradient of TA across the pycnocline ($\delta_{\text{TA\_70m\_10m}}/\delta z$, where δz represents the depth interval) and the eddy diffusion coefficient (E) according to:

$$F_{TA\_70m\_10m} = -E \frac{\delta_{TA\_70m\_10m}}{\delta z} \qquad (4)$$

$F_{\text{TA\_upwelling}}$ was computed from the TA at 70 m (TA_{70m}) and the upwelling flow ($Q_{\text{upwelling}}$), according to:

$$F_{TA\_upwelling} = TA_{70m} Q_{upwelling} \qquad (5)$$

A DIC mass balance was constructed assuming a steady-state, according to:

$$F_{DIC\_river} + F_{DIC\_70m\_10m} + F_{DIC\_upwelling} = F_{DIC\_Ruzizi} + F_{CO2} \atop + F_{CaCO3} + F_{POC} \qquad (6)$$

where $F_{\text{DIC\_river}}$ is the input of DIC from rivers, $F_{\text{DIC\_Ruzizi}}$ is the output of DIC by the Ruzizi river, $F_{\text{DIC\_70m\_10m}}$ is the flux of DIC from the monimolimnion to the mixolimnion by eddy diffusion, $F_{\text{DIC\_upwelling}}$ is the flux of DIC from the monimolimnion to the mixolimnion by upwelling, F_{CO2} is the exchange of CO_2 with the atmosphere, F_{CaCO3} is the precipitation and subsequent export to depth of $CaCO_3$, and F_{POC} is the closing term and represents the export of POC from surface to depth.

$F_{\text{DIC\_river}}$ was computed from discharge-weighted average DIC in the 12 sampled rivers draining into Lake Kivu (DIC_{river}), and Q_{river}, according to:

$$F_{DIC\_river} = DIC_{river} Q_{river} \qquad (7)$$

$F_{\text{DIC\_Ruzizi}}$ was computed from DIC measured in the Ruzizi River (DIC_{Ruzizi}), and Q_{Ruzizi}, according to:

$$F_{DIC\_Ruzizi} = DIC_{Ruzizi} Q_{Ruzizi} \qquad (8)$$

$F_{\text{DIC\_70m\_10m}}$ was computed from the gradient of DIC across the metalimnion ($\delta_{\text{DIC\_70m\_10m}}/\delta z$) and E according to:

$$F_{DIC\_70m\_10m} = -E \frac{\delta_{DIC\_70m\_10m}}{\delta z} \qquad (9)$$

$F_{\text{DIC\_upwelling}}$ was computed from the DIC at 70 m (DIC_{70m}) and $Q_{\text{upwelling}}$, according to:

$$F_{DIC\_upwelling} = DIC_{70m} Q_{upwelling} \qquad (10)$$

F_{CO2} was computed according to:

$$F_{CO2} = k \alpha \Delta pCO_2 \qquad (11)$$

where k is the gas transfer velocity, α is the CO_2 solubility coefficient, and ΔpCO_2 is the air-water gradient of pCO_2 computed from water pCO_2 (1 m depth) and an atmospheric pCO_2 value ranging from $\sim$372 ppm to $\sim$376 ppm (depending on the cruise), corresponding to the monthly average at Mount Kenya (Kenya, 0.05°S 37.30°E) obtained from GLOBALVIEW-CO2 (Carbon Cycle Greenhouse Gases Group of the National Oceanic and Atmospheric Administration, Earth System Research Laboratory), and converted into wet air using the water vapour algorithm of Weiss and Price [39].

α was computed from temperature and salinity using the algorithm of Weiss [40], k was computed from wind speed using the parameterization of Cole and Caraco [41] and the Schmidt number of CO_2 in fresh water according to the algorithm given by Wanninkhof [42]. Wind speed data were acquired with a Davis Instruments meteorological station in Bukavu (2.51°S 28.86°E). The wind speed data were adjusted to be representative of wind conditions over the lake by adding 2 m s^{-1} according to Thiery et al. [20]. F_{CO2} was computed with daily wind speed averages for a time period of one month centred on the date of the middle of each field cruise. Such an approach allows to account for the day-to-day variability of wind speed, and to provide F_{CO2} values that are seasonally representative.

F_{CaCO3} was computed according to:

$$F_{CaCO3} = \frac{F_{TA*}}{2} \qquad (12)$$

The average value of Q_{river} (76.1 m^3 s^{-1}) was given by Muvundja et al. [43], the average Q_{Ruzizi} for 2007–2009 (87.8 m^3 s^{-1}) measured at Ruzizi I Hydropower Plant was provided by the Société Nationale d'Electricité. A value of $Q_{\text{upwelling}}$ of 42 m^3 s^{-1} and a value of E of 0.06 cm^2 s^{-1} were given by Schmid et al. [19].

2.3 δ^{13}C-DIC mass balance model

The combination of the DIC and δ^{13}C-DIC budget for the mixed layer allows to estimate independently the total DIC vertical input by upwelling and by eddy diffusion ($F_{\text{DIC\_upward}}$) and F_{POC} [44]. At steady-state, DIC and δ^{13}C-DIC mass balances are given by the following equations:

$$F_{DIC\_river} + F_{upward} = F_{DIC\_Ruzizi} + F_{CO2} + F_{CaCO3} + F_{POC} \qquad (13)$$

$$F_{DIC\_river} (^{13}C/_{12}C)_{DIC\_river} + F_{DIC\_Ruzizi} (^{13}C/_{12}C)_{DIC\_lake}$$
$$+ F_{DIC\_upward} (^{13}C/_{12}C)_{DIC\_upward}$$
$$+ F_{CO2} (^{13}C/_{12}C)_{CO2} + F_{CaCO3} (^{13}C/_{12}C)_{CaCO3}$$
$$+ F_{POC} (^{13}C/_{12}C)_{POC} = 0 \qquad (14)$$

$$F_{upward} = F_{DIC\_70m\_10m} + F_{DIC\_upwelling} \qquad (15)$$

The $(^{13}C/^{12}C)$ in the equation (14) represents the ^{13}C to ^{12}C ratio of net C fluxes, and can be expressed using the classical $\delta^{13}C$ notation [45]. 10 out of the 12 different terms in equations (13) and (14) were measured or can be computed from measured variables, and then the two equations can be solved in order to estimate the F_{DIC_upward} and F_{POC} fluxes. F_{DIC_river}, F_{DIC_Ruzizi}, F_{CaCO3} and F_{CO2} were calculated as described above. $(^{13}C/^{12}C)_{DIC\_river}$, $(^{13}C/^{12}C)_{DIC\_lake}$ and $(^{13}C/^{12}C)_{POC}$ were measured during the 4 field surveys. $(^{13}C/^{12}C)_{CaCO3}$ was computed from the measured $\delta^{13}C_{DIC}$ in surface and the fractionation factor $\varepsilon\_CaCO3\text{-}HCO3$ of 0.88 ‰ [46].

The $(^{13}C/^{12}C)_{DIC\_upward}$ which represents the $\delta^{13}C$ signature of the net upward DIC input, was estimated from the $\delta^{13}C$-DIC vertical gradient as follows:

$$(^{13}C/_{12}C)_{DIC\_upward} = \frac{DIC_{z+1}(^{13}C/_{12}C)_{DIC\_z+1} - DIC_z(^{13}C/_{12}C)_{DIC\_z}}{DIC_{z+1} - DIC_z} \qquad (16)$$

where z is the depth, DIC_z and DIC_{z+1} are the DIC concentration at the depth z and z+1, $(^{13}C/^{12}C)_{DIC\_z}$ and $(^{13}C/^{12}C)_{DIC\_z+1}$ are the $\delta^{13}C$ signature of DIC at the depth z and z+1.

The $\delta^{13}C$ signature of the net flux of CO_2 at the air-water interface was calculated from the $\delta^{13}C$ signature of the different DIC species in surface water and the atmospheric CO_2 (−8.0‰) according to:

$$(^{13}C/_{12}C)_{CO2} = \alpha_{am}\alpha_{sol} \frac{pCO_{2atm}(^{13}C/_{12}C)_{CO2atm} - pCO_{2w}(^{13}C/_{12}C)_{DIC\_lake}\alpha_{DIC\text{-}g}}{pCO_{2atm} - pCO_{2w}} \qquad (17)$$

where $(^{13}C/^{12}C)_{CO2atm}$ and $(^{13}C/^{12}C)_{DIC\_lake}$ are the $\delta^{13}C$ signature of atmospheric CO_2 and lake surface DIC, respectively, α_{am} and α_{sol} are respectively the kinetic fractionation effect during CO_2 gas transfer, and the equilibrium fractionation during CO_2 dissolution measured by Zhang et al. [47] in distilled water. $\alpha_{DIC\text{-}g}$ is the equilibrium fractionation factor between aqueous DIC and gaseous CO_2 and is defined by:

$$\alpha_{DIC-g} = \frac{(^{13}C/_{12}C)_{CO2am} - (^{13}C/_{12}C)_{diseq}}{(^{13}C/_{12}C)_{DIC\_lake}} \qquad (18)$$

where $(^{13}C/^{12}C)_{diseq}$ is the air-water $\delta^{13}C$ disequilibrium, that is the difference between the $\delta^{13}C$-DIC expected at equilibrium with atmosphere CO_2 minus the measured $\delta^{13}C$-DIC in surface water of the lake.

2.4 Bulk DIC mixing models

A mixing model was developed to compute the theoretical evolution of TA, DIC, and pCO_2 between March 2007 and September 2007, when the mixed layer deepened. The aim of this model is to compare theoretical evolution considering conservative mixing (no biology or other in/outputs) with observational data to infer the importance of certain processes. The model was computed by daily time steps assuming the conservative mixing (no biological activity) of surface waters with deep waters for TA and DIC. At each time step, pCO_2 was calculated from TA, DIC,

salinity and temperature, allowing the computation of F_{CO2} and correcting DIC for F_{CO2}. The mixing model was also run without correcting DIC for F_{CO2}. The MLD, salinity and temperature were interpolated linearly between March and September 2007.

At each time step, TA was computed according to:

$$TA_{i+1\_ML} = \frac{TA_{deep}(MLD_{i+1} - MLD_i) + TA_{i\_ML}MLD_i}{MLD_{i+1}} \qquad (19)$$

where TA_{i_ML} is TA in the mixed layer at time step i, TA_{i+1_ML} is TA in the mixed layer at time step $i+1$, MLD_i is the MLD at time step i, MLD_{i+1} is the MLD at time step $i+1$, and TA_{deep} is TA in the deep waters.

At each time step, DIC corrected for F_{CO2} was computed according to:

$$DIC_{i+1\_ML} = \frac{DIC_{deep}(MLD_{i+1} - MLD_i) + DIC_{i\_ML}MLD_i - F_{CO2i}\delta_t}{MLD_{i+1}} \qquad (20)$$

where DIC_{i_ML} is DIC in the mixed layer at time step i, DIC_{i+1_ML} is DIC in the mixed layer at time step $i+1$, F_{CO2i} is F_{CO2} at time step i, δt is the time interval between each time step (1 d), and DIC_{deep} is DIC in the deep waters.

At each time step, DIC not corrected for F_{CO2} was computed according to:

$$DIC_{i+1\_ML} = \frac{DIC_{deep}(MLD_{i+1} - MLD_i) + DIC_{i\_ML}MLD_i}{MLD_{i+1}} \qquad (21)$$

Results

In surface waters (1 m depth) of the main basin of Lake Kivu (excluding Kabuno Bay), pCO_2 values were systematically above atmospheric equilibrium (~372 ppm to ~376 ppm depending on the cruise), and varied within narrow ranges of 534–605 ppm in March 2007, 701–781 ppm in September 2007, 597–640 ppm in June 2008, and 583–711 ppm in April 2009 (Fig. 2). The most prominent feature of the spatial variations was the much higher pCO_2 values in Kabuno Bay, ranging between 11213 ppm and 14213 ppm (i.e., between 18 and 26 times higher than in the main basin). Wind speed showed little seasonal variability (ranging between 3.2 and 3.6 m s^{-1}), hence, the seasonal variations of the CO_2 emission rates followed those of ΔpCO_2 with higher F_{CO2} values in September 2007 (14.2 mmol m^{-2} d^{-1}) and lowest F_{CO2} in March 2007 (8.0 mmol m^{-2} d^{-1}) in the main basin (Table 1). In Kabuno Bay, the F_{CO2} values ranged between 414.2 and 547.7 mmol m^{-2} d^{-1}, and were on average ~46 times higher than in the main basin.

Compared to the main basin, surface and deep waters of Kabuno Bay were characterized by higher salinity, DIC and TA values and by lower pH and $\delta^{13}C$-DIC values (Fig. 3). Comparison of DIC and TA profiles shows that the relative contribution of CO_2 to DIC was more important in Kabuno Bay than in the main lake, since TA is mainly as HCO$_3^-$, and if the CO_2 contribution to DIC is low, then DIC and TA should be numerically close. At 60 m depth, CO_2 contributes ~30% to DIC in Kabuno Bay, and only ~1% in the main basin. Kabuno Bay was also characterized by a very stable chemocline (salinity, pH) and oxycline at ~11 m irrespective of the sampling period [28]. In the main basin of Lake Kivu, the oxycline varied seasonally between ~35 m in March and September 2007 and $\sim$60 m in June 2008 [28]. The deepening of the mixed layer and entrainment of deeper waters to the surface mixed layer was shown to be main driver of the seasonal variations of CH_4 [28]. The positive correlations between pCO_2 and CH_4 and between pCO_2 and the MLD also show that the mixing of deep and surface waters was a major driver of the seasonal variability of pCO_2 (Fig. 4). This is also consistent with the negative relation between pCO_2 and δ^{13}C-DIC (Fig. 4), as DIC in deeper waters is more ^{13}C-depleted than that in surface waters (Fig. 3).

DIC concentrations in surface waters averaged 13.0 mmol L^{-1} and 16.9 mmol L^{-1} in the main basin of Lake Kivu and in Kabuno Bay, respectively, but were much lower in the inflowing rivers (on average $\sim$0.5 mmol L^{-1}). The comparison with the lake values shows that the δ^{13}C-DIC were always more negative in rivers (mean -7.0 ± 2.1‰) than in the main basin (mean 3.4 ± 0.5‰) and Kabuno Bay (mean 0.8 ± 0.5‰) (Fig. 5). This difference suggests that the DIC in surface waters of Lake Kivu originates from a different source than that in the rivers. POC concentrations in surface waters of the main basin averaged 32 μmol L^{-1} in June 2008, 24 μmol L^{-1} in April 2009 and 42 μmol L^{-1} in October 2010. In the rivers, POC concentration was higher, 358 μmol L^{-1} in June 2008 and 499 μmol L^{-1} in April 2009. However, POC in the rivers never contributed more than 4.4% of TSM. δ^{13}C-POC and δ^{13}C-DIC signatures appeared uncoupled in rivers (Fig. 6), but a positive relationship between δ^{13}C-DIC and δ^{13}C-POC was found in the lake when combining the data from the main lake and Kabuno Bay (model I linear regression, $p<0.001$, $r^2=0.71$, n = 15). Furthermore, the δ^{13}C-POC in the main basin and Kabuno Bay (mean -24.1 ± 2.0‰, n = 15) was significantly lower than the δ^{13}C-POC in rivers (mean $= -22.9\pm1.5$‰, n = 21) (t-test; p<0.05), but the δ^{13}C-DOC in the main lake (mean -23.1 ± 1.1‰, n = 15) did not differ from the δ^{13}C-DOC in rivers (mean -23.9±1.4‰, n = 21) (Fig. 7) and was vertically uncoupled from δ^{13}C-POC (Fig. 8).

In order to test if vertical mixing was the only driver of seasonal variations of pCO_2, we applied the mixing model to the March 2007 data in order to predict the evolution of TA, DIC, and pCO_2 up to September 2007 and we compared the predicted values to the actual data obtained at that period (Fig. 9). The TA value predicted by the mixing model was higher than the observations in September 2007 by 108 μmol L^{-1}. We assumed that the process removing TA was $CaCO_3$ precipitation in the mixolimnion and subsequent export to depth (F_{CaCO3}). In order to account for the difference between the mixing model prediction and the observations, F_{CaCO3} was estimated to be 14.2 mmol m^{-2} d^{-1} between March and September 2007.

The DIC value predicted by the mixing model was higher than the observations in September 2007 by 108 μmol L^{-1}. The emission of CO_2 to the atmosphere only accounted for 19% of the DIC removal. We assumed that the remaining DIC was removed by the combination of F_{CaCO3} and POC production in the epilimnion and export to depth (F_{POC}) that was estimated to be 11.5 mmol m^{-2} d^{-1}, using the F_{CaCO3} value estimated above from the TA data. The modeled pCO_2 was above the observed pCO_2 and the CO_2 emission only accounted for 27% of the difference. This implies that the decrease of pCO_2 was mainly related to F_{POC}.

To further investigate the drivers of CO_2 dynamics in Lake Kivu, we computed the TA and DIC whole-lake (bulk concentration) mass balances based on averages for the cruises (Fig. 10). The major flux of TA was the vertical input from deeper waters (50.9 mmol m^{-2} d^{-1}) and the outflow by the Ruzizi (42.6 mmol m^{-2} d^{-1}), which was higher than the inputs from rivers by one

order of magnitude (1.2 mmol m^{-2} d^{-1}). The closing term of the TA mass balance was 9.5 mmol m^{-2} d^{-1}. We assume that this was related to F_{CaCO3} (4.7 mmol m^{-2} d^{-1}). Similarly, the major fluxes of DIC were the vertical input (63.5 mmol m^{-2} d^{-1}) and the outflow of the Ruzizi (39.3 mmol m^{-2} d^{-1}), which were higher than the inputs from rivers by one order of magnitude (1.3 mmol m^{-2} d^{-1}), and than the emission of CO_2 to the atmosphere (10.8 mmol m^{-2} d^{-1}). The closing term of the DIC mass balance was 14.8 mmol m^{-2} d^{-1}. We assume that this was related to the sum of F_{CaCO3} and F_{POC}, allowing to compute F_{POC} using the F_{CaCO3} values computed from the TA mass balance. The estimated F_{POC} values was 10.0 mmol m^{-2} d^{-1}. The whole-lake DIC stable isotope mass balance provided a F_{POC} value of 25.4 mmol m^{-2} d^{-1}, and vertical inputs of DIC of 78.0 mmol m^{-2} d^{-1}.

Planktonic metabolic rates in the epilimnion (PNPP and BP) were measured during each cruise (Table 2). The PNPP values ranged from 14.2 to 49.7 mmol m^{-2} d^{-1}, and were relatively similar in March 2007, September 2007 and June 2008, but distinctly lower in April 2009. The BP values ranged from 3.9 to 49.8 mmol m^{-2} d^{-1}. This range encompasses the one reported for BP in the euphotic layer ($\sim$40 m) of Lake Tanganyika (3.0 to 20.0 mmol m^{-2} d^{-1}) [16]. The BR values estimated from BP ranged from 13.6 to 61.2 mmol C m^{-2} d^{-1}. PNPP was markedly in excess of BR only in June 2008. In March 2007 and September 2007, BR was balanced by PNPP or slightly in excess of PNPP. In April 2009, BR was markedly in excess of PNPP.

Discussion

The amplitude of the seasonal variations of mean pCO_2 across the main basin of Lake Kivu was 163 ppm (between 579±23 ppm on average in March 2007 and 742±28 ppm on average in September 2007). Such pCO_2 seasonal amplitude is low compared to temperate and boreal lakes, where it is usually between $\sim$500 ppm [48] and >1000 ppm [49],[50],[51],[52],[53],[54], and even up to $\sim$10,000 ppm in small bog lakes [53]. The lower amplitude of seasonal variations of the pCO_2 in Lake Kivu might be related to the tropical climate leading only to small surface water temperature seasonal variations (from 23.6°C in September 2007 to 24.6°C in March 2007 on average), and also for relatively modest variations in mixing (MLD changed from 20 m to 70 m). Hence, compared to temperate and boreal lakes, the seasonal variations of biological activity are less marked (due to relatively constant temperature and light, and modest changes in mixing), and also there is an absence of large episodic CO_2 inputs to surface waters such as those occurring in temperate or boreal systems during lake overturns or of CO_2 accumulation during ice covered periods.

The spatial variations of pCO_2 in the main basin of Lake Kivu were also low. The coefficient of variation of pCO_2 in surface waters of the main basin ranged for each cruise between 3% and 6%, below the range reported by Kelly et al. [54] in five large boreal lakes (range 5% to 40%). The relative horizontal homogeneity of pCO_2 could be in part related to the absence of extensive shallow littoral zones, owing to the steep shores [18], and also due to very small influence of C inputs from rivers in the overall DIC budget (Fig. 10). The most prominent spatial feature in Lake Kivu was the much larger pCO_2 values in surface waters of Kabuno Bay compared to the main basin. Furthermore, surface and deep waters of Kabuno Bay were characterized by higher salinity, DIC and TA values and by lower pH and δ^{13}C-DIC values. These vertical patterns indicate that there is a much larger contribution of subaquatic springs to the whole water column

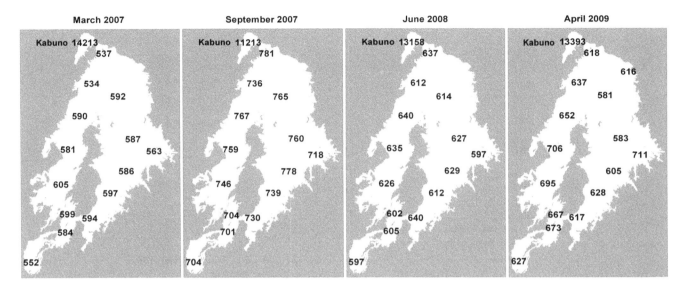

Figure 2. Spatial distribution of the partial pressure of CO$_2$ (pCO$_2$, ppm) in the surface waters of Lake Kivu (1 m depth) in March 2007, September 2007, June 2008 and April 2009.

including surface waters in Kabuno Bay than in the main basin of Lake Kivu relative to their respective volumes. This is related to the different geomorphology, since Kabuno Bay is shallower than the main basin (maximum depth of 110 m versus 485 m) and exchanges little water with the main basin (narrow connection ~10 m deep). Also, Kabuno Bay is smaller (~48 km^2) than the main basin (~2322 km^2). Hence, there is a stronger fetch limitation of wind induced turbulence that also contributes to the stability of the vertical water column structure in Kabuno Bay irrespective of the season [28].

The overall average of pCO$_2$ for the 4 cruises in the main basin of Lake Kivu (646 ppm) is lower than the average of 41 large lakes (>500 km^2) of the world (850 ppm) [5], than the global average for freshwater lakes (1287 ppm) [12], than the average of tropical freshwater lakes (1804 ppm) [55], and than the average for tropical African freshwater lakes (2296 ppm) [49]. Lake Kivu

corresponds to a saline lake according to the definition of Duarte et al. [56] (specific conductivity>1000 µS cm^{-1}; salinity>0.68) that collectively have a global average pCO$_2$ of 1900 ppm (derived from carbonic acid dissociation constants for freshwater) or 3040 ppm (derived from carbonic acid dissociation constants for seawater). Kabuno Bay, in contrast, was characterized by an exceptionally high average pCO$_2$ value (12994 ppm) compared to other freshwater lakes, tropical (African) freshwater lakes, and saline lakes globally.

The average F_{CO2} of the 4 cruises in the main basin of Lake Kivu was 10.8 mmol m^{-2} d^{-1}, which is lower than the global average for freshwater lakes of 16.0 mmol m^{-2} d^{-1} reported by Cole et al. [49], and the average for saline lakes ranging between 81 and 105 mmol m^{-2} d^{-1} reported by Duarte et al. [56]. The average F_{CO2} in Kabuno Bay (500.7 mmol m^{-2} d^{-1}) is distinctly higher than the F_{CO2} global averages for freshwater and saline

Table 1. Average wind speed (m s^{-1}), air-water gradient of the partial pressure of CO$_2$ (ΔpCO$_2$, ppm), and air-water CO$_2$ flux (F_{CO2}, mmol m^{-2} d^{-1}) in the main basin of Lake Kivu and Kabuno Bay in March 2007, September 2007, June 2008, and April 2009.

	wind speed (m s^{-1})	ΔpCO$_2$ (ppm)	F_{CO2} (mmol m^{-2} d^{-1})
March 2007			
Main basin	3.3±0.4	207±22	8.0±1.3
Kabuno Bay		13841	536.4±61.8
September 2007			
Main basin	3.2±0.4	370±27	14.2±2.0
Kabuno Bay		10841	547.7±38.7
June 2008			
Main basin	3.6±0.2	245±15	10.5±1.0
Kabuno Bay		12783	547.7±38.7
April 2009			
Main basin	3.3±0.2	267±41	10.3±1.8
Kabuno Bay		13016	504.5±36.7

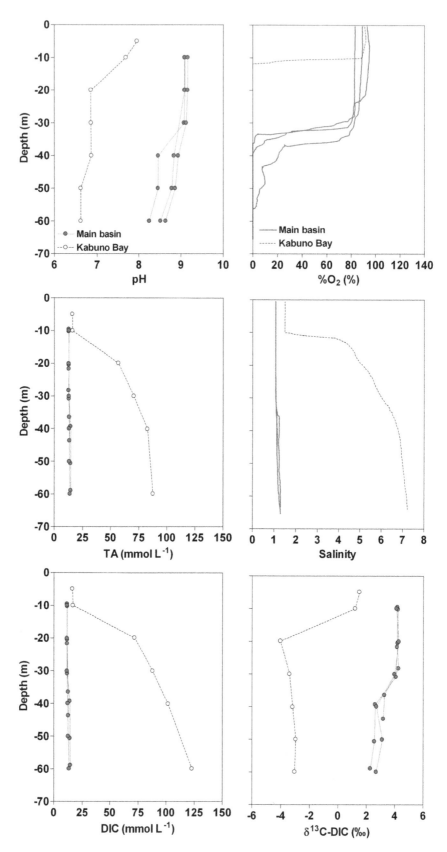

Figure 3. Vertical profiles in March 2007 of pH, oxygen saturation level (%O₂, %), total alkalinity (TA, mmol L⁻¹), salinity, dissolved inorganic carbon (DIC, mmol L⁻¹), δ¹³C signature of DIC (δ¹³C-DIC, ‰) in Kabuno Bay and in the three northernmost stations in the main basin of Lake Kivu.

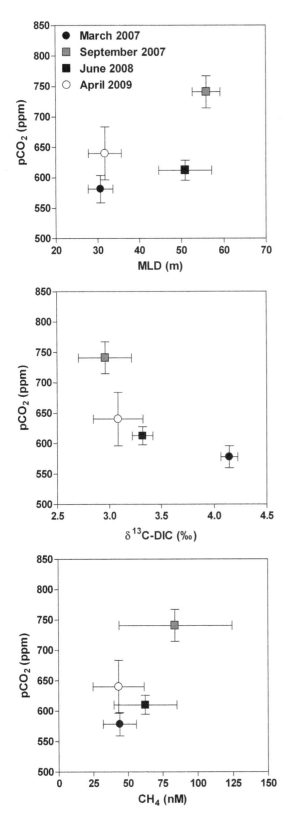

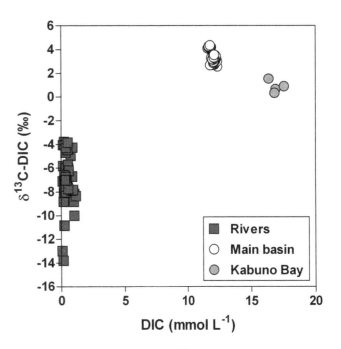

Figure 5. Relation between $\delta^{13}C$ signature of dissolved inorganic carbon (DIC) ($\delta^{13}C$-DIC, ‰) and DIC concentration (mmol L^{-1}), in the mixed layer of the main basin of Lake Kivu, Kabuno Bay, and various inflowing rivers, in March 2007, September 2007, June 2008, and April 2009.

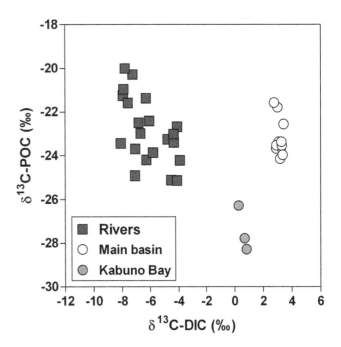

Figure 4. Average partial pressure of CO_2 (pCO_2, ppm) in the surface waters of the main basin of Lake Kivu (1 m depth) versus mixed layer depth (MLD, m), $\delta^{13}C$ signature of dissolved inorganic carbon (DIC) ($\delta^{13}C$-DIC, ‰), and methane concentration (CH_4, nmol L^{-1}) in March 2007, September 2007, June 2008, and April 2009. Vertical and horizontal bars represent standard deviations.

Figure 6. Relation between $\delta^{13}C$ signature of particulate organic carbon (POC) ($\delta^{13}C$-POC, ‰) and $\delta^{13}C$ signature of dissolved inorganic carbon (DIC) ($\delta^{13}C$-DIC, ‰), in the mixed layer of the main basin of Lake Kivu, Kabuno Bay and various inflowing rivers, in June 2008, April 2009, and October 2010.

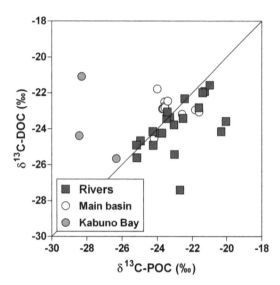

Figure 7. Relation between δ¹³C signature of dissolved organic carbon (DOC) (δ¹³C-DOC, ‰) and δ¹³C signature of particulate organic carbon (POC) (δ¹³C-POC, ‰), in the mixed layer of the main basin of Lake Kivu, Kabuno Bay and various inflowing rivers, in June 2008, April 2009, and October 2010. Solid line is the 1:1 line.

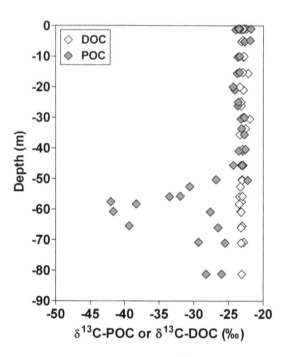

Figure 8. Vertical profiles of δ¹³C signature of dissolved organic carbon (DOC) (δ¹³C-DOC, ‰) and δ¹³C signature of particulate organic carbon (POC) (δ¹³C-POC, ‰) in the main basin of Lake Kivu in October 2010.

lakes. However, the average F_{CO2} in Kabuno Bay is equivalent to average of F_{CO2} value of alkaline volcanic lakes (458 mmol m^{-2} d^{-1}) but lower than average of F_{CO2} of acid volcanic lakes (51183 mmol m^{-2} d^{-1}) reported by Pérez et al. [57].

Cross system regional analyses show a general negative relationship between pCO₂ and lake surface area [5], [54], [58], [59], [60] and a positive relationship between pCO₂ and DOC [61] (and reference therein). The low pCO₂ and F_{CO2} values in Lake Kivu are consistent with these general patterns, since this is a large (>2000 km²) and organic poor (DOC ~0.2 mmol L^{-1}) system. However, the low seasonal amplitude of pCO₂ and relative horizontal homogeneity of pCO₂ in Lake Kivu are not necessarily linked to its large size. Indeed, spatial and temporal variability of pCO₂ within a single lake have been found to be no greater nor smaller in larger lakes than in smaller lakes, in cross system analyses in Northwest Ontario [54] and northern Québec [60].

Borges et al. [28] reported diffusive CH₄ emissions of 0.04 mmol m^{-2} d^{-1} and 0.11 mmol m^{-2} d^{-1} for the main basin of Lake Kivu and Kabuno Bay, respectively. Using a global warming potential of 72 for a time horizon of 20 yr [62], the CH₄ diffusive emissions in CO₂ equivalents correspond to 0.26 mmol m^{-2} d^{-1} and 0.77 mmol m^{-2} d^{-1} for the main basin of Lake Kivu and Kabuno Bay, respectively, hence 41 to 650 times lower than the actual F_{CO2} values.

DIC concentrations in surface waters of the main basin of Lake Kivu and Kabuno Bay averaged 13.0 and 16.8 mmol L^{-1}, respectively, and were well within the range of DIC reported for saline lakes by Duarte et al. [56], which range from 0.1 to 2140 mmol L^{-1}, but are lower than the global average for saline lakes of 59.5 mmol L^{-1}. DOC averaged in surface waters 0.15 mmol L^{-1} and 0.20 mmol L^{-1} in the main basin of Lake Kivu and in Kabuno Bay, respectively. Hence, DIC strongly dominated the dissolved C pool, with DIC:DOC ratios of 82 and 87 in the main basin of Lake Kivu and in Kabuno Bay, respectively. These DIC:DOC ratios are higher than those in 6 hard-water lakes of the northern Great Plains ranging from 3 to 6 [63], and higher than those in boreal lakes where DOC is the

dominant form of the dissolved C pool, with DIC:DOC ratios ranging from 0.01 to 0.68 e.g. [64], [65], this range reflecting both seasonal changes [59] and differences in catchment characteristics [66]. Unlike the 6 hard-water lakes of the northern Great Plains, where the high DIC concentrations are due to river inputs [67], the high DIC concentrations in Lake Kivu were related to vertical inputs of DIC from deep waters that were on average 49 times larger than the DIC inputs from rivers (Fig. 10), as confirmed by δ¹³C-DIC values clearly more positive in the lake than in the rivers (Fig. 5). The difference in C stable isotope composition of POC between the lake and rivers indicates that these two pools of organic C do not share the same origin. In the small, turbid rivers flowing to Lake Kivu, we expect the POC and DOC pools to be derived from terrestrial inputs, as reflected by the low contribution of POC to TSM e.g. [68]. In contrast, the positive relationship between the δ¹³C-DIC and δ¹³C-POC in surface waters (Fig. 6) suggests that DIC is the main C source for POC in surface waters of Lake Kivu, implying that the whole microbial food web could be supported by autochthonous organic C. However, the δ¹³C data indicate a surprising difference between the origin of DOC and POC in the lake (Figs. 7, 8). The δ¹³C-POC signatures were constant from the surface to the oxic-anoxic interface, then showed a local and abrupt excursion to values as low as −40‰, reflecting the incorporation of a ¹³C-depleted source in the POC (Fig. 8). Indeed, while the large pool of DIC is the main C source for POC in surface waters, it appears that CH₄ with a δ¹³C signature of approximately −60‰ (own data not shown) contributes significantly to C fixation at the oxic-anoxic interface, as also shown in Lake Lugano [69]. In contrast, the δ¹³C signature of the DOC pool in the mixolimnion showed little seasonal and spatial variations and appeared to be uncoupled from the POC pool (Figs. 7, 8). Heterotrophic bacteria quickly mineralized the labile autochthonous DOC that reflects the δ¹³C signature of POC, produced by cell lysis, grazing, or phytoplankton excretion

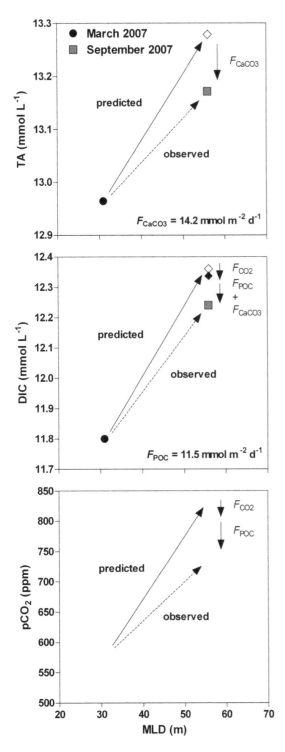

Figure 9. Observed data (circles and squares) and predicted values from a mixing model (diamonds) from March 2007 to September 2007 of total alkalinity (TA, mmol L^{-1}), dissolved inorganic carbon (DIC, mmol L^{-1}), and the partial pressure of CO$_2$ (pCO$_2$, ppm) as a function of mixed layer depth (MLD, m) in the main basin of Lake Kivu. F_{CO_2} = air-water CO$_2$ flux; F_{POC} = export of particulate organic carbon to depth; F_{CaCO_3} = export of CaCO$_3$ to depth.

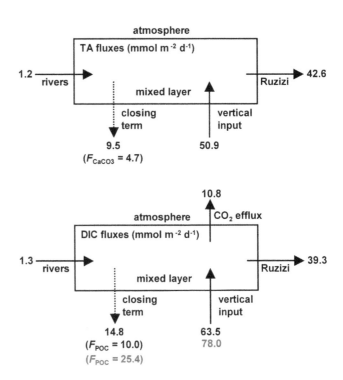

Figure 10. Average mass balance of total alkalinity (TA) and dissolved inorganic carbon (DIC) in the mixed layer of the main basin of Lake Kivu based on data collected in March 2007, September 2007, June 2008 and April 2009. F_{POC} = export of particulate organic carbon to depth; F_{CaCO_3} = export of CaCO$_3$ to depth. Numbers in black correspond to the mass balance based on bulk concentrations, and numbers in red correspond to the mass balance based on DIC stable isotopes. All fluxes are expressed in mmol m^{-2} d^{-1}.

[70]. Hence, standing stocks of autochthonous DOC are small, and older refractory compounds constitute the major part of the DOC pool.

The F_{CaCO_3} value computed from the whole-lake TA budget was 4.7 mmol m^{-2} d^{-1}, higher than the total inorganic C (TIC) annual average fluxes in sediment traps ranging from 0.3 mmol m^{-2} d^{-1} (at 50 m depth) to 0.5 mmol m^{-2} d^{-1} (at 130 m depth) reported by Pasche et al. [71] at Ishungu. The maximum individual monthly TIC flux from sediment traps at Ishungu reported by Pasche et al. [71] was 4.0 mmol m^{-2} d^{-1}. However, the F_{CaCO_3} value was closer to the TIC deposition fluxes in the top cm of sediment cores ranging from 1.4 mmol m^{-2} d^{-1} (at Ishungu) to 4.8 mmol m^{-2} d^{-1} (at Gisenyi) also reported by Pasche et al. [71]. The F_{POC} computed from the DIC whole-lake budget was 10.0 mmol m^{-2} d^{-1} close to the total organic C (TOC) average fluxes in sediment traps ranging from 8.7 mmol m^{-2} d^{-1} (at 50 m depth) to 9.8 mmol m^{-2} d^{-1} (at 172 m depth) reported by Pasche et al. [71] at Ishungu.

Due to thermodynamic equilibria of the dissolved carbonate system, CaCO$_3$ precipitation leads to a shift from the HCO$_3^-$ to the CO$_2$ pool according to:

$$Ca^{2+} + 2HCO_3^- \rightleftharpoons CaCO_3 + H_2O \qquad (22)$$

However, CaCO$_3$ precipitation has frequently been reported in lakes as biologically mediated by primary producers [50], [72], [73], whereby the CO$_2$ produced by the precipitation of CaCO$_3$ is fixed into organic matter by photosynthesis and does not accumulate in the water [50], [74].

Table 2. Photic depth (Ze, m), chlorophyll-a concentration in the mixed layer (Chl-a, mg m^{-2}), particulate net primary production (PNPP, mmol m^{-2} d^{-1}), bacterial production integrated over Ze (BP, mmol m^{-2} d^{-1}), bacterial respiration (BR, mmol m^{-2} d^{-1}), and percent of extracellular release (PER, %), at two stations in the main basin of Lake Kivu (Kibuye, Ishungu) in March 2007, September 2007, June 2008, and April 2009.

		Ze (m)	Chl-a (mg m^{-2})	PNPP (mmol m^{-2} d^{-1})	BP (mmol m^{-2} d^{-1})	BR (mmol m^{-2} d^{-1})	PNPP-BR (mmol m^{-2} d^{-1})	PER (%)
March 2007								
15/03/2007	Kibuye	18	38.3	27.0	23.2	35.7	−8.7	54
17/03/2007	Kibuye	20	48.4	42.5	25.8	39.9	2.6	33
23/03/2007	Ishungu	17	36.1	49.7	40.5	49.6	0.1	32
September 2007								
09/09/2007	Kibuye	19	56.4	42.9	35.9	47.9	−5.0	42
12/09/2007	Kibuye	18	55.1	45.9	34.0	45.9	0.1	36
04/09/2007	Ishungu	20	48.2	n.d.	16.0	29.9	n.d.	n.d.
June 2008								
23/06/2008	Kibuye	24	42.8	46.0	7.7	21.6	24.4	3
11/07/2008	Kibuye	20	37.8	42.0	11.1	24.3	17.7	17
03/07/2008	Ishungu	19	28.1	40.7	3.9	13.6	27.1	−4
April 2009								
04/05/2009	Kibuye	21	22.9	14.2	49.8	61.2	−47.0	82
21/04/2009	Ishungu	24	39.3	24.5	43.5	58.9	−34.4	68

PNPP and BP were derived from experimental measurements. BR was computed from BP (see material and methods), and PER was computed from PNPP, BR, river inputs and vertical export of organic matter according to equation (24).

Precipitation and preservation of $CaCO_3$ in lakes are not considered in global compilations of C fluxes in lakes, that focus exclusively on organic C and CO_2 fluxes e.g. [4], [7]. However, in Lake Kivu, F_{CaCO3} was found to be a major flux term in the C budgets, 3.6 times larger than the DIC inputs from rivers, and comparable to the emission of CO_2 to the atmosphere and F_{POC} (~2 times lower). The $F_{POC}{:}F_{CaCO3}$ ratio in the main basin of Lake Kivu was 2.1, which is consistent with the values reported in 6 hard-water lakes of the northern Great Plains, ranging from 1.0 to 4.0 [63], and with the values in Lake Malawi ranging from 0.2 to 7.3 (on average 2.5) [75]. As a comparison, the average $F_{POC}{:}F_{CaCO3}$ in the ocean (so called rain ratio) has been estimated from models of varying complexity to be 4.0 [76], 3.5 to 7.5 [77], and 11.0 [78].

The export ratio (ER) is the fraction of PP that is exported from surface waters to depth and is an important metric of the net metabolism and overall C fluxes in aquatic systems. We computed ER as defined by Baines et al. [79], according to:

$$ER = \frac{F_{POC}}{PNPP_i} 100 \qquad (23)$$

where F_{POC} is derived from the DIC mass balance (Fig. 10), and $PNPP_i$ is the average PNPP for a given cruise i measured by incubations (Table 2).

ER was 25%, 23%, 23%, and 52% in March 2007, September 2007, June 2008, and April 2009, respectively. These values are consistent with the fact that the ER in lakes is negatively related to lake primary production based on the analysis of Baines et al. [79]. These authors reported ER values as high as 50% for oligotrophic lakes such as Lake Kivu.

The general agreement between the F_{CaCO3} computed from the TA budget and the TIC deposition fluxes derived from sediment cores reported by Pasche et al. [71], and the F_{POC} computed from the DIC budget and TOC average fluxes in sediment traps reported by Pasche et al. [71], give confidence on the overall robustness of the TA and DIC whole-lake budget we computed. Also, the F_{CaCO3} and F_{POC} values computed from whole-lake budget are consistent with those derived independently from a mixing model based on the March and September 2007 data (Fig. 9). The whole-lake DIC stable isotope mass balance budgets give F_{POC} and upward DIC inputs estimates that are of same order of magnitude as those derived from whole-lake bulk DIC mass balance budget. The difference in the two approaches is that in the DIC stable isotope mass balance the upward DIC inputs were computed from vertical distributions of DIC and δ^{13}C-DIC while in the bulk DIC mass balance they are computed from the E and $Q_{upwelling}$ values from the model of Schmid et al. [19] and the DIC vertical distribution. This can explain the mismatch between both approaches in the upward DIC input estimates (difference of 23%) that propagated into a relatively larger mismatch in the F_{POC} estimates (difference of 61%) computed as a closing term in both approaches.

Based on the POC and DOC data acquired during the June 2008 and April 2009 cruises in 12 rivers flowing into Lake Kivu (Fig. 1), we computed an overall TOC input from rivers of 0.7 mmol m^{-2} d^{-1} and 3.3 mmol m^{-2} d^{-1}, respectively. The F_{POC} was 10 mmol m^{-2} d^{-1}, implying a net organic C production in the epilimnion (net autotrophic community metabolic status). This would mean that the fraction of PNPP that does not sediment out of the epilimnion cannot meet BR, and that BR must then rely on other organic C sources. We assume these other organic C sources to be dissolved primary production (DPP), that was estimated assuming steady-state, according to:

$$DPP = F_{POC} + BR - PNPP - F_{TOC\_river} \qquad (24)$$

where F_{TOC_river} is the input of TOC from rivers that was computed from the discharge weighted average TOC concentrations from the June 2008 and April 2009 cruises.

The percent of extracellular release (PER) allows to determine the relative importance of DPP in overall C flows in an aquatic system. PER as defined by Baines and Pace [80] was computed according:

$$PER = \frac{DPP}{DPP + PNPP} 100 \qquad (25)$$

In June 2008, for two stations, the sum of organic C inputs (PNPP + F_{TOC_river}) exceeded the sum of organic C outputs (F_{POC} + BR), leading to negative DPP and PER estimates. If we exclude these values, PER estimates ranged from 3% to 80% (Table 2) encompassing the range reported by Baines and Pace [80] for freshwater lakes from ~0% to ~75%. PNPP in April 2009 was distinctly lower than during the other cruises, leading to high PER estimates. The average PER for all cruises was 32%, and if the April 2009 data are excluded, the average PER was 19%. During the April 2009 field survey, we carried 6 h incubations using the ^{14}C incorporation method [81] in light controlled (200 μE m^{-2} s^{-1}) conditions, allowing to measure PNPP and to compute DPP using the model of Moran et al. [82]. Experimentally-derived PER estimates were 57% at Ishungu, 62% at Kibuye, and 50% in Kabuno Bay [70]. These experimentally determined PER values are within the range of those determined from the mass balance (3% to 80%) and above the average for all cruises (32%). This confirms that a substantial part of the BR is subsidised by DPP, that part of the PNPP is available for export to depth, and consequently that the epilimnion of Lake Kivu is net autotrophic, although a source of CO_2 to the atmosphere.

Conclusions

Surface waters of 93% of the lakes in the compilation of Sobek et al. [12] were over-saturated in CO_2 with respect to the atmospheric equilibrium. Hence, the overwhelming majority of lakes globally act as a CO_2 source to the atmosphere. These emissions to the atmosphere have been frequently explained by the net heterotrophic nature of lakes sustained by terrestrial organic C inputs mainly as DOC [1], [6], [7], [12], [49], [83], [84], [85], [86], [87]. While this paradigm undoubtedly holds true for boreal humic lakes, several exceptions have been put forward in the literature. For instance, Balmer and Downing [88] showed that the majority (60%) of eutrophic agriculturally impacted lakes are net autotrophic and CO_2 sinks. Karim et al. [89] showed that surface waters of very large lakes such as the Laurentian Great Lakes are at equilibrium with atmospheric CO_2 and O_2. This is consistent with the negative relationship between pCO_2 and lake size reported in several regional analyses [5], [54], [58], [59], [60], and with the positive relationship between pCO_2 and catchment area: lake area reported for Northern Wisconsin lakes [90]. Also, in some lakes among which hard-water lakes, the magnitude of CO_2 emissions to the atmosphere seems to depend mainly on hydrological inputs of DIC from rivers and streams [67], [91], [92], [93] or ground-water [52], [94], [95], rather than on lake metabolism. Some of these hard-water lakes were actually found to be net autotrophic, despite acting as a source of CO_2 to the atmosphere [67], [91], [93].

Here, we demonstrate that Lake Kivu represents an example of a large, oligotrophic, tropical lake acting as a source of CO_2 to the atmosphere, despite having a net autotrophic epilimnion. The river inputs of TOC were modest, and were on average 11 times lower than the export of POC to depth. This is probably related to the very low ratio of catchment surface area: lake surface area ($5100:2370$ $km^2:km^2$) that is among the lowest in lakes globally [96]. We showed that BR was in part subsidized by DPP, based on mass balance considerations and incubations. Since the epilimnion of Lake Kivu is net autotrophic, the CO_2 emission to the atmosphere must be sustained by DIC inputs. The river DIC inputs are also low owing to very low ratio of catchment surface area: lake surface area, and cannot sustain the CO_2 emission to the atmosphere unlike the hard water lakes studied by Finlay et al. [67] and Stets et al. [91]. In Lake Kivu, the CO_2 emission is sustained by DIC inputs from depth, and this DIC is mainly geogenic [24] and originates from deep geothermal springs [19].

Carbonate chemistry in surface waters of Lake Kivu is unique from other points of view. The dissolved C pool is largely dominated by DIC, with DIC:DOC ratios distinctly higher than in hard-water lakes and humic lakes. The high DIC content in surface water results in $CaCO_3$ over-saturation, in turn leading to $CaCO_3$ precipitation and export to depth. This flux was found to be significant, being 4 times larger than the river inputs of DIC and of similar magnitude than the CO_2 emission to the atmosphere.

Supporting Information

Table S1 Data-set of depth (m), water temperature (°C), specific conductivity at 25°C (μS cm^{-1}), oxygen saturation level (%O_2, %), $\delta^{13}C$ signature of dissolved inorganic carbon (DIC) ($\delta^{13}C$-DIC, ‰), total alkalinity (TA, mmol L^{-1}), pH, partial pressure of CO_2 (pCO$_2$, ppm), dissolved methane concentration (CH_4, nmol L^{-1}), particulate organic carbon (POC, mg L^{-1}), $\delta^{13}C$ signature of POC ($\delta^{13}C$-POC, ‰), dissolved organic carbon (DOC, mg L^{-1}), $\delta^{13}C$ signature of DOC ($\delta^{13}C$-DOC, ‰), total suspended matter (TSM, mg L^{-1}) in Lake Kivu and 12 rivers flowing into Lake Kivu, in March 2007, September 2007, June 2008, April 2009 and October 2010.

Acknowledgments

We are grateful to Pascal Isumbisho Mwapu (Institut Supérieur Pédagogique, Bukavu, République Démocratique du Congo) and Laetitia Nyinawamwiza (National University of Rwanda, Butare, Rwanda) and their respective teams for logistical support during the cruises, to Bruno Delille, Gilles Lepoint, Bruno Leporcq, and Marc-Vincent Commarieu for help in field sampling, to an anonymous reviewer and Pirkko Kortelainen (reviewer) for constructive comments on a previous version of the manuscript.

Author Contributions

Conceived and designed the experiments: AVB SB PS JPD FD. Performed the experiments: AVB CM FD. Analyzed the data: AVB CM SB PS JPD FD. Contributed reagents/materials/analysis tools: AVB CM SB PS JPD FD. Wrote the paper: AVB CM SB PS JPD FD.

References

1. Cole JJ, Caraco NF (2001) Carbon in catchments: connecting terrestrial carbon losses with aquatic metabolism. Mar Fresh Res 52: 101–110.
2. Kempe S (1984) Sinks of the anthropogenically enhanced carbon cycle in surface fresh waters. J Geophys Res 89: 4657–4676.
3. Richey JE, Melack JM, Aufdenkampe AK, Ballester VM, Hess LL (2202) Outgassing from Amazonian rivers and wetlands as a large tropical source of atmospheric CO_2. Nature 416: 617–620.
4. Cole JJ, Prairie YT, Caraco NF, McDowell WH, Tranvik LJ, et al. (2007) Plumbing the global carbon cycle: Integrating inland waters into the terrestrial carbon budget. Ecosystems 10: 171–184.
5. Alin SR, Johnson TC (2007) Carbon cycling in large lakes of the world: A synthesis of production, burial, and lake-atmosphere exchange estimates. Global Biogeochem Cycles 21(GB3002): doi:10.1029/2006GB002881.
6. Battin TJ, Kaplan LA, Findlay S, Hopkinson CS, Marti E, et al. (2008) Biophysical controls on organic carbon fluxes in fluvial networks. Nature Geosc 1: 95–100.
7. Tranvik LJ, Downing JA, Cotner JB, Loiselle SA, Striegl RG, et al. (2009) Lakes and reservoirs as regulators of carbon cycling and climate. Limnol Oceanogr 54: 2298–2314.
8. Aufdenkampe AK, Mayorga E, Raymond PA, Melack JM, Doney SC, et al. (2011) Riverine coupling of biogeochemical cycles between land, oceans, and atmosphere, Front Ecol Environ 9: 53–60.
9. Butman D, Raymond PA (2011) Significant efflux of carbon dioxide from streams and rivers in the United States. Nature Geosc 4: 839–842.
10. Raymond PA, Hartmann J, Lauerwald R, Sobek S, McDonald C, et al. (2013) Global carbon dioxide emissions from inland waters. Nature 503: 355–359.
11. Takahashi T, Sutherland SC, Wanninkhof R, Sweeney C, Feely RA, et al. (2009) Climatological mean and decadal change in surface ocean pCO$_2$, and net sea-air CO_2 flux over the global oceans. Deep-Sea Res. II 56: 554–577.
12. Sobek S, Tranvik LJ, Cole JJ (2005) Temperature independence of carbon dioxide supersaturation in global lakes. Global Biogeochem Cycles 19(GB3003): doi:10.1029/2004GB002264.
13. Ludwig W, Probst JL, Kempe S (1996) Predicting the oceanic input of organic carbon by continental erosion. Global Biogeochem Cycles 10: 23–41.
14. Lehner B, Döll P (2004) Development and validation of a global database of lakes, reservoirs and wetlands. J Hydrol 296: 1–22.
15. Darchambeau F, Sarmento H, Descy J-P (2014) Primary production in a tropical large lake: The role of phytoplankton composition. Sci Total Environ 473–474: 178–188.
16. Stenuite S, Pirlot S, Tarbe AL, Sarmento H, Lecomte M, et al. (2009) Abundance and production of bacteria, and relationship to phytoplankton

production, in a large tropical lake (Lake Tanganyika). Freshwater Biol 54: 1300–1311.
17. Damas H (1937) La stratification thermique et chimique des lacs Kivu, Edouard et Ndalaga (Congo Belge). Verh Internat Verein Theor Angew Limnol 8: 51–68.
18. Degens ET, vos Herzes RP, Wosg H-K, Deuser WG, Jannasch HW (1973) Lake Kivu: Structure, chemistry and biology of an East African rift lake. Geol Rundsch 62: 245–277.
19. Schmid M, Halbwachs M, Wehrli B, Wüest A (2005) Weak mixing in Lake Kivu: new insights indicate increasing risk of uncontrolled gas eruption. Geochem Geophys Geosyst 6(Q07009): doi:07010.01029/02004GC000892.
20. Thiery W, Stepanenko VM, Fang X, Jöhnk KD, Li Z, et al. (2014) LakeMIP Kivu: Evaluating the representation of a large, deep tropical lake by a set of one-dimensional lake models. Tellus A 66(21390): doi:10.3402/tellusa.v66.21390.
21. Sarmento H, Isumbisho M, Descy J-P (2006) Phytoplankton ecology of Lake Kivu (Eastern Africa). J Plankton Res 28: 815–829.
22. Sarmento H, Darchambeau F, Descy J-P (2012) Phytoplankton of Lake Kivu. In: Lake Kivu - Limnology and biogeochemistry of a tropical great lake: Springer. pp. 67–83.
23. Schmid M, Busbridge M, Wüest A (2010) Double-diffusive convection in Lake Kivu. Limnol Oceanogr 55: 225–238.
24. Schoell M, Tietze K, Schoberth SM (1988) Origin of methane in Lake Kivu (East-Central Africa). Chem Geol 71: 257–265.
25. Schmid M, Tietze K, Halbwachs M, Lorke A, McGinnis D, et al. (2004) How hazardous is the gas accumulation in Lake Kivu? Arguments for a risk assessment in light of the Nyiragongo Volcano eruption of 2002. Acta Vulcanol 14/15: 115–121.
26. Nayar A (2009) A lakeful of trouble. Nature 460: 321–323.
27. Wüest A, Jarc L, Bürgmann H, Pasche N, Schmid M (2012) Methane Formation and Future Extraction in Lake Kivu. In: Lake Kivu - Limnology and biogeochemistry of a tropical great lake: Springer. pp. 165–180.
28. Borges AV, Abril G, Delille B, Descy J-P, Darchambeau F (2011) Diffusive methane emissions to the atmosphere from Lake Kivu (Eastern Africa) J Geophys Res 116(G03032): doi:10.1029/2011JG001673.
29. Schmid M, Wüest A (2012) Stratification, Mixing and Transport Processes in Lake Kivu. In: Lake Kivu - Limnology and biogeochemistry of a tropical great: Springer. pp. 13–29.
30. Frankignoulle M, Borges A, Biondo R (2001) A new design of equilibrator to monitor carbon dioxide in highly dynamic and turbid environments. Water Res 35: 1344–1347.
31. Frankignoulle M, Borges AV (2001) Direct and indirect pCO$_2$ measurements in a wide range of pCO$_2$ and salinity values (the Scheldt estuary). Aquat Geochem 7: 267–273.

32. Gran G (1952) Determination of the equivalence point in potentiometric titrations of seawater with hydrochloric acid. Oceanol Acta 5: 209–218.

33. Millero FJ, Graham TB, Huang F, Bustos-Serrano H, Pierrot D (2006) Dissociation constants of carbonic acid in sea water as a function of salinity and temperature. Mar Chem 100: 80–94.

34. Miyajima T, Yamada Y, Hanba YT, Yoshii K, Koitabashi T, et al. (1995) Determining the stable-isotope ratio of total dissolved inorganic carbon in lake water by GC/C/IRMS. Limnol Oceanogr 40: 994–1000.

35. Steemann-Nielsen E (1951) Measurement of production of organic matter in sea by means of carbon-14. Nature 267: 684–685.

36. Descy J-P, Higgins HW, Mackey DJ, Hurley JP, Frost TM (2000) Pigment ratios and phytoplankton assessment in northern Wisconsin lakes. J Phycol 36: 274–286.

37. Fuhrman JA, Azam F (1982) Thymidine incorporation as a measure of heterotrophic bacterioplankton production in marine surface waters: Evaluation and field results. Mar Biol 66: 109–120.

38. Del Giorgio PA, Cole JJ (1998) Bacterial growth efficiency in natural aquatic systems. Annu Rev Ecol Syst 29: 503–41.

39. Weiss RF, Price BA (1980) Nitrous oxide solubility in water and seawater. Mar Chem 8: 347–359.

40. Weiss RF (1974) Carbon dioxide in water and seawater: the solubility of a non-ideal gas. Mar Chem 2: 203–215.

41. Cole JJ, Caraco NF (1998) Atmospheric exchange of carbon dioxide in a low-wind oligotrophic lake measured by the addition of SF_6. Limnol Oceanogr 43: 647–656.

42. Wanninkhof R (1992) Relationship between wind speed and gas exchange over the ocean. J Geophys Res 97: 7373–7382.

43. Muvundja FA, Pasche N, Bugenyi FWB, Isumbisho M, Müller B, et al. (2009) Balancing nutrient inputs to Lake Kivu. J Great Lakes Res 35: 406–418.

44. Quay PD, Stutsman J, Feely RA, Juranek W (2009) Net community production rates across the subtropical and equatorial Pacific Ocean estimated from air-sea $\delta^{13}C$ disequilibrium. Global Biogeochem Cycles 23 (GB2006): doi:10.1029/2008GB003193.

45. Quay PD, Stutsman J (2003) Surface layer carbon budget for the subtropical N. Pacific: $\delta^{13}C$ constraints at station ALOHA, Deep-Sea Res I 50: 1045–1061.

46. Emrich K, Ehhalt DH, Vogel JC (1970) Carbon isotope fractionation during the precipitation of calcium carbonate. Earth Planet Sci Lett 8: 363–371.

47. Zhang J, Quay PD, Wilbur DO (1995) Carbon isotope fractionation during gas-water exchange and dissolution of CO_2. Geochim Cosmochim Acta 59: 107–114.

48. Atilla N, McKinley GA, Bennington V, Baehr M, Urban N, et al. (2011) Observed variability of Lake Superior pCO_2. Limnol Oceanogr 56: 775–786.

49. Cole JJ, Caraco NF, Kling GW, Kratz TK (1994) Carbon dioxide supersaturation in the surface waters of lakes. Science 265: 1568–1570.

50. McConnaughey TA, LaBaugh JW, Rosenberry DO, Striegl RG, Reddy MM, et al. (1994) Carbon budget for a groundwater-fed lake: Calcification supports summer photosynthesis. Limnol Oceanogr 39: 1319–1332.

51. Gelbrecht J, Fait M, Dittrich M, Steinberg C (1998) Use of GC and equilibrium calculations of CO_2 saturation index to indicate whether freshwater bodies in north-eastern Germany are net sources or sinks for atmospheric CO_2. Fresenius J Anal Chem 361: 47–53.

52. Striegl RG, Michmerhuizen CM (1998) Hydrologic influence on methane and carbon dioxide dynamics at two north-central Minnesota lakes. Limnol Oceanogr 43: 1519–1529.

53. Riera JL, Schindler JE, Kratz TK (1999) Seasonal dynamics of carbon dioxide and methane in two clear-water lakes and two bogs lakes in northern Wisconsin, U.S.A. Can J Fish Aquat Sci 56: 265–274.

54. Kelly CA, Fee E, Ramlal PS, Rudd JWM, Hesslein RH, et al. (2001) Natural variability of carbon dioxide and net epilimnetic production in the surface waters of boreal lakes of different sizes. Limnol Oceanogr 46: 1054–1064.

55. Marotta H, Duarte CM, Sobek S, Enrich-Prast A (2009) Large CO_2 disequilibria in tropical lakes. Global Biogeochem Cycles, 23(GB4022):doi:10.1029/2008GB003434.

56. Duarte CM, Prairie YT, Montes C, Cole JJ, Striegl R, et al. (2008) CO_2 emissions from saline lakes: A global estimate of a surprisingly large flux. J Geophys Res 113 (G04041): doi:10.1029/2007JG000637.

57. Pérez NM, Hernández PA, Padilla G, Nolasco D, Barrancos J, et al. (2011) Global CO_2 emission from volcanic lakes. Geology 39: 235–238.

58. Kortelainen P, Rantakari M, Huttunen JT, Mattsson T, Alm J, et al. (2006) Sediment respiration and lake trophic state are important predictors of large CO_2 evasion from small boreal lakes. Global Change Biol 12: 1554–1567.

59. Kortelainen P, Rantakari M, Pajunen H, Huttunen JT, Mattsson T, et al. (2013) Carbon evasion/accumulation ratio in boreal lakes is linked to nitrogen. Global Biogeochem Cycles 27: 363–374.

60. Roehm CL, Prairie YT, del Giorgio PA (2009) The pCO_2 dynamics in lakes in the boreal region of northern Québec, Canada, Global Biogeochem Cycles 23(GB3013): doi:10.1029/2008GB003297.

61. Lapierre J-F, del Giorgio PA (2012) Geographical and environmental drivers of regional differences in lake pCO_2 versus DOC relationship across northern landscapes. J Geophys Res 117(G03015): doi:10.1029/2012JG001945.

62. IPCC (2007) Climate Change 2007: The Physical Science Basis. Contribution of Working Group I In: Fourth Assessment Report of the Intergovernmental Panel on Climate: Cambridge University Press. pp. 1–996

63. Finlay K, Leavitt PR, Wissel B, Prairie YT (2009) Regulation of spatial and temporal variability of carbon flux in six hard-water lakes of the northern Great Plains. Limnol Oceanogr 54: 2553–2564.

64. Whitfield PH, van der Kamp G, St-Hilaire A (2009) Predicting the partial pressure of carbon dioxide in boreal lakes, Can Water Resour J 34: 303–310.

65. Einola E, Rantakari M, Kankaala P, Kortelainen P, Ojala A, et al. (2011) Carbon pools and fluxes in a chain of five boreal lakes: A dry and wet year comparison. J Geophys Res 116(G03009): doi:10.1029/2010JG001636.

66. Rantakari M, Kortelainen P (2008) Controls of total organic and inorganic carbon in randomly selected Boreal lakes in varied catchments. Biogeochemistry 91: 151–162.

67. Finlay K, Leavitt PR, Patoine A, Wissel B (2010), Magnitudes and controls of organic and inorganic carbon flux through a chain of hardwater lakes on the northern Great Plains. Limnol Oceanogr 55: 1551–15640.

68. Tamooh F, Van den Meersche K, Meysman F, Marwick TR, Borges AV, et al. (2012) Distribution and origin of suspended matter and organic carbon pools in the Tana River Basin, Kenya. Biogeosciences 9: 2905–2920.

69. Lehmann MF, Bernasconi SM, McKenzie JA, Barbieri A, Simona M, et al. (2004) Seasonal variation of the $\delta^{13}C$ and $\delta^{15}N$ of particulate and dissolved carbon and nitrogen in Lake Lugano: Constraints on biogeochemical cycling in a eutrophic lake. Limnol Oceanogr 49: 415–429.

70. Morana C, Sarmento H, Descy J-P, Gasol JM, Borges AV, et al. (2014) Production of dissolved organic matter by phytoplankton and its uptake by heterotrophic prokaryotes in large tropical lakes. Limnol Oceanogr 59: 1364–1375.

71. Pasche N, Alunga G, Mills K, Muvundja F, Ryves DB, et al. (2010) Abrupt onset of carbonate deposition in Lake Kivu during the 1960s: response to recent environmental changes. J Paleolimnol 44: 931–946.

72. Dittrich M, Obst M (2004) Are picoplankton responsible for calcite precipitation in lakes? Ambio 33: 559–564.

73. Obst M, Wehrli B, Dittrich M (2009) $CaCO_3$ nucleation by cyanobacteria: laboratory evidence for a passive, surface-induced mechanism. Geobiology 7: 324–347.

74. Nimick DA, Gammons CH, Parker SR (2011) Diel biogeochemical processes and their effect on the aqueous chemistry of streams: A review. Chem Geol 283: 3–17.

75. Pilskaln CH (2004) Seasonal and interannual particle export in an African rift valley lake: A 5-yr record from Lake Malawi, southern East Africa. Limnol Oceanogr 49: 964–977.

76. Broecker WS, Peng TH (1982) Tracers in the sea. Eldigio Press.

77. Shaffer G (1993) Effects of the marine carbon biota on global carbon cycling. In: The Global Carbon Cycle. Berlin: Springer Verlag. pp. 431–455.

78. Yamanaka Y, Tajika E (1996) The role of the vertical fluxes of particulate organic matter and calcite in the oceanic carbon cycle: Studies using an ocean biogeochemical general circulation model. Global Biogeochem Cycles 10: 361–382.

79. Baines SB, Pace ML, Karl DM (1994) Why does the relationship between sinking flux and planktonic primary production differ between lakes and the ocean? Limnol Oceanogr 39: 213–226.

80. Baines SB, Pace ML (1991) The production of dissolved organic matter by phytoplankton and its importance to bacteria: patterns across marine and freshwater systems. Limnol Oceanogr 36: 1078–1090.

81. Moran XAG, Gasol JM, Pedros-Alio C, Estrada M (2001) Dissolved and particulate primary production and bacterial production in offshore Antarctic waters during austral summer: coupled or uncoupled? Mar Ecol Prog Ser 222: 25–30.

82. Moran XAG, Estrada M, Gasol JM, Pedros-Alio C (2002) Dissolved primary production and the strength of phytoplankton-bacterioplankton coupling in constraring marine regions. Microb Ecol 44: 217–223.

83. Del Giorgio PA, Cole JJ, Caraco NF, Peters RH (1999) Linking planktonic biomass and metabolism to net gas fluxes in northern temperate lakes. Ecology 80: 1422–1431.

84. Prairie YT, Bird DF, Cole JJ (2002) The summer metabolic balance in the epilimnion of southeastern Quebec lakes. Limnol Oceanogr 47: 316–321.

85. Algesten G, Sobek S, Bergstrom A-K, Agren A, Tranvik LJ, et al. (2004) Role of lakes for organic carbon cycling in the boreal zone. Global Change Biol 10: 141–147.

86. Hanson PC, Bade DL, Carpenter SR, Kratz TK (2003) Lake metabolism: Relationships with dissolved organic carbon and phosphorus. Limnol Oceanogr 48: 1112–1119.

87. Kosten S, Roland F, Da Motta Marques DML, Van Nes EH, Mazzeo N, et al. (2010) Climate-dependent CO_2 emissions from lakes. Global Biogeochem Cycles 24(GB2007): doi:10.1029/2009GB003618.

88. Balmer MB, Downing JA (2011) Carbon dioxide concentrations in eutrophic lakes: undersaturation implies atmospheric uptake. Inland Waters 1: 125–132.

89. Karim A, Dubois K, Veizer J (2011) Carbon and oxygen dynamics in the Laurentian Great Lakes: Implications for the CO_2 flux from terrestrial aquatic systems to the atmosphere. Chem Geol 281: 133–141.

90. Hope D, Kratz TK, Riera JL (1996) The relationship between pCO_2 and dissolved organic carbon in the surface waters of 27 northern Wisconsin lakes. J Environ Qual 49: 1442–1445.

91. Stets EG, Striegl RG, Aiken GR, Rosenberry DO, Winter TC (2009) Hydrologic support of carbon dioxide flux revealed by whole-lake carbon budgets. J Geophys Res 114(G01008): doi:10.1029/2008JG000783.

92. Maberly SC, Barker PA, Stott AW, De Ville MM (2012) Catchment productivity controls CO_2 emissions from lakes. Nat Clim Chang doi:10.1038/NCLIMATE1748.

93. McDonald CP, Stets EG, Striegl RG, Butman D (2013) Inorganic carbon loading as a primary driver of dissolved carbon dioxide concentrations in the lakes and reservoirs of the contiguous United States, Global Biogeochem Cycles 27: doi:10.1002/gbc.20032.

94. Dubois K, Carignan R, Veizer J (2009) Can pelagic net heterotrophy account for carbon fluxes from eastern Canadian lakes? Appl Geochem 24: 988–998.

95. Humborg C, Mörth C-M, Sundbom M, Borg H, Blenckner T, et al. (2010) CO_2 supersaturation along the aquatic conduit in Swedish watersheds as constrained by terrestrial respiration, aquatic respiration and weathering, Global Change Biol 16: 1966–1978.

96. Spigel RH, Coulter GW (1996) Comparison of hydrology and physical limnology of the East African Great Lakes: Tanganyika, Malawi, Victoria, Kivu and Turkana (with reference to some North American Great Lakes). In: Johnson TC, Odada EO, editors. The limnology, climatology and paleoclimatology of the East African lakes. Boca Raton, FL: Gordon and Breach Publishers. pp. 103-139.

13

Gaseous Elemental Mercury (GEM) Emissions from Snow Surfaces in Northern New York

J. Alexander Maxwell[1], Thomas M. Holsen[2]*, Sumona Mondal[3]

1 Institute for a Sustainable Environment, Clarkson University, Potsdam, New York, United States of America, **2** Department of Civil and Environmental Engineering, Clarkson University, Potsdam, New York, United States of America, **3** Department of Mathematics, Clarkson University, Potsdam, New York, United States of America

Abstract

Snow surface-to-air exchange of gaseous elemental mercury (GEM) was measured using a modified Teflon fluorinated ethylene propylene (FEP) dynamic flux chamber (DFC) in a remote, open site in Potsdam, New York. Sampling was conducted during the winter months of 2011. The inlet and outlet of the DFC were coupled with a Tekran Model 2537A mercury (Hg) vapor analyzer using a Tekran Model 1110 two port synchronized sampler. The surface GEM flux ranged from -4.47 ng m^{-2} hr^{-1} to 9.89 ng m^{-2} hr^{-1}. For most sample periods, daytime GEM flux was strongly correlated with solar radiation. The average nighttime GEM flux was slightly negative and was not well correlated with any of the measured meteorological variables. Preliminary, empirical models were developed to estimate GEM emissions from snow surfaces in northern New York. These models suggest that most, if not all, of the Hg deposited with and to snow is reemitted to the atmosphere.

Editor: Stephen J. Johnson, University of Kansas, United States of America

Funding: New York State Energy Research and Development Authority (NYSERDA http://www.nyserda.ny.gov/) financially supported this research (Charles Driscoll, Syracuse University PI). It has not been subject to the NYSERDA's peer and policy review and, therefore, does not necessarily reflect the views of NYSERDA and no official endorsement should be inferred. The funders had no role in study design, data collection and analysis, decision to publish, or preparation of the manuscript.

Competing Interests: The authors have declared that no competing interests exist.

* E-mail: holsen@clarkson.edu

Introduction

Hg is a potent neurotoxin and regulated by the U.S. EPA [1], European Union Restriction of Hazardous Substances Directive (RoHS) [2], and other government agencies worldwide as a hazardous pollutant. In the form of monomethylmercury (MeHg) it can adversely impact the development and health of both humans and wildlife [3]. Gaseous elemental mercury (GEM) is emitted into the atmosphere from both natural and anthropogenic sources, and has an atmospheric residence time of 0.5–2 years, allowing it to be transported over great distances [4–6]. Anthropogenic sources can also emit Hg in the form of gaseous oxidized Hg (GOM) and particulate bound Hg (PBM), which have shorter atmospheric lifetimes on the order of days to weeks [4]. GOM is fairly soluble in water, thus allowing it to be readily deposited to terrestrial surfaces through wet deposition, including snow [4–6]. The Hg deposited with snow is then either quickly revolatilized back into the atmosphere or incorporated into the snowpack. Newly deposited Hg has been shown to preferentially revolatilize, depending on the deposition surface, in a process known as prompt recycling [7].

While the role of snow surfaces in Hg cycling has been widely studied in arctic regions [5,8–11], much less is known about its importance in more temperate climates [12–14]. Hg is deposited to snowpacks through both wet (snow) and dry deposition. Once deposited on the snowpack surface, it has been shown that >50% of the Hg deposited is reemitted within the first 24 hours [8,12]. This process is believed to be governed by photoinduced reduction of GOM to GEM. Hg in the snowpack is mainly found in the form

of GOM dissolved in snow grains, while <1% remains trapped in the interstitial air as GEM [8]. Hg concentrations are known to decrease with depth [12] with the higher concentrations up to 1.5 ng m^{-3} (GEM) remaining on the surface [8].

In the arctic, the snow surface-to-air flux of Hg is mainly the result of a diurnal pattern of GEM production in the interstitial air near the surface of the snowpack during the daytime ($\sim$15–50 ng m^{-2} hr^{-1}), with little contribution from deeper snow layers [15,16]. However, internal production of GEM increases slightly with higher temperatures and snowmelt [8]. Since this process has not been well studied in temperate climates, measurements of snow surface-to-air fluxes were made over the 2011 winter season in Potsdam, NY.

Materials and Methods

Site Description, Methods, and Materials

Flux measurements were conducted at an open field site located at the Potsdam Municipal Airport (Damon Field) in Potsdam, NY (44°40.41N, −74°57.06′W) near the Clarkson University Observatory. This site remains largely undisturbed throughout the year and has served as a background site for the New York State particulate matter (PM) monitoring network. Sampling periods were determined based on access to the site and snow conditions. Special considerations were made to ensure that the chamber was never buried in snow and that all inlets and outlets remained above the snow during sampling. Measurements were conducted on a concrete slab, isolating the snowpack from the soil surface.

Concentrations of GEM were measured using a DFC with a method previously described in Choi & Holsen (2009). Briefly, the ambient sampling line (inlet) and chamber sampling line (outlet) of the DFC (described below) were coupled with a Tekran Model 2537A Hg vapor analyzer operated at room temperature in a field shed (Tekran Corporation, Inc., Toronto, Ontario, Canada) using a Tekran Model 1110 two-port synchronized sampler. The Tekran 1110 unit allowed for alternating five minute sampling pairs to be made between the inlet and outlet sample lines every 20 minutes (trap A inlet, trap B inlet; trap A outlet, trap B outlet). During inlet sampling, outlet air is bypassed at the same 1 L min^{-1} flow rate as the Tekran Model 2537A to maintain a constant turnover time (TOT) of 0.78 minutes and an optimized flushing flow rate (FFR) of 5 L min^{-1}[17] through the flux chamber. The inlet and outlet openings were placed next to each other at the same height, roughly 2 cm above the snow surface. Four, 1 cm diameter holes were evenly distributed around the perimeter of the chamber wall to insure the chamber was well-mixed. Although a standard method for the use of DFCs does not exist and this method has not been used in other snow studies, this sampling approach is similar to methods used in past studies over soil surfaces [17–19]. The 5 L min^{-1} FFR and 0.78 minute TOT are also similar to those used in a study by Eckley, et al. (2010).

Modified Teflon fluorinated ethylene propylene (FEP) chambers were used in the study. The modified Teflon chamber was constructed using a polycarbonate (PC) chamber frame and thin, 25 μm thick Teflon FEP film (CS Hyde Company, Lake Villa, IL) to cover the top and side windows (Figure 1). In previous studies [20,21], Teflon film was shown to allow better UV permeability, up to 85±11% of light for wavelengths between 260 and 970 nm.

Each DFC had a chamber volume of 3.9 L with a 18 cm diameter opening covering an area of approximately 254 cm^2 of the snow surface.

Manual spike Hg recovery tests were conducted at the start of each sampling period by injecting 20 μL of Hg at roughly 13.23 pg μL^{-1} (20°C) or approximately 0.26 ng into an operating chamber using a calibrated (ANSI/NCSL Z540-1-1994) Hamilton Digital Syringe (Hamilton Company, Reno, NV) and a Tekran Model 2505 Hg vapor calibration unit (Tekran Corporation, Inc., Toronto, Ontario, Canada). The recorded Hg concentrations after each manual spike test were roughly 9 ng m^{-3}, on the same order as the average daytime Hg concentrations around 2 ng m^{-3}. The recovery was 97.5±3.8%. Flow rates were calibrated using a Bios Definer 220 volumetric flow meter (Bios International Corporation, Butler, NJ) at the beginning of each sampling period.

Prior to all field measurements, the Tekran Model 2537A was calibrated with an internal permeation source to ensure acceptable response factors (>6,000,000) and that the concentration difference between the inlet and outlet samples was less than 5%. In addition, all soda-lime traps and 0.2 μm polytetrafluoroethylene (PTFE) membrane filters were replaced at the start of each sampling period.

Meteorological data was collected using a weather station (Vantage Pro 2 Weather Station, Davis Instruments, Hayward, CA) located 1–2 m away from the chamber. The weather station measured ambient air temperature (°C), relative humidity (%), and solar radiation (W m^{-2}) at a 10 minute time resolution.

Sampling Analysis and Calculations

The GEM flux from the snow under the chamber was calculated using the following mass balance equation:

$$F = (C_{outlet} - C_{inlet}) \times (Q/A) \qquad (1)$$

Where F is flux (ng GEM m^{-2} h^{-1}); C_{outlet} and C_{inlet} are the concentrations of GEM (ng GEM m^{-3}) at the outlet and inlet, respectively; Q is the FFR (m^3 h^{-1}) through the chamber; and A is the surface area (m^2) of the snow exposed in the chamber. When fluxes were negative (-), Hg was being deposited on the snow surface, and when fluxes were positive (+) Hg was being emitted from the snow surface. All flux data was then smoothed using a Savitzky-Golay smoothing filter [22], (Eqn 2), to account for random error/noise while also preserving the quantitative information and trends.

$$F_4^* = [(-2 \times F_1) + (3 \times F_2) + (6 \times F_3) + (7 \times F_4) + \\ (6 \times F_5) + (3 \times F_6) + (-2 \times F_7)]/21 \qquad (2)$$

Where F_4^* is the smoothed flux (ng GEM m^{-2} h^{-1}), F_{1-7} are the range of measured abscissa flux values (ng GEM m^{-2} h^{-1}), and 21 is the normalizing factor.

Histograms of the GEM flux and the three meteorological predictor variables (temperature, solar radiation, and relative humidity) showed that none of the variables were normally distributed. Figure 2 provides histograms and residual plots of daytime GEM flux when compared to solar radiation. Similar plots were constructed for each individual variable, temperature, solar radiation, and relative humidity. Shapiro-Wilk normality tests [23] were then employed to confirm that the data deviated from normality. Non-parametric Pearson product-moment tests

Figure 1. Modified Teflon Fluorinated Ethylene Propylene (FEP) Chamber with Polycarbonate (PC) Frame and 25 μm Teflon FEP Film Top and Side Windows.

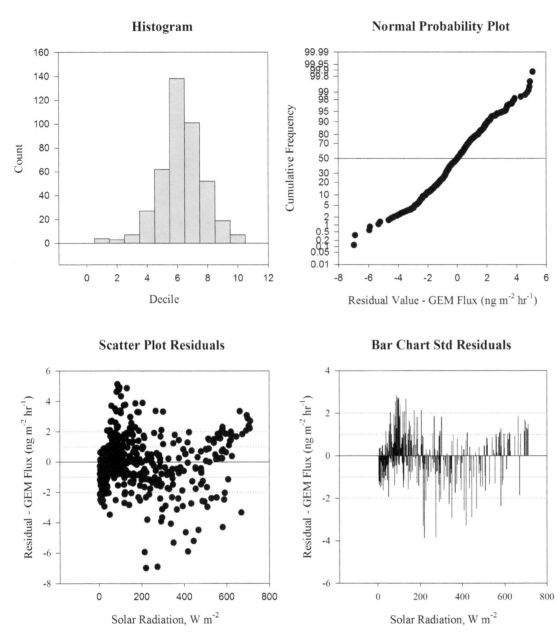

Figure 2. Histograms and Residual Plots of Daytime GEM Flux and Solar Radiation During Winter 2011.

Table 1. Measured Daytime And Nighttime GEM Flux Over 5 Winter 2011 Sampling Periods.

Date	Diurnal Period	GEM Flux (ng m^{-2} hr^{-1})					
		Mean	Std. Dev.	Range	Max	Min	Median
21–24 Jan	Daytime	1.13	1.37	5.60	4.44	−1.16	0.94
	Nighttime	−0.44	0.29	1.47	0.14	−1.33	−0.43
26–31 Jan	Daytime	2.65	1.92	8.20	6.57	−1.63	2.69
	Nighttime	−0.21	0.42	1.95	0.69	−1.26	−0.22
15–18 Feb	Daytime	1.88	2.96	12.56	8.09	−4.47	1.30
	Nighttime	−0.57	0.45	2.10	0.39	−1.71	−0.50
23–24 Feb	Daytime	3.03	2.60	9.38	8.13	−1.25	2.46
	Nighttime	−0.29	0.23	1.00	0.25	−0.75	−0.30
08–09 Mar	Daytime	3.50	3.08	9.97	9.89	−0.08	2.29
	Nighttime	−0.08	0.22	0.87	0.27	−0.60	−0.10
Overall	Daytime	2.37	2.48	14.36	9.89	−4.47	1.89
	Nighttime	−0.35	0.41	2.41	0.69	−1.71	−0.33

Table 2. Pearson Product-Moment Correlation Coefficients and P-Values For Correlations Between GEM Flux and Temperature, Relative Humidity, and Solar Radiation.

Date	Diurnal Period	Temperature (°C)		Relative Humidity (%)		Solar Radiation (W m^{-2})	
		Coefficient	P Value	Coefficient	P Value	Coefficient	P Value
21–24 Jan	Daytime	0.562	0.000	−0.494	0.000	0.546	0.000
	Nighttime	−0.004	0.959	−0.192	0.026	−	−
26–31 Jan	Daytime	0.189	0.022	−0.046	0.585	0.304	0.000
	Nighttime	−0.183	0.009	−0.319	0.000	−	−
15–18 Feb	Daytime	−0.518	0.000	0.673	0.000	0.820	0.000
	Nighttime	−0.553	0.000	−0.600	0.000	−	−
23–24 Feb	Daytime	0.300	0.027	−0.629	0.000	0.875	0.000
	Nighttime	0.000	0.997	−0.053	0.745	−	−
08–09 Mar	Daytime	0.446	0.001	−0.787	0.000	0.942	0.000
	Nighttime	−0.518	0.000	0.251	0.129	−	−
Overall	Daytime	0.103	0.035	−0.385	0.000	0.684	0.000
	Nighttime	−0.222	0.000	−0.132	0.002	−	−

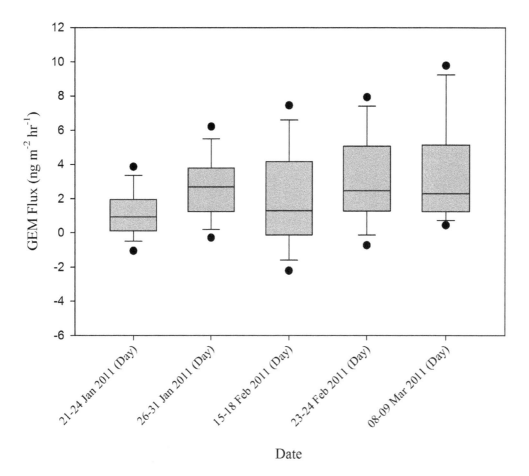

Figure 3. Average Daytime GEM Flux Measurements Made For Each Sampling Conducted During Winter 2011.

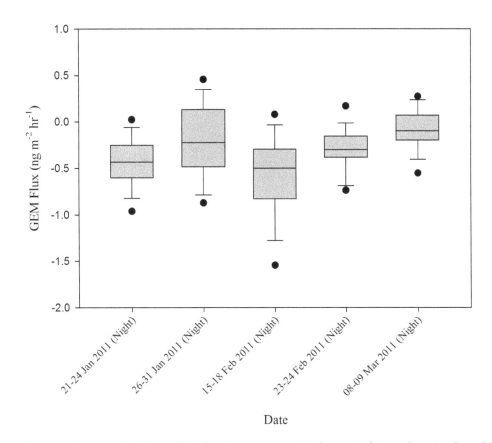

Figure 4. Average Nighttime GEM Flux Measurements Made For Each Sampling Conducted During Winter 2011.

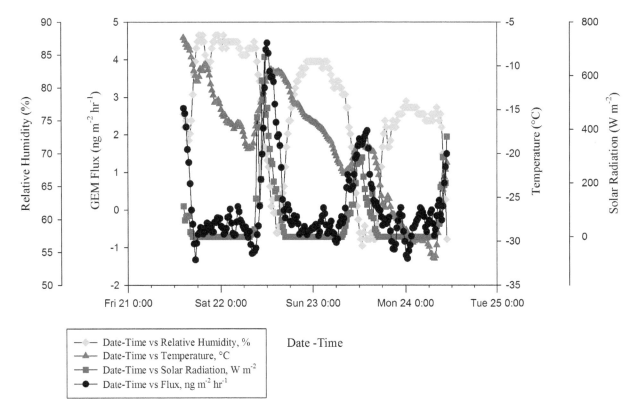

Figure 5. Diurnal Pattern Of GEM Flux For 21–24 January 2011 Sampling With Temperature, Relative Humidity, And Solar Radiation.

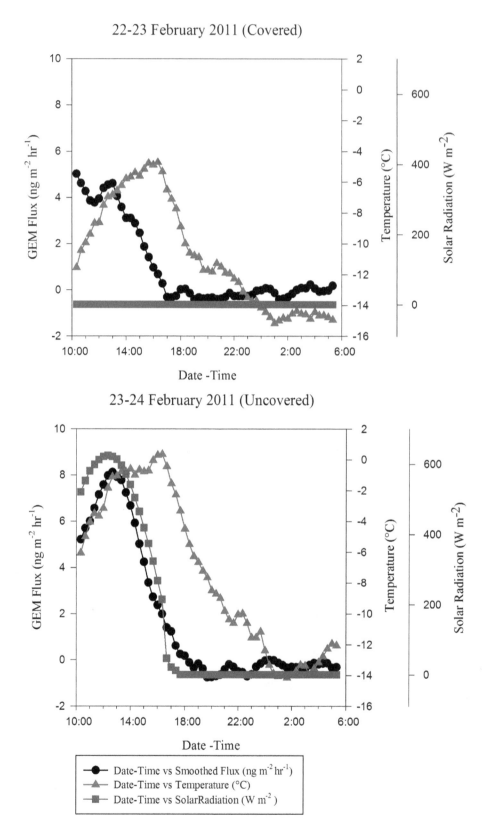

Figure 6. GEM Fluxes Measured Using Covered And Uncovered Chambers To Determine The Impact Of Solar Radiation On GEM Flux.

Table 3. PPMCs For Covered And Uncovered Chamber Tests For Impact Of Solar Radiation On GEM Flux.

Date	Diurnal Period	Temperature (°C)		Relative Humidity (%)		Solar Radiation (W m^{-2})	
		Coefficient	P Value	Coefficient	P Value	Coefficient	P Value
22–23 Feb (Covered)	Daytime	−0.624	0.002	0.489	0.021	−	−
	Nighttime	−0.374	0.025	0.141	0.412	−	−
23–24 Feb (Uncovered)	Daytime	0.300	0.027	−0.629	0.000	0.875	0.000
	Nighttime	0.000	0.997	−0.053	0.745	−	−

were used to determine the correlation coefficients (PPMC) between the variables [24].

Results and Discussion

Flux Measurements

During the 2011 winter sampling season, the flux was measured over five sampling periods, each lasting from one to six days. The measured flux ranged from a minimum −4.47 to a maximum 9.89 ng GEM m^{-2} h^{-1} (Table 1). The average daytime flux was 2.37±2.48 ng GEM m^{-2} h^{-1} and the average nighttime flux was −0.35±0.41 ng GEM m^{-2} h^{-1}. Measured nighttime Hg emission fluxes from other snowpack studies have been ≈0 ng GEM m^{-2} h^{-1} [8], while daytime fluxes have been shown to be much higher, ≈30–50 ng GEM m^{-2} h^{-1} [8]. Daytime fluxes were strongly correlated with solar radiation (PPMC value = 0.684, p-value = 0.000<0.050) and to a lesser extent temperature and relative humidity (PPMC value = 0.103, p-value = 0.035<0.050 and PPMC value = −0.385, p-value = 0.000<0.050 respectively) (Table 2). This strong correlation with solar radiation suggests that the daytime Hg emissions from the snow surface are a result of the photoreduction of GOM associated with the snow to GEM. Similar results have been reported in Ferrari, et al. (2005), where it was also reported that GEM emissions from the snowpack were negligible in comparison to emissions caused by solar irradiation at the surface. Nighttime fluxes were only weakly correlated with both temperature (PPMC value = −0.222, p-value = 0.000<0.050) and relative humidity (PPMC value = −0.132, p-value = 0.002<0.000) (Table 2) and showed a statistically significant difference from zero.

Overall, peak fluxes tended to increase later in the sampling season (Figure 3 and Figure 4). Emissions were highest during the last sampling period, 08–09 March, corresponding with the highest solar radiation peak (Max: 712 W m^{-2}). Fluxes also tended to follow a diurnal pattern (Figure 5) with peaks occurring during the day following increased exposure to solar radiation, and deposition occurring at night, similar to patterns reported in other literature [15,16].

Impact of Solar Radiation

To test the impact of solar radiation on GEM fluxes, the chamber was covered with aluminum foil to simulate zero UV conditions. The uncovered measurements were made on 23–24 February 2011, while the covered measurements were made on 22–23 February 2011 (Figure 6). During the uncovered and covered tests, the average GEM fluxes were 1.76±3.06 and 0.99±1.81 ng GEM m^{-2} h^{-1} respectively. The covered DFC daytime measurements were negatively correlated with temperature (PPMC coefficient = −0.624, p-value = 0.000<0.050) (Table 3). The slow decline in GEM flux after covering the

chamber is likely a result of diffusion of GEM from the interstitial air in the snowpack into the DFC. The uncovered DFC daytime measurements were positively correlated with solar radiation, and to a lesser degree, temperature (PPMC coefficient = 0.875, p-value = 0.000<0.050 and PPMC coefficient = 0.300, p-value = 0.027<0.050) (Table 3), similar to what has been reported in other arctic studies [8]. Overall, solar radiation had the highest positive impact on GEM emissions, and though temperature and relative humidity were correlated to GEM flux, their correlation with solar radiation (PPMC coefficient = 0.711 & −0.686 respectively, p-value = 0.000<0.050) indicate that their influence was likely a result of their codependence on solar radiation.

Modeling

In the past, empirical models have been developed using meteorological data in order to estimate surface GEM flux from soils in temperate regions of eastern North America [17,21,25]. However, no model exists to estimate GEM flux from snow in the temperate climate of northern New York. Previous models for this region [17] excluded winter fluxes from snow surfaces. In order to better model GEM flux throughout the winter season, two multiple linear regression models were developed based on aggregated seasonal flux data:

Winter 2011 (Daytime): ($\text{R}^2 = 0.481$)

$$F = 0.722 + 0.0358(T) + 0.00906(SR)$$

Winter 2011 (Nighttime): ($\text{R}^2 = 0.0616$)

$$F = -0.167 - 0.00939(T) - 0.00344(RH)$$

where F is GEM flux in ng m^{-2}hr^{-1}, T is ambient temperature in °C, RH is relative humidity in %, and SR is solar radiation in W m^{-2}. Fluxes predicted by this model for the 22–24 January sampling period are shown in Figure 7.

Several nonlinear polynomial and power equation fits and variable transformations were conducted using SigmaPlot, ver. 12 in order to develop a more precise correlative model structure. However, the dynamic fits showed little improvement.

Using the multiple linear regression models in conjunction with 5-year winter (December-March, 2005–2010) EPA Clean Air Status and Trends Network (CASTNET) meteorological data from the National Atmospheric Deposition Program (NADP) site, NY20, it is estimated that the average snow surface emissions from the open Huntington Wildlife Forest (HWF) site range from −0.10±0.07 ng m^{-2} hr^{-1} (nighttime) to 1.53±1.69 ng m^{-2} hr^{-1} (daytime) or ~17.22 ng m^{-2} year^{-1}. During the same time period Mercury Deposition Network (MDN) data from the same site yield

21-24 Jan 2011 Daytime GEM Flux Model Comparison

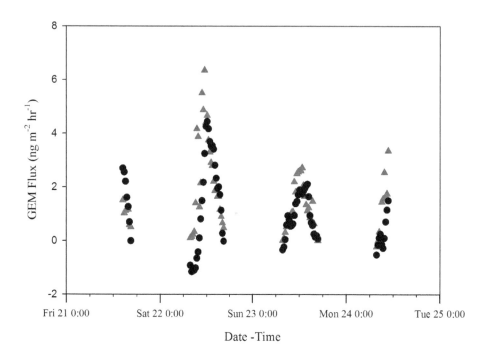

21-24 Jan 2011 Nighttime GEM Flux Model Comparison

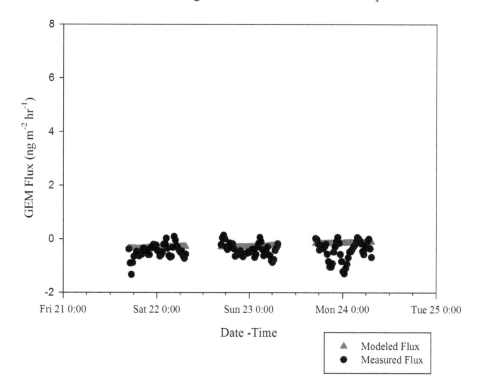

Figure 7. Daytime And Nighttime GEM Flux Model Comparison For 21–24 January Sampling Period.

a similar value with an average deposition flux of 0.48±0.41 ng m^{-2} hr^{-1} or 11.52±9.84 ng m^{-2} $year^{-1}$. The reason for the slightly higher modeled flux compared to the measured flux is likely due to the fact that some of the measurements used to make the empirical model were made after fresh snowfall when GEM fluxes would be at their maximum values.

Overall, these models suggest that most if not all the Hg deposited to snow surfaces is promptly recycled. Similar reemission

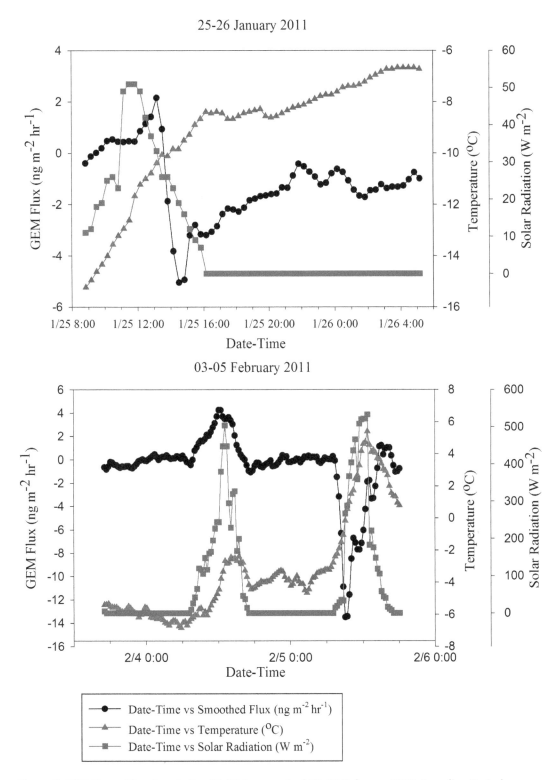

Figure 8. GEM Deposition Events For 25–26 January And 03–05 February 2011 Sampling Period.

phenomena have been reported by other research groups [8,12] with mean emission fluxes of 2–5 ng m^{-2} hr^{-1}, zero change in surface snow Hg concentration after deposition events, and up to 54% GEM reemission during the first 24 hours after a snowfall.

Deposition Events

Two unique deposition events with fluxes as high as -14 ng m^{-2} hr^{-1} occurred during separate sampling periods, one on 25 January and one on 05 February. Both of these event followed snowfalls ≥3 cm (Table 4) and melting also occurred during the 03–05 February sampling (Figure 8). During both of these events,

Table 4. Field Observations Made During Various Measurement Periods Throughout the 2011 Winter Sampling Season.

Date	Observation
21–24 Jan.	No recent snow
26–27 Jan.	Fresh snow (dusting, <2.5 cm) prior to sampling
27–30 Jan.	Fresh snow (dusting, <2.5 cm) and intermittent light snowing throughout sampling (no enough to cover inlets of chamber)
30–31 Jan.	No recent snow
15–18 Feb.	No recent snow
23–24 Feb.	No recent snow
08–09 Mar.	Fresh snow ($\approx$7.5 cm); during end of season melt

fluxes were negatively correlated with temperature (PPMC coefficient $= -0.421$ and -0.439 respectively, p-value $= 0.000 < 0.050$) and displayed patterns opposite to the diurnal patterns typically seen. This sudden deposition event is similar to atmospheric Hg depletion events (AMDEs) witnessed in arctic regions during polar sunrise [9–11].

During an AMDE, rapid oxidation of GEM forms GOM that is subsequently deposited to the snow surface. Arctic AMDEs are springtime phenomenon that occur as a result of reactions with ozone and other halogen compounds, especially bromine oxides [26]. Though the cause of the two deposition events seen in Potsdam is unclear, they could coincide with sudden increases in atmospheric oxidant concentrations including free halogens.

Conclusions

Snow surface-to-air exchange of gaseous elemental Hg (GEM) was measured using a modified Teflon fluorinated ethylene propylene (FEP) dynamic flux chamber (DFC) in a remote, open site in Potsdam, New York during the winter months of 2011. The surface GEM flux ranged from -4.47 ng m^{-2} hr^{-1} to 9.89 ng m^{-2} hr^{-1}. For most sample periods, the daytime GEM flux was

strongly correlated with solar radiation. The average nighttime GEM flux was slightly negative and was weakly correlated with all of the measured meteorological variables. Overall, preliminary models indicate that most if not all the Hg being deposited to snow surfaces is being reemitted back into the atmosphere. Two unique deposition events with fluxes as high as -14 ng m^{-2} hr^{-1} occurred during separate sampling periods following snowfalls $\geq$3 cm. During both of these events, fluxes were negatively correlated (PPMC coefficient $= -0.421$ and -0.439 respectively, p-value $= 0.000 < 0.050$) with temperature and displayed patterns opposite to the diurnal patterns typically seen.

Acknowledgments

A special thank you to Dr. Andrea Ferro, Dr. Stephan Grimberg, and Dr. Jiaoyan Huang for their help in reviewing this work, and James Laing for assistance with sampling.

Author Contributions

Conceived and designed the experiments: JAM TMH. Performed the experiments: JAM. Analyzed the data: JAM SM. Contributed reagents/materials/analysis tools: JAM TMH SM. Wrote the paper: JAM.

References

1. EPA (2013) Laws and Regulations | Mercury | US EPA. Available: http://www.epa.gov/hg/regs.htm. Accessed 14 December 2012.
2. EU (2011) DIRECTIVE 2011/65/EU of the European Parliament and of the Council of 8 June 2011 on the restriction of the use of certain hazardous substances in electrical and electronic equipment (recast): 88–110. Available: http://eur-lex.europa.eu/LexUriServ/LexUriServ.do?uri = OJ:L:2011:174:0088:0110:EN:PDF.
3. Mergler D, Anderson HA, Chan LHM, Mahaffey KR, Murray M, et al. (2007) Methylmercury Exposure and Health Effects in Humans: A Worldwide Concern. AMBIO: A Journal of the Human Environment 36: 3–11. Available: http://dx.doi.org/10.1579/0044-7447(2007)36[3:MEAHEI]2.0.CO;2. Accessed 14 December 2012.
4. Lin C-J, Pehkonen SO (1999) The chemistry of atmospheric mercury: a review. Atmospheric Environment 33: 2067–2079. Available: http://dx.doi.org/10.1016/S1352-2310(98)00387-2. Accessed 14 December 2012.
5. Schroeder WH, Munthe J (1998) Atmospheric mercury–An overview. Atmospheric Environment 32: 809–822. Available: http://dx.doi.org/10.1016/S1352-2310(97)00293-8. Accessed 14 December 2012.
6. Lindqvist O, Rodhe H (1985) Atmospheric mercury-a review. Tellus B 37B: 136–159. Available: http://www.tellusb.net/index.php/tellusb/article/view/15010. Accessed 14 December 2012.
7. Selin NE, Jacob DJ, Yantosca RM, Strode S, Jaegle L, et al. (2008) Global 3-D Land-Ocean-Atmosphere Model for Mercury: Present- Day Versus Preindustrial Cycles and Anthropogenic Enrichment Factors for Deposition. Global Biogeochemical Cycles 22: 13.
8. Ferrari C, Gauchard P, Aspmo K, Dommergue A, Magand O, et al. (2005) Snow-to-air exchanges of mercury in an Arctic seasonal snow pack in Ny-Ålesund, Svalbard. Atmospheric Environment 39: 7633–7645. Available: http://dx.doi.org/10.1016/j.atmosenv.2005.06.058. Accessed 14 December 2012.

9. Ariya PA, Dastoor AP, Amyot M, Schroeder WH, Barrie L, et al. (2004) The Arctic: a sink for mercury. Tellus B 56: 397–403. Available: http://www.tellusb.net/index.php/tellusb/article/view/16458. Accessed 14 December 2012.
10. Skov H, Christensen JH, Goodsite ME, Heidam NZ, Jensen B, et al. (2004) Fate of Elemental Mercury in the Arctic during Atmospheric Mercury Depletion Episodes and the Load of Atmospheric Mercury to the Arctic. Environmental Science & Technology 38: 2373–2382. Available: http://dx.doi.org/10.1021/es030080h. Accessed 14 December 2012.
11. Lindberg SE, Brooks S, Lin C-J, Scott KJ, Landis MS, et al. (2002) Dynamic Oxidation of Gaseous Mercury in the Arctic Troposphere at Polar Sunrise. Environmental Science & Technology 36: 1245–1256. Available: http://dx.doi.org/10.1021/es0111941. Accessed 14 December 2012.
12. Lalonde JD, Poulain AJ, Amyot M (2002) The Role of Mercury Redox Reactions in Snow on Snow-to-Air Mercury Transfer. Environmental Science & Technology 36: 174–178. Available: http://dx.doi.org/10.1021/es010786g. Accessed 14 December 2012.
13. Lalonde JD, Amyot M, Doyon M-R, Auclair J-C (2003) Photo-induced Hg(II) reduction in snow from the remote and temperate Experimental Lakes Area (Ontario, Canada). Journal of Geophysical Research: Atmospheres (1984–2012) 108: 4200.
14. Nelson SJ, Fernandez IJ, Kahl JS (2010) A review of mercury concentration and deposition in snow in eastern temperate North America. Hydrological Processes 24: 1971–1980.
15. Faïn X, Grangeon S, Bahlmann E, Fritsche J, Obrist D, et al. (2007) Diurnal production of gaseous mercury in the alpine snowpack before snowmelt. Journal of Geophysical Research 112: D21311. Available: http://www.agu.org/pubs/crossref/2007/2007JD008520.shtml. Accessed 14 December 2012.
16. Dommergue A, Ferrari CP, Poissant L, Gauchard P-A, Boutron CF (2003) Diurnal Cycles of Gaseous Mercury within the Snowpack at Kuujjuarapik/Whapmagoostui, Québec, Canada. Environmental Science & Technology 37:

3289–3297. Available: http://dx.doi.org/10.1021/es026242b. Accessed 14 December 2012.

17. Choi H-D, Holsen TM (2009) Gaseous mercury fluxes from the forest floor of the Adirondacks. Environmental pollution (Barking, Essex□: 1987) 157: 592–600. Available: http://dx.doi.org/10.1016/j.envpol.2008.08.020. Accessed 14 December 2012.

18. Zhang H, Lindberg SE, Barnett MO, Vette AF, Gustin MS (2002) Dynamic flux chamber measurement of gaseous mercury emission fluxes over soils. Part 1: simulation of gaseous mercury emissions from soils using a two-resistance exchange interface model. Atmospheric Environment 36: 835–846. Available: http://dx.doi.org/10.1016/S1352-2310(01)00501-5. Accessed 14 December 2012.

19. Lindberg SE, Zhang H, Vette AF, Gustin MS, Barnett MO, et al. (2002) Dynamic flux chamber measurement of gaseous mercury emission fluxes over soils: Part 2–effect of flushing flow rate and verification of a two-resistance exchange interface simulation model. Atmospheric Environment 36: 847–859. Available: http://dx.doi.org/10.1016/S1352-2310(01)00502-7. Accessed 14 December 2012.

20. Eckley CS, Gustin M, Lin C-J, Li X, Miller MB (2010) The influence of dynamic chamber design and operating parameters on calculated surface-to-air mercury fluxes. Atmospheric Environment 44: 194–203. Available: http://dx.doi.org/10.1016/j.atmosenv.2009.10.013. Accessed 14 December 2012.

21. Carpi A, Frei A, Cocris D, McCloskey R, Contreras E, et al. (2007) Analytical artifacts produced by a polycarbonate chamber compared to a Teflon chamber for measuring surface mercury fluxes. Analytical and bioanalytical chemistry 388: 361–365. Available: http://www.ncbi.nlm.nih.gov/pubmed/17260134. Accessed 14 December 2012.

22. Savitzky A, Golay MJE (1964) Smoothing and Differentiation of Data by Simplified Least Squares Procedures. Analytical Chemistry 36: 1627–1639. Available: http://dx.doi.org/10.1021/ac60214a047. Accessed 31 October 2012.

23. Shapiro SS, Wilk MB (1965) An Analysis of Variance Test for Normality (Complete Samples). Biometrika 52: 591–611. Available: http://www.jstor.org/stable/10.2307/2333709.

24. Pearson K (1895) No Title. Royal Society Proceedings. 241.

25. Gbor P, Wen D, Meng F, Yang F, Zhang B, et al. (2006) Improved model for mercury emission, transport and deposition. Atmospheric Environment 40: 973–983. Available: http://dx.doi.org/10.1016/j.atmosenv.2005.10.040. Accessed 14 December 2012.

26. Steffen A, Douglas T, Amyot M, Ariya P, Aspmo K, et al. (2008) A synthesis of atmospheric mercury depletion event chemistry in the atmosphere and snow. Atmospheric Chemistry and Physics 8.

14

Large Carbon Dioxide Fluxes from Headwater Boreal and Sub-Boreal Streams

Jason J. Venkiteswaran[1,2]*, Sherry L. Schiff[1], Marcus B. Wallin[3]

1 Department of Earth and Environmental Sciences, University of Waterloo, Waterloo, Ontario, Canada, **2** Department of Geography and Environmental Studies, Wilfrid Laurier University, Waterloo, Ontario, Canada, **3** Department of Ecology and Genetics/Limnology, Uppsala University, Uppsala, Sweden

Abstract

Half of the world's forest is in boreal and sub-boreal ecozones, containing large carbon stores and fluxes. Carbon lost from headwater streams in these forests is underestimated. We apply a simple stable carbon isotope idea for quantifying the CO_2 loss from these small streams; it is based only on in-stream samples and integrates over a significant distance upstream. We demonstrate that conventional methods of determining CO_2 loss from streams necessarily underestimate the CO_2 loss with results from two catchments. Dissolved carbon export from headwater catchments is similar to CO_2 loss from stream surfaces. Most of the CO_2 originating in high CO_2 groundwaters has been lost before typical in-stream sampling occurs. In the Harp Lake catchment in Canada, headwater streams account for 10% of catchment net CO_2 uptake. In the Krycklan catchment in Sweden, this more than doubles the CO_2 loss from the catchment. Thus, even when corrected for aquatic CO_2 loss measured by conventional methods, boreal and sub-boreal forest carbon budgets currently overestimate carbon sequestration on the landscape.

Editor: Ben Bond-Lamberty, DOE Pacific Northwest National Laboratory, United States of America

Funding: A Natural Sciences and Engineering Research Council grant and Swedish Research Council grant funded this research. The funders had no role in study design, data collection and analysis, decision to publish, or preparation of the manuscript.

Competing Interests: The authors have declared that no competing interests exist.

* Email: jvenkiteswaran@wlu.ca

Introduction

Boreal and sub-boreal ecozones are large (about 17 million km^2) and about half the world's forest cover [1]. Forests in these zones contain large carbon stores and contribute significant fluxes in the global carbon budget [2]. Part of the carbon fixed from the atmosphere by forests is returned to the atmosphere via aquatic surfaces [3–6]. In the past, this flux has been ignored in the construction of carbon budgets for forested watersheds and is now predicted to be higher than expected [7]. In Sweden, annual estimates of carbon sequestration in forests are 10% lower if aquatic carbon losses are included [8]; similar amounts are expected for Norway [9]. In the USA, stream and river CO_2 loss may be five times greater than previously thought [10]. The difficulty in accounting for the aquatic loss lies in inadequate information on the enormous number of small headwater streams with high CO_2 concentrations. In boreal Sweden, 90% of total stream length is in catchments less than 15 km^2 [11].

Neglecting the aquatic export of terrestrially fixed carbon (via dissolved inorganic carbon (DIC), CH_4, particulate organic carbon (POC), and dissolved organic carbon (DOC)) in landscape carbon budgets results in overestimating net ecosystem exchange by 25% to 70% [12–14]. The active pipe model for inland waters [15] increased estimates of total terrestrial export by two-fold while concluding that CO_2 loss to the atmosphere was necessarily underestimated because small streams were entirely excluded for lack of emission and distribution data [3–7,10]. Subsequently, 'in-stream heterotrophy' was added to the stream portion of the budget [5] but other DIC sources, e.g., DIC-rich groundwaters, are poorly known and therefore excluded. The knowledge gap

between the significant contribution of headwater streams in a basin and the small amount of data about them has been termed *aqua incognita* [11].

All streams in an extensively studied 4th order Swedish boreal catchment were supersaturated in CO_2 [16]. There, CO_2 loss from stream surfaces directly to the atmosphere was 30%–50% of the dissolved carbon (DIC+DOC) exported from the forest (2.9 $gC/m^2/yr$ of 9.8 $gC/m^2/yr$) [17,18]. In an Alaskan headwater stream network, greatest variability and mean fluxes of CO_2 were in the first order streams [19] with 9.0 $gC/m^2/yr$ lost from the entire Yukon River basin [20].

In sub-boreal catchments in Canada [21] and a peat catchment in northeast Scotland [13], CO_2 loss from stream surfaces was 36% and 34% of the dissolved carbon exported from the catchments. These studies noted that gas loss from stream surfaces to the atmosphere is significant, but all neglected the higher CO_2 evasion rates from stream surfaces upstream of their sampling locations. Without good estimates of CO_2 fluxes, ecosystem-scale metabolism calculations also contain this uncertainty and bias. DIC in headwater streams is a net result of a number of processes, including: dissolution of carbonate and weathering of some minerals in soils and bedrock, in-stream biotic respiration and fixation, exchange with atmospheric CO_2, and shallow groundwater input [17,22–25]. Although it is commonly assumed that CO_2 saturation is controlled by mineral weathering and in-stream biotic activity in carbonate systems, boreal forests are, in many cases, characterized by high-DIC shallow groundwaters derived from terrestrially respired organic carbon [8,24].

Here, we focus on silicate bedrock catchments because they constitute 45% of boreal catchments [1] and we avoid the complications of the contribution of significant amounts of DIC from carbonate [26]. Silicate-dominated boreal landscapes contribute significantly to global carbon fluxes [27]. Headwater streams in these catchments have low to moderate pH values and moderate to high concentrations of CO_2 that is modern in radiocarbon terms [28,29]. Thus, DIC from shallow groundwater is directly related to soil and root respiration instead of long-term mineral weathering.

In undisturbed, shaded, nutrient-poor, silicate-bedrock, boreal headwater streams, shallow groundwater is the major source of DIC since in-stream respiration is low relative to gas exchange with the atmosphere [30–33]. Furthermore, in silicate terrain, shallow groundwater retains the $\delta^{13}C$ value of the DIC input from DOC and soil OC decomposition and exhibits modern radiocarbon ages [28,29] indicating weathering is unimportant at this scale. Thus for small streams of up to a few hundred metres length in this landscape, soil and root respiration is the key $\delta^{13}C$-DIC source.

Degassing of CO_2 from these streams causes the $\delta^{13}C$-DIC to increase due to well known equilibrium isotopic fractionation between CO_2, HCO_3^-, and CO_3^{2-} and during CO_2 loss to the atmosphere — since the $\delta^{13}C$-CO_2 value is less than the bulk $\delta^{13}C$-DIC value, the loss of CO_2 will cause the remaining $\delta^{13}C$-DIC value to increase [34,35]. Additionally, since CO_2 loss does not affect carbonate alkalinity, there is a concomitant increase in pH as degassing occurs along the stream [36]. This CO_2 loss also includes CO_2 lost as groundwater transits the riparian zone adjacent to the stream since these DIC and $\delta^{13}C$-DIC changes are indistinguishable from those in the stream proper. Recent work presented a moderately complex modelled formulation of the degassing effects on $\delta^{13}C$-DIC and focused on several French streams [35].

At a much broader level, we can use the characteristic degassing trajectories of decreasing DIC vs increasing $\delta^{13}C$-DIC to estimate (1) how far along the degassing curve a particular sampling site in a stream is and (2) how much CO_2 loss there must have been upstream of the sampling site. This gives us the opportunity to use discrete samples from a stream to estimate the upstream CO_2 loss without direct knowledge of the DIC concentration of the groundwater end-member. Since groundwater $\delta^{13}C$-DIC can be easily constrained based on C3 plants and soil organic matter formation [22,28,37] in silicate-dominated catchments, the average groundwater DIC concentration can then be determined as a result of the modelling process. Discrete samples are available from the mouths of streams where whole-lake mass balances have been completed [38], along stream networks [16,39], and potentially from archived samples and data sets [40]. Because only in-stream samples are needed, this technique makes it possible to cover large catchment areas. Logistically, headwater stream CO_2 fluxes can be rapidly assessed without the large investment required to study only one catchment in detail.

Given the (1) poor knowledge of the extent of headwater stream surface area (lengths and widths [11,24], but see recent methods using digital elevation models [18,20] and branching theory [41]) and (2) narrow constraint on $\delta^{13}C$-DIC in groundwater, we suggest these small streams can be studied with archived and new samples. We place the results in a context of landscape-scale degassing trajectories where $\delta^{13}C$-DIC of groundwater, stream, and small lakes all differ. We focus on improving CO_2 loss estimates from headwater stream surfaces in two catchments: (1)

Harp Lake, Canada where stream CO_2 loss is situated in a catchment C budget and (2) Krycklan, Sweden where stream CO_2 losses are higher than previously estimated.

Materials and Methods

Site description

Two field sites of contrasting vegetation and location were selected: Harp Lake catchment, Canada ($45°22'$ N, $79°08'$ W) and Krycklan catchment, Sweden ($64°14'$ N, $19°50'$ E). Three headwater streams in the Harp Lake catchment were studied. All are underlain by the gneissic bedrock of the Canadian Shield [42] with mixed forests of sugar maple (*Acer saccharum*), beech (*Fagus grandifolia*), yellow birch (*Betula alleghaniensis*), white pine (*Pinus strobus*), and aspen (*Populous tremuloides*). Harp Lake catchment receives 1033 mm/yr precipitation. Mean annual air temperature is 5°C. Stream catchments ranged from 3.7 km^2 to 21.7 km^2 with stream lengths from 170 m to 760 m. All streams were first or second order and catchments had varying amount of wetland cover (<0.3% to 8.5%) [43,44].

Stream samples from around Harp Lake were collected from July 1989 to November 1991 across various flow conditions near the mouth of each stream ($n = 111$). DIC samples were collected in Pyrex culture tubes with polycone caps and analysed colorimetrically as per standard methods in the Ontario Ministry of the Environment lab at Dorset, Ontario, Canada. Samples for $\delta^{13}C$-DIC analysis were collected in 500 mL glass bottles with polycone caps and analysed by standard off-line dual-inlet IRMS techniques. Groundwater was sampled from piezometers near the middle of the Harp 4–21 catchment ($n = 8$) [28,29]. An automated titrator (PC-Titrate Man-Tech Associates) was used to measure pH. Stream temperature was measured in the field. Stream DOC concentrations varied with stream flow with a median concentration of 5.1 mgC/L (range of 1.1 mgC/L and 17.2 mgC/L) and flow-weighted annual average around 10 mgC/L [45,46]. Harp Lake epilimnion was sampled from March 1990 to November 1991 for DIC concentration and $\delta^{13}C$-DIC ($n = 22$).

This work was initially conducted to study the origin, transport, cycling, and fate of organic carbon from soils to DOC and DIC in a soft water catchment by combining $\delta^{13}C$ and $\Delta^{14}C$ analyses [28,29,42,43,45,46]. DOC turnover rates in streams, lakes, and wetlands are fast, <40 yr, and on the same time scale as acid deposition in the area. There is extensive DOC cycling in upper soil layers and the source of DOC to headwater streams changes seasonally with water table level. Soil CO_2 and $\delta^{13}C$-CO_2 profiles show characteristic patterns of root respiration, decomposition of SOM and diffusional gradients along with modern radiocarbon ages.

The Krycklan catchment is underlain by metagreywacke bedrock and has forests of Norway spruce (*Picea abies*) and Scots pine (*Pinus sylvestris*), but deciduous trees are commonly found in the riparian zone of larger streams. It receives 600 mm/yr precipitation. Mean annual air temperature is 1°C. Stream catchment areas ranged from 0.03 km^2 to 67 km^2 and stream lengths from 0.02 km to 96.5 km. Streams were first to fourth order and had varying amounts of wetland cover (0% to 40%) [16].

Stream samples were collected in June, August, and November 2006 ($n = 43$). All samples were collected in septum-capped, screw-topped glass vials and analysed via headspace equilibration by gas chromatography or on-line GC-IRMS. Shallow groundwater was sampled from suction lysimeters along a transect parallel to the lateral flow paths towards one of the headwater streams ($n = 12$).

Stream temperature was measured in the field. Typical annual pH range of was 3.7 to 6.3 in headwaters and 5.7 to 7.4 in 4th order streams. First-order-stream DOC concentrations were 5.0 mgC/L to 40.0 mgC/L and in 4th order streams were 5.0 mgC/L to 15.0 mgC/L [47].

Research at Krycklan catchment is focused on integrating water quality, hydrology, and ecology in flowing waters [48]. Some recent work includes characterizing the loss of CO_2 from stream surfaces by stream size and season [16,47]. Streams were always supersaturated in CO_2, greatest in the headwaters, and negatively correlated with pH [16]. Owing to the importance of the gas exchange coefficient in controlling CO_2 loss rates and its requirement for scaling across landscapes, it was independently measured across streams in Krycklan catchment [47]. The source of the excess CO_2 was largely explained as respired carbon being exported from catchment soils [49]. As part of this work, δ^{13}C-DIC values can be used to identify CO_2 degassing, compared with the labour-intensive flux measurements, scaled flux estimates, and to asses the upstream CO_2 loss that is demonstrated to be large and important [18].

Calculations

We built a simple, parsimonious degassing model in Matlab (MathWorks, Natick MA USA) for high-DIC, low-pH, silicate headwater streams. Only DIC concentration, pH, δ^{13}C-DIC and temperature are required to be measured in the stream since all other carbonate species concentrations and δ^{13}C values can be calculated. The model partitioned total DIC into CO_2, HCO_3^-, and CO_3^{2-} according to pH and temperature-dependent acid dissociation constants and equilibrium [50,51] and kinetic [52] fraction factors as summarized by [34]. The calculations here are similar to those employed in CO2sys [53], seacarb [54], and streamCO2-DEGAS [35] but do not require as many input parameters, many of which are not available for our catchments. In small headwater streams, this simple approach may be

appropriate to survey a large area. Larger, more detailed models may be required for larger and longer streams and rivers, but it becomes more difficult to measure all variables required to ground-truth such models.

Key assumptions to this approach are that streams are nutrient-poor, low pH, with low community metabolic rates relative to stream velocity and gas exchange, and catchments have C3 vegetation, silicate bedrock, with high DIC shallow groundwater reflective of the vegetation and soils. Both Harp Lake catchment [21,28,29,42–46,55] and Krycklan catchment [16,24,47,56–58] meet these assumptions. Both have low nutrient concentrations, pH values, and associated metabolic rates under undisturbed forest canopies. Stream lengths and segments are on the order of hundreds of metres and water travel times are less than an hour. Forests in these catchments contain only C3 vegetation, are sub-boreal and boreal, and are underlain by crystalline silicate bedrock. Groundwaters are acidic and DIC-rich, as outlined below. This approach is generalizable to the plethora of small streams that meet these criteria across the silicate bedrock areas of the sub-boreal and boreal forests such as the headwaters of catchments on the Canadian and Fennoscandian/Baltic Shields. Source code is provided so that additional components can be added to address the complexities of carbon cycling as needed if the basic assumptions for field sites are not met, such as larger streams with longer travel times, large diel temperature variability, the confluence of several stream sections, measured rates of in-stream metabolism, known patterns of groundwater discharge, or the influence of carbonate-rich bedrock following portions of the streamCO2-DEGAS model [35], which has been used in larger rivers. Further, model sensitivity to variability in input parameters may be easily assessed with Monte Carlo approaches.

The chemical equilibrium calculations can be easily recreated in any modelling language. Since ionic strength of these soft waters is very low (conductivity is typically 15–40 μS), the activity coefficients were assumed to be approaching unity. This time-forward model differs from [35] in that: (a) carbonate dissolution

Table 1. Typical model input and calculated parameters.

Parameter	Typical Value	Unit	Notes	Citation
Typical model input parameters (Fig. 2 and raw data)				
DIC	1200	μmol/L	Typical measured value	
pH	5.5		Typical measured value	
Temperature	4	°C	Typical measured value	
δ^{13}C-DIC	−26	‰	Typical measured value	
Calculated parameters based on model inputs above				
Alk	103	μeq/L	Carbonate alkalinity held constant	
CO_2	1098	μmol/L	Calculated from DIC, pH, $\log K_a$, $\log K_b$	
CO_{2eq}	25	μmol/L	Calculated from $\log k_H$ and p_{CO_2atm}	
ϵ_b	−10.3	‰	CO_2^*–HCO_3^-, function of temperature	[34, 50]
ϵ_c	3.3	‰	HCO_3^-–CO_3^{2-}, function of temperature	[34, 52]
ϵ_g	−1.2	‰	CO_2– CO_2, function of temperature	[34, 50]
$\log K_a$	−6.53		CO_2–HCO_3^-, function of temperature	[68]
$\log K_b$	−10.6		HCO_3^-–CO_3^{2-}, function of temperature	[69]
$\log k_H$	−1.18	log(mol/L/atm)	CO_{2atm}– CO_2, function of temperature	[50]
p_{CO_2atm}	380	ppmv	Partial pressure of CO_2 in the atmosphere	
pK_w	14.8		Function of temperature	[70, 71]

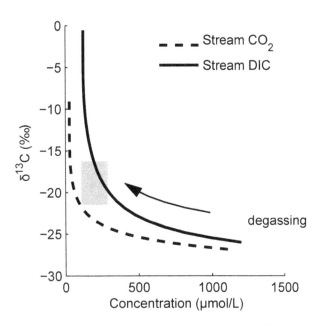

Figure 1. DIC and CO$_2$ degassing trajectories show that as CO$_2$ is lost from streams, δ^{13}C values increase at different rates for DIC and CO$_2$. δ^{13}C-DIC is typically measured, not δ^{13}C- CO$_2$, but the δ^{13}C of both DIC and CO$_2$ during CO$_2$ loss can be modelled using well known isotope fractionation factors. For this example, a typical groundwater end-member was chosen with initial DIC of 1200 μmol/L, δ^{13}C-DIC of -26‰, pH of 5.5, and thus carbonate alkalinity of 103 μC- CO$_2$, but the δ^{13}C of both DIC and CO$_2$ during CO$_2$ loss can be modelled using well known isotope fractionation factors. For this example, a typical groundwater end-member was chosen with initial DIC of 1200 μmol/L, δ^{13}C-DIC of -26‰, pH of 5.5, and thus carbonate alkalinity of 103 μeq/L. In watersheds, variation in the groundwater end-member results in a 'family' of curves with similar trajectories. The grey box represents the 25th and 75th percentiles of DIC and δ^{13}C-DIC collected from three different first- and second-order streams draining into Harp Lake, Ontario, Canada. Samples were collected at the mouth of each stream from July 1989 to November 1991 ($n = 111$). Significant CO$_2$ loss (60% to 80%) must have occurred by the sampling point in each stream in order for the box to fall where it does on the degassing curve.

was not included since our catchments are on silicate bedrock; (b) total alkalinity calculations differ slightly; (c) DIC concentration is the basis for calculations rather than total alkalinity; (d) the model starts with a small plug of shallow groundwater discharged to the surface; (e) groundwater is continuously added to the modelled stream if there are corresponding flow measurements; and (f) the iterative process to find a best fit between measured stream values and modelled results.

Though there is some contribution of organic acids to total alkalinity [59], loss or gain of CO$_2$ does not change organic or carbonate alkalinity. Here we model alkalinity based on the carbonate system since in high-DIC, low-pH waters the main cause of pH change is CO$_2$ loss and assume that the contribution of organic alkalinity to pH is relatively constant and minor.

In practice, to calculate the fraction of CO$_2$ lost by the sampling point in the stream, DIC concentration, pH, and δ^{13}C-DIC measurements in the stream are required. An inverse modelling approach is used as the time-forward model takes initial constrained estimates of groundwater DIC, pH, and known δ^{13}C-DIC, and allows CO$_2$ to degas with time. The resulting values of stream DIC, pH, δ^{13}C-DIC, and δ^{13}C- CO$_2$ with time are outputs. The model was iteratively re-run with Matlab's *fminsearch* function to reduce the sum of squared errors between measured stream DIC, pH, and δ^{13}C-DIC and modelled values. The resulting best-fits then provide the average groundwater DIC concentration and the amount of CO$_2$ that was lost upstream of the sampling point. In this way, an average groundwater DIC concentration can be determined. Additionally, the groundwater δ^{13}C-DIC must be determined but this value is highly constrained by plant and soil δ^{13}C values, as above, and can be confirmed with

a number of piezometers [28,29]. Matlab code to run the model forward and inversely is available from the corresponding author and https://github.com/jjvenky/CO2-from-headwater-streams.

At each time step, a small portion of CO$_2$ was removed from the modelled aquatic system via gas exchange:

$$\frac{dCO_2}{dt} = k \times \left(CO_{2eq} - CO_2 \right) \tag{1}$$

where k is the gas exchange coefficient, CO$_{2eq}$ is the CO$_2$ equilibrium concentration determined with the Henry's constant [50], and CO$_2$ is the solvated CO$_2$ concentration. This necessarily reduced the DIC concentration since:

$$\frac{dDIC}{dt} = \frac{dCO_2}{dt} \tag{2}$$

$$DIC = CO_2 + HCO_3^- + CO_3^{2-} \tag{3}$$

The remaining DIC was then re-apportioned to its constituent species by determining the pH (as H$^+$) required to hold the carbonate-based alkalinity constant [36,60]:

$$H^+ = HCO_3^- + 2 \times CO_3^{2-} + OH^- - \text{Alkalinity} \tag{4}$$

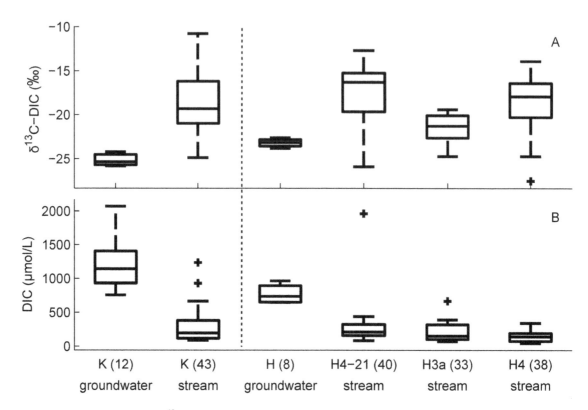

Figure 2. Groundwater and stream (a) δ^{13}**C-DIC values and (b) DIC concentrations from headwater catchments in Sweden (left) and Canada (right).** Groundwater and stream data are from the headwater stream network in Krycklan catchment, Sweden (K) and three first- or second-order streams draining into Harp Lake, Ontario, Canada (H with stream numbers; [43,44]). The boxes represent 25th and 75th percentiles, mid-line the median, whiskers the most extreme datum not outside 1.5 × inter-quartile range (IRQ), and + the data outside 1.5 × IQR. Number of samples is indicated in parentheses. At these two sites where we have both groundwater and stream data, there is a clear increase in δ^{13}C-DIC values from groundwater to stream sampling points along with a large decrease in DIC concentration.

The δ^{13}C portion of the model employed well-known temperature-dependent isotopic fractionation factors between DIC species, and solvated CO_2 and atmospheric CO_2 [34]: at 4°C, these are kinetic gas exchange fractionation ($-1.3‰$) since the k for $^{13}CO_2$ is slightly slower than for $^{12}CO_2$, equilibrium fractionation ($-1.2‰$) between atmospheric CO_2 and solvated CO_2, CO_2–HCO_3^- fractionation ($-10.3‰$), and HCO_3^-–CO_3^{2-} fractionation ($+3.3‰$). At each time step, the δ^{13}C value of each species was calculated. Typical values of input parameters and calculated parameters are summarized in Table 1.

The effects of continuous groundwater input to the modelled stream that significantly changes the stream volume over the reach of interest can also be included by adding an amount of water with known or estimated initial groundwater values (DIC, δ^{13}C-DIC, pH, alkalinity) at each time step. The amount can be determined by stream length and discharge measured at the catchment outlet or as the difference in flow between two sites. This comprised an additional input parameter used during best-fit modelling of the H4-21 data since this is a small first-order stream, the stream length is known, and the discharge was measured at the stream mouth along with DIC, δ^{13}C-DIC, and pH. Here, groundwater input was assumed to be constant down the reach. In each case, the CO_2 lost was multiplied by the measured daily discharge and reported relative to the catchment area.

Permissions

Headwater streams in the Harp Lake catchment were studied under the auspices of pre-existing Ontario Ministry of the

Environment research program. The Krycklan catchment is a part of the Svartberget LTER site run by the Swedish University of Agricultural Sciences and the area is developed for scientific purposes. Specific permissions for these activities were not required. No endangered or protected species were involved in either site.

Results and Discussion

As CO_2 is initially lost from a stream, there is little associated change in δ^{13}C-DIC (Fig. 1). Only after about half of the DIC has been lost is there an observable change of about 2‰ in δ^{13}C-DIC. In streams where measured δ^{13}C-DIC values are significantly different than shallow groundwater, a large amount of CO_2 must have already been lost from the stream by the time the sample was collected (Fig. 1). Thus, CO_2 flux measurements obtained with data from an individual sampling site where the δ^{13}C-DIC value is several per mille greater than the groundwater δ^{13}C-DIC value must underestimate stream and catchment CO_2 loss rates since the stream surface with higher CO_2 concentrations and higher flux rates is upstream of such a sampling site. The degree of underestimation needs to be better quantified because these results suggest that net ecosystem exchange and C storage may be lower than assumed in boreal and sub-boreal ecozones.

We applied these ideas to a suite of headwater stream data from Krycklan catchment in Sweden and Harp Lake in Canada for which both groundwater and stream measurements were available. Stream DIC concentrations were lower and δ^{13}C-DIC and

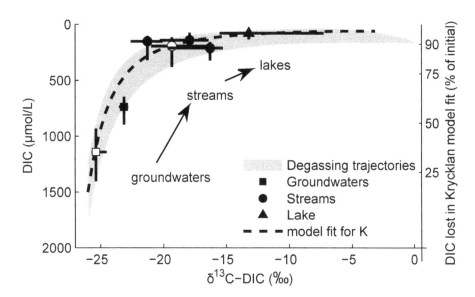

Figure 3. Typical DIC and CO$_2$ degassing trajectories (grey band) show that small initial increases in δ^{13}C-DIC values signify large losses in CO$_2$ from groundwater to streams. The δ^{13}C progress along this trajectory is a natural result of CO$_2$ degassing, not other processes (e.g., primary production, mineral weathering, etc.). The grey band represents the combined trajectories of degassing curves with a combination of initial DIC concentrations (1200 to 1800 μmol/L), δ^{13}C-DIC (-26‰) and pH values (4.0 to 5.5). The grey band also describes the landscape-scale trajectory of DIC concentrations and δ^{13}C-DIC values from shallow groundwater discharging into headwater streams and degassing while flowing downhill to small lakes. Groundwater, stream, and lake data from the two study sites are displayed as 25th and 75th percentiles (lines) with medians (symbols). Harp Lake data are from the epilimnion from March 1990 to November 1991 ($n = 22$). The line represents the degassing trajectory that was fit to the median of the Krycklan (open symbols) dataset. The secondary y-axis indicates the amount of DIC lost along the degassing model fit for the Krycklan site.

pH values greater than their respective groundwater values in both catchments (Fig. 2). The range of δ^{13}C-DIC in shallow groundwater, -26‰ to -24‰ is tightly constrained by the narrow range of δ^{13}C in C3 plants and soil. This is comparable to measured δ^{13}C-DOC values [28] so in-stream respiration would produce DIC with the same δ^{13}C-DIC value as shallow groundwater. Surface water was allowed to lose CO$_2$ via gas exchange and the chemical and isotopic equilibria were adjusted accordingly (see Methods).

There are many groundwater discharge patterns possible. In small streams, the difference in discharge measured at two locations may be the only way to estimate groundwater inputs. As such, it may be difficult to parameterize [43]. In a stream network in Alaska, high resolution CO$_2$ and discharge data suggest variable and varying groundwater inputs [19] along the lengths of the streams. The net effect of groundwater input to small streams is to increase stream DIC concentration, reduce the δ^{13}C-DIC value, decrease pH, and make the stream appear closer to the initial shallow groundwater values (Figs 1 and 2).

While shallow groundwaters in boreal forest catchments vary, to some extent in DIC concentration, δ^{13}C-DIC, pH, and temperature, the effect of changing these variables does not alter the fundamental relationship between δ^{13}C-DIC and DIC during CO$_2$ degassing. Increased initial DIC concentration, lower pH, and lower alkalinity cause the sigmoid-like relationship to approach its plateau δ^{13}C-DIC value more quickly and thus requires a greater loss of CO$_2$ before δ^{13}C-DIC values will increase appreciably. This is demonstrated in the range of landscape-scale degassing trajectories (grey area in Fig. 3). This range is akin to a sensitivity analysis by confirming the shape of the DIC vs δ^{13}C-DIC relationship across a wide range of DIC concentrations and pH values. Buffered groundwaters discharging

into headwater streams exhibit a shallower curve than acidic, poorly buffered waters. The degassing trajectories in DIC vs δ^{13}C-DIC space describe the CO$_2$ loss from groundwaters to streams and ultimately to headwater lakes as decreasing DIC, increasing δ^{13}C-DIC, and increasing pH (Fig. 3).

Scaling-up fluxes estimated via these methods to catchment- and landscape-scale requires areas or lengths of headwater streams, catchment areas, or groundwater discharge areas. Here, we avoid using stream length and surface area, parameters not easily obtained in headwater catchments, but see [18,20], and instead report stream losses relative to their catchment area. In this manner, the CO$_2$ loss from aquatic surfaces is easily compared in the same units used to assess net ecosystem exchange and productivity at the catchment scale.

One of the study catchments, a small first-order upland stream in a sub-boreal temperate forest (H4-21) [61], lost a median 15 mgC/m^2/d between its source and mouth ($n = 21$, range 3–100 mgC/m^2/d). This is a flux-weighted loss of 5 gC/m^2/yr (range 1–40 gC/m^2/yr) and is comparable to the annual DOC export (1–8 gC/m^2/yr) from the catchment (Fig. 4). It is also around 10% of catchment net CO$_2$ uptake [62]. Unlike DOC, this CO$_2$ loss occurred from the small stream surface and was lost directly to the atmosphere before the stream outlet. Furthermore, this CO$_2$ loss would not have been quantified at the mouth of the stream using conventional measures of CO$_2$ and DIC fluxes [38]. Previous estimates of CO$_2$ flux rates from the mouth of Harp Lake streams [21] are much smaller than our estimate. This highlights the fact that measured CO$_2$ concentrations decline down the length of streams and CO$_2$ loss is underestimated when it is calculated from the sampling points of lowest measured concentrations.

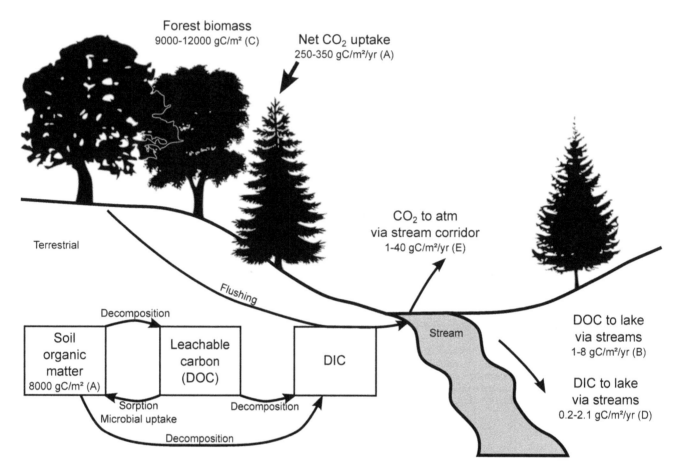

Figure 4. Schematic drawing showing the size of some of the organic carbon pools, carbon fluxes, and important processes affecting DOC, DIC and CO₂ in terrestrial catchments and aquatic surfaces. Decomposition from soil organic matter includes microbial exudates. Production of DOC, decomposition, microbial uptake/sorption, root respiration, mineral weathering, and flushing are competing processes affecting the export of DOC and DIC via streams in forested catchments. Loss of CO₂ directly to the atmosphere from the surfaces in the stream corridor can exceed the sum of DIC and DOC export to downstream lakes. Total losses of carbon lost by CO₂ emissions and dissolved carbon export (DIC+DOC) can be important relative to net CO₂ uptake by forests from the atmosphere. All rates are per catchment area. (Figure after [42]; A. [62]; B. [38]; C. range for boreal and temperature forests [66]; D. [38,67]; E. this study.)

The other study catchment, a typical stream network in Krycklan, Sweden [18], lost upwards of 13–28 mgC/m²/d. This estimate is necessarily larger than the CO₂ concentration based estimated loss of 5.0 gC/m²/yr. It is also larger than the DOC+ DIC export of 4.6 gC/m²/yr and is 14–29% of estimated the net ecosystem exchange of 96 gC/m²/yr [18].

This method of using measured in-stream δ^{13}C-DIC values shows that the CO₂ loss from aquatic surfaces in these boreal and sub-boreal catchments is large and under-estimated. The areal flux rates we present for Harp Lake and Krycklan catchments are similar to those arrived at using much more intensive sampling and different sets of assumptions. This supports the use of a parsimonious model in similar locations. The pattern of measured δ^{13}C-DIC values in small streams being much higher than expected shallow groundwater δ^{13}C-DIC values, is present in other datasets [23,25,40] and indicates there is a large flux of CO₂ from headwater streams that has not been included in our continental carbon budgets.

With a measurement at one point in a small stream, the CO₂ loss can be estimated by this δ^{13}C approach for a certain distance upstream. In H4-21, using a first-order-loss-rate expression ($3U/k$,

where U is stream velocity and k is gas transfer coefficient, [63]), typical gas transfer coefficients (7–25 d^{-1}, [21]), typical measured mean discharge (1–20 L/s), typical cross-sectional areas (250–1000 cm²), the upstream distance over which the CO₂ degassing occurred was 375–1375 m.

Grab samples can be used to demonstrate the degree of degassing that has occurred at a given location with a parsimonious model — the simplicity of the DIC–alkalinity–pH–δ^{13}C-DIC calculations means this idea can be easily incorporated into site-specific calculations. Further, stream CO₂ loss, normalized to catchment area, is likely to be larger than CO₂ loss from lake surfaces since lake CO₂ loss is typically less than that half of the DOC input [5,38,64].

Headwater stream CO₂ loss is a reduction in the net ecosystem productivity of boreal and sub-boreal forests. It is a globally important flux since boreal and sub-boreal ecozones are so large [1]. Here, we have shown it can be easily estimated at the catchment scale by combining δ^{13}C-DIC measurements with in-stream and groundwater samples. Ultimately, C loss via stream degassing may be required to be integrated into northern hemispheric CO₂ uptake and loss rates [2,7,18,65].

Acknowledgments

We thank R. J. Elgood for extensive field work and preparation of off-line δ^{13}C-DIC samples from the Harp Lake catchment. We thank the Dorset Environmental Science Centre and the Ontario Ministry of the Environment for geochemical analyses and field support. We thank the crew of the Krycklan Catchment Study (KCS) for logistics and great field support. We thank K. M. Chomicki and J. Parks for assistance with calculating forest and geologic areas and K. Bishop for assistance at Krycklan. Raw geochemical and isotopic data and Matlab code are available from the corresponding author and https://github.com/jjvenky/CO2-from-headwater-streams.

Author Contributions

Conceived and designed the experiments: JJV SLS MBW. Performed the experiments: JJV SLS MBW. Analyzed the data: JJV SLS MBW. Contributed reagents/materials/analysis tools: JJV SLS MBW. Wrote the paper: JJV SLS MBW.

References

1. FAO (2001) Global forest resources assessment 2000. Technical report, FAO. URL http://www.fao.org/DOCREP/004/Y1997E/Y1997E00.HTM.

2. Stephens BB, Gurney KR, Tans PP, Sweeney C, Peters W, et al. (2007) Weak northern and strong tropical land carbon uptake from vertical profiles of atmospheric CO_2. Science 316: 1732–1735.

3. Battin TJ, Kaplan LA, Findlay S, Hopkinson CS, Marti E, et al. (2008) Biophysical controls on organic carbon fluxes in fluvial networks. Nature Geoscience 1: 95–100.

4. Battin TJ, Luyssaert S, Kaplan LA, Aufdenkampe AK, Richter A, et al. (2009) The boundless carbon cycle. Nature Geoscience 2: 598–600.

5. Tranvik LJ, Downing JA, Cotner JB, Loiselle SA, Striegl RG, et al. (2009) Lakes and reservoirs as regulators of carbon cycling and climate. Limnology and Oceanography 54: 2298–2314.

6. Aufdenkampe AK, Mayaorga E, Raymond PA, Melack JM, Doney SC, et al. (2011) Riverine coupling of biogeochemical cycles between land, oceans, and atmosphere. Frontiers in Ecology and the Environment 9: 53–60.

7. Raymond PA, Hartmann J, Lauerwald R, Sobek S, McDonald C, et al. (2013) Global carbon dioxide emissions from inland waters. Nature 503: 355–359.

8. Humborg C, Morth CM, Sundbom M, Borg H, Blenckner T, et al. (2010) CO_2 supersaturation along the aquatic conduit in Swedish watersheds as constrained by terrestrial respiration, aquatic respiration and weathering. Global Change Biology 16: 1966–1978.

9. de Wit HA, Austnes K, Dalsgaard L, Hylen G (2013) A carbon budget of Norway: integration of terrestrial and aquatic C fluxes. Geophysical Research Abstracts 15: EGU2013–5158.

10. Butman D, Raymond PA (2011) Significant efflux of carbon dioxide from streams and rivers in the United States. Nature Geoscience 4: 839–842.

11. Bishop K, Buffam I, Erlandsson M, Folster J, Laudon H, et al. (2008) Aqua incognita: the unknown headwaters. Hydrological Processes 22: 1239–1242.

12. Kling GW, Kipphut GW, Miller MC (2001) Arctic lakes and streams as gas conduits to the atmosphere: implications for tundra carbon budgets. Science 251: 298–301.

13. Hope D, Palmer SM, Billett MF, Dawson JJC (2001) Carbon dioxide and methane evasion from a temperate peatland stream. Limnology and Oceanography 46: 847–857.

14. Richey JE, Melack JM, Aufdenkampe AK, Ballester VM, Hess LL (2002) Outgassing from Amazonian rivers and wetlands as a large tropical source of atmospheric CO_2. Nature 416: 617–620.

15. Cole JJ, Prairie YT, Caraco NF, McDowell WH, Tranvik LJ, et al. (2007) Plumbing the global carbon cycle: Integrating inland waters into the terrestrial carbon budget. Ecosystems 10: 171–184.

16. Wallin M, Buffam I, Öquist MG, Laudon H, Bishop K (2010) Temporal and spatial variability of dissolved inorganic carbon in a boreal stream network: Concentrations and downstream fluxes. Journal of Geophysical Research: Biogeosciences 115: G02014.

17. Öquist MG, Wallin M, Seibert J, Bishop K, Laudon H (2009) Dissolved inorganic carbon export across the soil/stream interface and its fate in a boreal headwater stream. Environmental Science & Technology 43: 7364–7369.

18. Wallin MB, Grabs T, Buffam I, Laudon H, Ågren A, et al. (2013) Evasion of CO_2 from streams – the dominant component of the carbon export through the aquatic conduit in a boreal landscape. Global Change Biology 19: 785–797.

19. Crawford JT, Striegl RG, Wickland KP, Dornblaser MM, Stanley EH (2013) Emissions of carbon dioxide and methane from a headwater stream network of interior Alaska. Journal of Geophysical Research: Biogeosciences 118: 482–494.

20. Striegl RG, Dornblaser MM, McDonald CP, Rover JR, Stets EG (2012) Carbon dioxide and methane emissions from the Yukon River system. Global Biogeochemical Cycles 26: GB0E05.

21. Koprivnjak JF, Dillon PJ, Molot LA (2010) Importance of CO_2 evasion from small boreal streams. Global Biogeochemical Cycles 24: GB4003.

22. Palmer SM, Hope D, Billett MF, Dawson JJC, Bryant CL (2001) Sources of organic and inorganic carbon in a headwater stream: Evidence from carbon isotope studies. Biogeochemistry 52: 321–338.

23. Hope D, Palmer SM, Billett MF, Dawson JJC (2004) Variations in dissolved CO_2 and CH_4 in a first-order stream and catchment: an investigation of soil-stream linkages. Hydrological Processes 18: 3255–3275.

24. Jonsson A, Algesten G, Bergstrom AK, Bishop K, Sobek S, et al. (2007) Integrating aquatic carbon fluxes in a boreal catchment carbon budget. Journal of Hydrology 334: 141–150.

25. Doctor DH, Kendall C, Sebestyen SD, Shanley JB, Ote N, et al. (2008) Carbon isotope fractionation of dissolved inorganic carbon (DIC) due to outgassing of carbon dioxide from a headwater stream. Hydrological Processes 22: 2410–2423.

26. Kortelainen P (1993) Content of total organic-carbon in Finnish lakes and its relationship to catchment characteristics. Canadian Journal of Fisheries and Aquatic Sciences 50: 1477–1483.

27. Houghton RA (2003) The contemporary carbon cycle. In: Holland HD, Turekian KK, editors, Treatise on Geochemistry. Volume 8: Biogeochemistry, Elsevier, chapter 8.10.

28. Schiff SL, Aravena R, Trumbore SE, Dillon PJ (1990) Dissolved organic-carbon cycling in forested watersheds — a carbon isotope approach. Water Resources Research 26: 2949–2957.

29. Aravena R, Schiff SL, Trumbore SE, Dillon PJ, Elgood R (1992) Evaluating dissolved inorganic carbon cycling in a forested lake watershed using carbon isotopes. Radiocarbon 34: 636–645.

30. Naiman RJ (1983) The annual pattern and spatial distribution of aquatic oxygen metabolism in boreal forest watersheds. Ecological Monographs 53: 73–94.

31. Lamberti GA, Steinman AD (1997) A comparison of primary production in stream ecosystems. Journal of the North American Benthological Society 16: 95–104.

32. Naiman RJ, Melillo JM, Lock MA, Ford TE, Reice SR (1987) Longitudinal patterns of ecosystem processes and community structure in a sub-arctic river continuum. Ecology 68: 1139–1156.

33. Roberts BJ, Mulholland PJ, Houser AN (2007) Effects of upland disturbance and instream restoration on hydrodynamics and ammonium uptake in headwater streams. Journal of the North American Benthological Society 26: 38–53.

34. Zeebe RE, Wolf-Gladrow D (2001) CO_2 in seawater: equilibrium, kinetics, isotopes. New York: Elsevier.

35. Polsenaere P, Abril G (2012) Modelling CO_2 degassing from small acidic rivers using water pCO_2, DIC and δ^{13}C-DIC data. Geochimica et Cosmochimica Acta 91: 220–239.

36. Di Toro DM (1976) Combining chemical equilibrium and phytoplankton models — a general methodology. In: Canale RP, editor, Modeling Biochemical Processes in Aquatic Ecosystems, Ann Arbor, MI, USA: Ann Arbor Science. pp. 233–255.

37. Cerling TE, Wang Y, Quade J (1993) Expansion of C4 ecosystems as an indicator of global ecological change in the Late Miocene. Nature 361: 344–345.

38. Dillon PJ, Molot LA (1997) Dissolved organic and inorganic carbon mass balances in central Ontario lakes. Biogeochemistry 36: 29–42.

39. Giesler R, Lyon SW, Mörth CM, Karlsson J, Karlsson EM, et al. (2014) Catchment-scale dissolved carbon concentrations and export estimates across six subarctic streams in northern Sweden. Biogeosciences 11: 525–537.

40. Waldron S, Scott EM, Soulsby C (2007) Stable isotope analysis reveals lower-order river dissolved inorganic carbon pools are highly dynamic. Environmental Science & Technology 41: 6156–6162.

41. Downing JA, Cole JJ, Duarte CA, Middelburg JJ, Melack JM, et al. (2012) Global abundance and size distribution of streams and rivers. Inland Waters 2: 229–236.

42. Schiff SL, Aravena R, Trumbore SE, Hinton MJ, Elgood R, et al. (1997) Export of DOC from forested catchments on the precambrian shield of central ontario: Clues from ^{13}C and ^{14}C. Biogeochemistry 36: 43–65.

43. Hinton MJ (1998) The role of groundwater flow in streamflow generation within two small forested watersheds of the Canadian Shield. Ph.D. thesis, University of Waterloo.

44. Dillon PJ, Molot LA, Scheider WA (1991) Phosphorus and nitrogen export from forested stream catchments in central Ontario. Journal of Environmental Quality 20: 857–864.

45. Hinton MJ, Schiff SL, English MC (1997) The significance of storms for the concentration and export of dissolved organic carbon from two Precambrian Shield catchments. Biogeochemistry 36: 67–88.

46. Hinton MJ, Schiff SL, English MC (1998) Sources and flowpaths of dissolved organic carbon during storms in two forested watersheds of the Precambrian Shield. Biogeochemistry 41: 175–197.

47. Wallin MB, Öquist MG, Buffam I, Billett MF, Nisell J, et al. (2011) Spatiotemporal variability of the gas transfer coefficient (K_{CO_2}) in boreal streams: Implications for large scale estimates of CO_2 evasion. Global Biogeochemical Cycles 25: GB3025.

48. Laudon H, Taberman I, Ågren A, Futter M, Ottosson-Löfvenius M, et al. (2013) The Krycklan Catchment Study — a flagship infrastructure for hydrology, biogeochemistry, and climate research in the boreal landscape. Water Resources Research 49: 7154–7158.

49. Wallin M (2011) Evasion of CO_2 from Streams: Quantifying a Carbon Component of the Aquatic Conduit in the Boreal Landscape. Ph.D. thesis, Swedish University of Agricultural Sciences. URL http://urn.kb.se/resolve?urn = urn:nbn:se:slu:epsilon-3150.

50. Mook WG, Bommerso JC, Staverman WH (1974) Carbon isotope fractionation between dissolved bicarbonate and gaseous carbon-dioxide. Earth and Planetary Science Letters 22: 169–176.

51. Vogel JC, Grootes PM, Mook WG (1970) Isotopic fractionation between gaseous and dissolved carbon dioxide. Zeitschrift für Physik 230: 225–238.

52. Zhang J, Quay PD, Wilbur DO (1995) Carbon-isotope fractionation during gas-water exchange and dissolution of CO_2. Geochimica et Cosmochimica Acta 59: 107–114.

53. van Heuven S, Pierrot D, Rae JWB, Lewis E, Wallace DWR (2011). MATLAB program developed for CO_2 system calculations. doi:10.3334/CDIAC/otg. CO2SYS_MATLAB_v1.1. ORNL/CDIAC-105b. Carbon Dioxide Information Analysis Center, Oak Ridge National Laboratory, U.S. Department of Energy, Oak Ridge, Tennessee.

54. Lavigne H, Epitalon JM, Gattuso JP (2014) seacarb: seawater carbonate chemistry with R. URL http://CRAN.R-project.org/package = seacarb.R package version 3.0.

55. Devito KJ, Dillon PJ, Lazerte BD (1989) Phosphorus and nitrogen retention in five Precambrian shield wetlands. Biogeochemistry 8: 185–204.

56. Berggren M, Laudon H, Jonsson A, Jansson M (2010) Nutrient constraints on metabolism affect the temperature regulation of aquatic bacterial growth efficiency. Microbial Ecology 60: 894–902.

57. Grabs T, Bishop K, Laudon H, Lyon SW, Seibert J (2012) Riparian zone hydrology and soil water total organic carbon (TOC): implications for spatial variability and upscaling of lateral riparian TOC exports. Biogeosciences 9: 3901–3916.

58. Kuglerová L, Jansson R, Ågren A, Laudon H, Malm-Renöfält B (2014) Groundwater discharge creates hotspots of riparian plant species richness in a boreal forest stream network. Ecology 95: 715–725.

59. Herczeg AL, Hesslein RH (1984) Determination of hydrogen-ion concentration in softwater lakes using carbon-dioxide equilibria. Geochimica et Cosmochimica Acta 48: 837–845.

60. Choi J, Hulseapple SM, Conklin MH, Harvey JW (1998) Modeling CO_2 degassing and pH in a stream-aquifer system. Journal of Hydrology 209: 297–310.

61. Hinton MJ, Schiff SL, English MC (1993) Physical properties governing groundwater flow in a glacial till catchment. Journal of Hydrology 142: 229–249.

62. Trumbore SE, Schiff SL, Aravena R, Elgood R (1992) Sources and transformation of dissolved organic carbon in the Harp Lake forested catchment: The role of soils. Radiocarbon 34: 626–635.

63. Chapra SC, Di Toro DM (1991) Delta method for estimating primary production, respiration, and reaeration in streams. Journal of Environmental Engineering 117: 640–655.

64. Hesslein RH, Broecker WS, Quay PD, Schindler DW (1980) Whole-lake radiocarbon experiment in an oligotrophic lake at the Experimental Lakes Area, Northwestern Ontario. Canadian Journal of Fisheries and Aquatic Sciences 37: 454–463.

65. Peters W, Jacobson AR, Sweeney C, Andrews AE, Conway TJ, et al. (2007) An atmospheric perspective on North American carbon dioxide exchange: CarbonTracker. Proceedings of the National Academy of Sciences of the United States of America 104: 18925–18930.

66. Hudson RJM, Gherini SA, Goldstein RA (1994) Modeling the global carbon-cycle: Nitrogen-fertilization of the terrestrial biosphere and the missing CO_2 sink. Global Biogeochemical Cycles 8: 307–333.

67. Molot LA, Dillon PJ (1997) Photolytic regulation of dissolved organic carbon in northern lakes. Global Biogeochemical Cycles 11: 357–365.

68. Harned HS, Davis R Jr (1943) The ionization constant of carbonic acid in water and the solubility of carbon dioxide in water and aqueous salt solutions from 0 to 50°. Journal of the American Chemical Society 65: 2030–2037.

69. Harned HS, Scholes SR Jr (1941) The ionization constant of HCO_3^- from 0 to 50°. Journal of the American Chemical Society 63: 1706–1709.

70. Dickson AG, Millero FJ (1987) A comparison of the equilibrium-constants for the dissociation of carbonic-acid in seawater media. Deep-Sea Research Part A – Oceanographic Research Papers 34: 1733–1743.

71. Millero FJ (1995) Thermodynamics of the carbon-dioxide system in the oceans. Geochimica et Cosmochimica Acta 59: 661–677.

15

Hailstones: A Window into the Microbial and Chemical Inventory of a Storm Cloud

Tina Šantl-Temkiv[1,2,3], Kai Finster[2,3], Thorsten Dittmar[4], Bjarne Munk Hansen[1], Runar Thyrhaug[†5], Niels Woetmann Nielsen[6], Ulrich Gosewinkel Karlson[1]*

1 Department of Environmental Science, Aarhus University, Roskilde, Denmark, **2** Microbiology Section, Department of Bioscience, Aarhus University, Aarhus, Denmark, **3** Stellar Astrophysics Centre, Department of Physics and Astronomy, Aarhus University, Aarhus, Denmark, **4** Max Planck Research Group for Marine Geochemistry, Institute for Chemistry and Biology of the Marine Environment, University of Oldenburg, Oldenburg, Germany, **5** Department of Biology, University of Bergen, Bergen, Norway, **6** Danish Meteorological Institute, Copenhagen, Denmark

Abstract

Storm clouds frequently form in the summer period in temperate climate zones. Studies on these inaccessible and short-lived atmospheric habitats have been scarce. We report here on the first comprehensive biogeochemical investigation of a storm cloud using hailstones as a natural stochastic sampling tool. A detailed molecular analysis of the dissolved organic matter in individual hailstones via ultra-high resolution mass spectrometry revealed the molecular formulae of almost 3000 different compounds. Only a small fraction of these compounds were rapidly biodegradable carbohydrates and lipids, suitable for microbial consumption during the lifetime of cloud droplets. However, as the cloud environment was characterized by a low bacterial density (Me = 1973 cells/ml) as well as high concentrations of both dissolved organic carbon (Me = 179 µM) and total dissolved nitrogen (Me = 30 µM), already trace amounts of easily degradable organic compounds suffice to support bacterial growth. The molecular fingerprints revealed a mainly soil origin of dissolved organic matter and a minor contribution of plant-surface compounds. In contrast, both the total and the cultivable bacterial community were skewed by bacterial groups (γ-Proteobacteria, Sphingobacteriales and Methylobacterium) that indicated the dominance of plant-surface bacteria. The enrichment of plant-associated bacterial groups points at a selection process of microbial genera in the course of cloud formation, which could affect the long-distance transport and spatial distribution of bacteria on Earth. Based on our results we hypothesize that plant-associated bacteria were more likely than soil bacteria (i) to survive the airborne state due to adaptations to life in the phyllosphere, which in many respects matches the demands encountered in the atmosphere and (ii) to grow on the suitable fraction of dissolved organic matter in clouds due to their ecological strategy. We conclude that storm clouds are among the most extreme habitats on Earth, where microbial life exists.

Editor: Stefan Bertilsson, Uppsala University, Sweden

Funding: TST was supported by a Ph.D. fellowship granted by the Danish Agency for Science, Technology and Innovation (Forsknings- og Innovationsstyrelsen). Funding for the Stellar Astrophysics Centre is provided by The Danish National Research Foundation. The research is supported by the ASTERISK project (ASTERoseismic Investigations with SONG and Kepler) funded by the European Research Council (Grant agreement no.: 267864). The funders had no role in study design, data collection and analysis, decision to publish, or preparation of the manuscript.

Competing Interests: The authors have declared that no competing interests exist.

* E-mail: uka@dmu.dk

† Deceased.

Introduction

Airborne bacteria have lately generated a lot of interest, due to their ubiquitous presence and the accumulating evidence of their activity in the atmosphere [1]. Previous studies indicate that terrestrial habitats, in particular soils and plant leaf surfaces, are the major sources of airborne bacteria, whereas marine environments are a less prominent source [2]. By performing a meta-analysis of the composition of the airborne community and of their potential source environments, Bowers et al [3] identified bacterial taxa indicative for soil and plant-surface origin. Generally, they found that the airborne community was more similar to plant-surface than to soil communities. Depending on the land-use type, however, either soil or plant-surface bacteria were found to dominate the community. As the atmospheric bacterial community was distinct from its source communities, which was driven by the different relative abundances of bacterial taxa, the existence of a microbial community characteristic for the atmosphere was implied [3].

Diverse bacterial communities have been described in the atmosphere [4] and in clouds [5], [6]. However, bacterial communities in cloud water may be distinct from bacterial communities in the dry atmosphere, as the chances of airborne bacteria to enter into cloud droplets are increased for those that can act as cloud condensation nuclei [7]. After entering cloud droplets, bacteria are thought to influence physical and chemical processes in the atmosphere [1]. They may do this both by the means of their outer membrane structures as well as their metabolic activity. During their residence time in clouds, a group of mainly epiphytic Gram-negative bacteria could influence patterns of precipitation by facilitating the formation of ice crystals [8]. The so-called ice nucleation active (INA) bacteria are among the most efficient described ice nucleators. By forming large aggregates of INA proteins, which are anchored in their outer membrane [9], INA bacteria substantially elevate the freezing temperature of water. Thus, they may be important in mixed phase clouds, where subzero temperatures are often too high for water to freeze in the absence of ice nucleators.

There is also growing evidence that some cloudborne bacteria proliferate in cloud droplets. It was observed for two cloud events that the majority (72% and 95%) of cloud bacteria were viable [10]. Also, Hill et al [11] showed that on average 76% of cloudborne bacteria from two clouds were metabolically active. A couple of studies confirmed that the indigenous bacterial communities from rain- and cloud water could grow on either naturally present or supplemented organic compounds [12], [13]. Several isolates from clouds were shown capable of metabolizing nutrients present in cloud water [14] at rates that make them competitive with photooxidation [15]. However, it remains unclear whether cloud bacteria are in fact active *in situ*.

Inside storm clouds water droplets can coalesce into hailstones. During their formation, hailstones collect cloud and rain droplets in a non-selective way as they circulate inside the cloud, following unpredictable individual paths. We have recently shown that hailstones, which preserve the samples by freezing in real time, are useful sampling tools of storm cloud water and, indirectly, of air from the atmospheric boundary layer that has been sucked up by the storm cloud [6]. The storm cloud bacterial community was diverse with the estimated total bacterial richness of 1800 operational taxonomic unites (OTUs) at the species level and with a medium species evenness as estimated from Lorenz curves [6]. We also suggested that the highly diverse community encompasses strains with opportunistic ecologic strategy, which may grow despite the short residence times in clouds. Although some of the isolates have been characterized as opportunists [6], it remains unclear, whether the pool of organic chemicals can support the metabolism of these bacteria and if selective enrichment of some bacterial groups actually occurs in clouds. We report here on a comprehensive biogeochemical study, analyzing large hailstones from the same hail event [6]. By performing a detailed molecular characterization of water-soluble organic matter in hailstones and by aligning the potential substrates with the characteristic bacterial genera present in the cloud, we investigate the possibility of microbial growth in the storm cloud.

Materials and Methods

Ethics Statement

All sampling sites were public property and non-protected areas. In addition, in Slovenia there is no legal requirement for obtaining permits for taking precipitation samples. Thus, there were no specific permits required for the described field studies. Endangered or protected species were not in any way affected by or involved in the sampling activity.

Collection and cleaning of the hailstones

Forty two hailstones were collected after a thunderstorm discharged over Ljubljana, Slovenia in the late afternoon of May 25th, 2009. Hailstones were collected into sterile bags within 5 minutes after they fell on ground and stored at $-20°C$. For molecular characterization and analysis of dissolved organic carbon (DOC) as well as total dissolved nitrogen (TDN), the surface of 18 hailstones was cleaned by rinsing with deionized water. Ice cubes of deionised water, with their surface contaminated by soil and grass, were treated in the same way as a control for the rinsing procedure. All plastic and glass lab ware was acid washed; all metal equipment used was treated by dry-heat-

sterilization (160°C, over night). The cleaning of hailstones was done under conditions minimizing contamination by organic vapour.

For microbiological analysis, the surface of 24 hailstones was sterilized under sterile conditions as previously described [6]. For flow cytometry analysis 1.8 ml each of 12 hailstones was fixed in 2% glutaraldehyde, the remainder of these 12 hailstones was either refrozen, stored at $-20°C$, or used for the enumeration of colony forming units (CFU) using R2A plates [16].

Determination of dissolved organic carbon (DOC) and total dissolved nitrogen (TDN)

DOC and TDN were analyzed in 18 hailstones by low-volume manual injection and catalytic high-temperature combustion on a Shimadzu TOC-V analyzer with a total nitrogen module (TNM-1) [17]. Samples were acidified to pH = 2 with HCl (p.a.) and purged for 10 minutes with synthetic air prior to analysis to remove inorganic carbon. The accuracy of the analysis was confirmed with deep-sea reference water samples provided by the University of Miami. The accuracy with respect to deep-sea water was within 5% relative error and detection limits were 5 µM for DOC and 1 µM for TDN. Procedural blanks did not yield detectable amounts of DOC and TDN. Eight controls for the cleaning procedure were analyzed in the same way as hailstones and showed significantly lower values than the hailstones (Mann–Whitney U test, $W = 144$, $p < 0.0001$ for both DOC and TDN). The negative controls were used for blank-correcting DOC and TDN concentrations. As the data were not normally distributed (Shapiro-Wilk normality test, $W = 0.4$, $p < 0.0001$ for both DOC and TDN), we report the median (Me) together with the quartile 1–quartile 3 values (Q1–Q3) and use a nonparametric test for the analysis of correlation (Spearman's rank correlation coefficient).

Characterization of dissolved organic matter (DOM)

On three individual hailstones, covering the DOC concentration range, a detailed molecular characterization was performed using ultrahigh-resolution mass spectrometry on a 15 Tesla Bruker Solarix electrospray ionization Fourier-transform ion cyclotron resonance mass spectrometer (ESI-FT-ICR-MS). For FT-ICR-MS analysis, DOM was isolated from the hailstones via solid phase extraction [18]. DOM was directly infused into the mass spectrometer in methanol:water (1:1). The samples were ionized by electrospray ionization (ESI) in negative and positive mode. This ionization technique produces singly charged ions and keeps covalent bonds intact. 500 scans were accumulated in broad band mode for each sample. Procedural blanks did not contain detectable impurities. The mass spectra were internally calibrated. A mass error of <20 ppb was achieved for each detected mass. Based on this ultrahigh precision, molecular formulae were calculated for each peak. Programs used for data analysis and interpretation were Bruker Solarix Control, Bruker Data Analysis, Microsoft Access, and Ocean Data View. The difference between the three analyzed hailstones was insignificant compared to triplicate analysis of the same sample; therefore we discuss the average of the three samples.

Total bacterial abundance

Total bacterial abundance was determined using a FacsCalibur flow cytometer (Becton Dickinson, Franklin Lakes, NJ) equipped with an air-cooled laser providing 15 mW at 488 nm employing a standard filter set-up. The samples were stained with SYBRGreen I (final concentration 0.02% of the stock solution, Molecular Probes Inc., Eugene, OR) for 15 min in the dark, at room temperature [19]. Fluorescent microspheres (Molecular Probes Inc.) with a diameter of 0.95 μm were analyzed as a standard. Sterilized ice cubes of deionized water were fixed and analyzed in the same way as the samples. The densities in the negative controls were significantly lower than the densities in hailstones (Mann–Whitney U test, $W = 47$, $p<0.005$). As the data were not normally distributed (Shapiro-Wilk normality test, $W = 0.5$, $p<0.0001$), we report the median (Me) and the quartile 1–quartile 3 values (Q1–Q3).

The analysis of bacterial sources

The 16S rRNA gene sequences of the clones and the isolates, which have been previously reported under GenBank accession numbers JQ896628–JQ897350 [6], were analyzed for the community composition of individual hailstones. Operational taxonomic units (OTUs) were created by 99% similarity using the CD-HIT Suite: Biological Sequence Clustering and Comparison [20]. The Ribosomal Database Project (RDP) classifier was used for naive Bayesian classification of sequences [21]. The taxa that were independently sampled by at least 3 hailstones were considered characteristic and the presence of taxa only sampled by 1 or 2 hailstones was regarded as coincidental. The cultivable community was investigated on the genus level, whereas the total community was analyzed on the phylum, class or order level.

Results and Discussion

The composition of dissolved organic matter (DOM) was determined in terms of quantity and quality. Our bulk analysis of dissolved organic carbon (DOC) and total dissolved nitrogen (TDN) in 18 hailstones revealed high concentrations of both DOC and TDN. Concentrations of DOC ranged between 90 and 1569 μM, with a median DOC concentration of 179 μM (Q1–Q3 = 132–220 μM). TDN concentrations ranged between 23 and 228 μM, with a median of 30 μM (Q1–Q3 = 27–35 μM). Similar DOC concentrations were previously reported for cloud water from orographic clouds [22] and rain [23]. The concentration of TDN was in the range of values reported for TDN in precipitation [24]. On average, more than two thirds of TDN was present as dissolved inorganic nitrogen (DIN) in the form of nitrate and ammonium [6]. It has previously been reported for precipitation in both rural and urban areas world-wide that inorganic nitrogen accounts for the major fraction of dissolved nitrogen [24]. Considering that the concentrations of DOC and TDN in storm clouds are within the same range as the concentrations measured in rivers, lakes and oceans [25], storm clouds can be classified as eutrophic environments. There was a significant correlation (Spearman's rank correlation coefficient, rho = 0.749, p<0.001, n = 18) between DOC and TDN concentrations (Figure 1), which suggests that carbon and nitrogen were derived from the same organic source, which either served as a condensation nucleus or got dissolved in cloud water. Subsequently, the source was diluted by deposition of water vapour or coalescence of other cloud droplets, causing the range of concentrations that we observed between individual hailstones. As most of the TDN in hailstones was inorganic, we assert that a mineralization process, involving photochemistry or biodegradation, took place after dissolution of the source organic compound into cloud droplets.

The ability of heterotrophic microorganisms to metabolize DOM is not only dependent on the quantity of DOM that is available, but also on the molecular composition of the DOM pool. Not all compounds may be equally degradable by the microorganisms that are co-occurring in the cloud droplets. Using ultrahigh-resolution mass spectrometry we characterized the

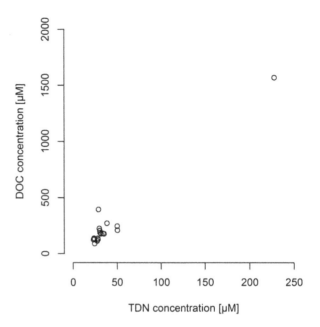

Figure 1. The correlation between DOC and TDN. Dissolved organic carbon (DOC) is presented as a function of dissolved total nitrogen (TDN). The DOC and TDN concentrations were significantly correlated (Spearman's rank correlation coefficient, rho = 0.749, p<0.001, n = 18).

molecular composition of DOM in three individual hailstones. FT-ICR-MS is the only method that allows obtaining molecular information on individual compounds in complex DOM mixtures. The method has been successfully used to get insights into the molecular composition of DOM in marine and in freshwater systems in hitherto unprecedented detail [26]. Here, we applied this advanced analytical technique for the first time on hailstones. Very small volatile organic compounds (<150 Da) escaped our analytical window. Their ubiquitous presence in the atmosphere has already been described elsewhere [22], thus we focused on molecules of the higher molecular mass range. The molecular formulae of 2839 compounds were identified. More than 99% of them were in the molecular mass range of 150–1000 Da. The median mass of all compounds was 354 Da. The median molecular formula was $C_{16}H_{22}O_7$, i.e. half of the detected compounds contained more, and half less, of the respective element. The large molecular diversity and the molecular mass range of DOM in hailstones were comparable to DOM in aquatic systems [26] as well as to water-soluble compounds in aerosols [27]. Forty-four percent of all identified compounds contained one or two nitrogen atoms. All nitrogen was associated to phenolic and unsaturated compounds, whereas peptides and proteins were not present in detectable concentrations. While the molecular diversity of nitrogen-containing compounds was high (1242 compounds contained nitrogen), their abundance in terms of relative concentration was low. Thirteen percent of all compounds contained one sulfur atom. Most sulfur containing compounds were sulfonic acids, some of which are common synthetic products.

The molecular composition of higher molecular mass range DOM is indicative of its history. A few compounds contained less than 10 carbon atoms (Figure 2) and were potentially volatile, but most compounds were too large to be volatile and must have reached the atmosphere as particles. As aromatic compounds in dissolved organic matter are very susceptible to photochemical

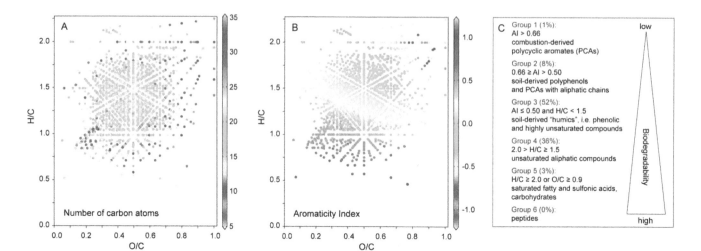

Figure 2. The molecular composition of dissolved organic matter in hail. The element ratio H/C is plotted as a function of O/C for each detected molecular formula detected by ultrahigh-resolution mass spectrometry (FT-ICR-MS) in at least one of the three hailstones. Each dot in these plots represents the molecular formula of an intact molecule. Panel A: The number of carbons in each molecular formula is displayed as a color code in the third dimension. Most compounds are large (C>10) and polar (O/C>0) and are likely not volatile. Panel B: The aromaticity index (AI-mod, [29]) of each molecular formula is displayed as a color code in the third dimension. An aromaticity index 0.5<AI is unambiguous evidence for aromatic compounds and an aromaticity index 0.66<AI is unambiguous evidence for condensed aromatics. Panel C: Compound groups were assigned to molecular formulae based on their aromaticity and element ratios [29], [39]. The biodegradability roughly increases from group 1 to group 6, e.g., polycyclic aromates are among the most stable compounds in the environment, whereas most peptides are quickly decomposed in the environment. Peptides (group 6) have the same characteristics as group 4, but contain nitrogen.

decay [28], the dominance of this compound class in the hailstones is indicative of a fast transfer from soil to atmosphere and hail on the time scale of less than one day.

The aromaticity of each molecular formula was assessed with help of the Aromaticity Index (AI) [29]. An aromaticity index 0.66<AI is an unambiguous criterion for polycyclic aromatic structures that are produced during combustion, but not by organisms. An aromaticity index 0.5<AI is an unambiguous criterion for the presence of aromatic compounds that are abundant in vascular plant debris (e.g. lignin). Thus compounds having an aromaticity index 0.5<AI<0.66 are plant derived material that has undergone microbial degradation in soils. By using AI and element ratios of molecular formulae (H/C and O/C), all detected molecules were grouped according to their molecular structure. The majority of the compounds (60%) were highly unsaturated or phenolic organic acids, typical for soil-derived organic matter in rivers (groups 2 and 3 in Figure 2) (e.g. [30], [31]). Less than 3% of the identified compounds were plant waxes, fatty acids or carbohydrates (group 5 in Figure 2), and less than 1% of the compounds were unambiguously combustion-derived (group 1 in Figure 2). As the majority of compounds were soil-derived (groups 2, 3 and 4 in Figure 2) and leaf surface compounds were present to a lesser degree (part of group 5 in Figure 2), the most likely scenario that explains the molecular composition of DOM in the hailstones is dissolution and desorption of organic matter from soil particles that were mobilized from a local source. A fraction of these particles most probably carried bacterial cells, which often are found aerosolized attached to particles [2]. Peptides were not present in detectable concentrations, thus the only highly biodegradable group of compounds were plant waxes, fatty acids and carbohydrates (group 5 in Figure 2), which represented less than 3% of the identified compounds. Due to the short lifetime of the storm cloud, from which the hailstones were obtained [6], only group 5 compounds are likely to be metabolized by bacteria during their residence time in the atmosphere. As 80% of the cloud water

remains airborne after the hail event, the hailstones do not represent the eventual fate of all cloud water, containing organic matter and bacteria. When the cloud dissipates and the droplets evaporate, the aerosols remain airborne and can serve as condensation nuclei for future cloud events. Thus, the detected trace amounts of easily degradable organic compounds, may serve as bacterial substrate even after the cloud has dissipated.

The analysis of 12 individual hailstones by flow cytometry revealed total bacterial numbers ranging from 778 to 21 321 cells per ml (Me = 1973, Q1–Q3 = 1485–2960, Figure 3). The bacterial densities in the storm cloud are in the lower range of previously reported cell numbers in cloud water, which ranged between 1500 [13] and 430 000 [11] bacteria per ml. Based on the average bacterial density in hailstones and an assumed initial cloud droplet diameter of 10 μm [32], we can calculate that on average only 1 out of 10^6 storm cloud droplets carried a bacterial cell. Thus, cloud droplets are sparsely populated environments, where competition for nutrients and space between bacterial cells is likely insignificant. In addition, we can conclude that cloud water is a nutrient-rich microbial environment, in which significant increase in cell numbers would be possible even if only 3% of the high molecular mass DOM is readily biodegradable.

The median cultivability of bacteria was 0.8% (Q1–Q3 = 0.2%–1.5%), with high variability characteristic for individual hailstones. The reported range of cultivable bacteria in the atmosphere is between 0.01% and 75% [2]. However, cloudborne bacterial communities have previously been found to be characterized by a lower cultivability of between <1% and 2% [10], [33]. Up to 10.5% of all storm cloud bacteria were cultivable on nutrient agar plates (Figure 3), a property that is consistent with an opportunistic ecologic strategy. Lower cultivability of cells from some hailstones was probably a result of stress factors that the cells were subjected to during hailstone formation. E.g. these cells could have been subjected to several cycles of freeze-thawing during hailstone formation, which may cause cultivable bacteria to die or develop into a viable but non-cultivable state [34]. The fact than an

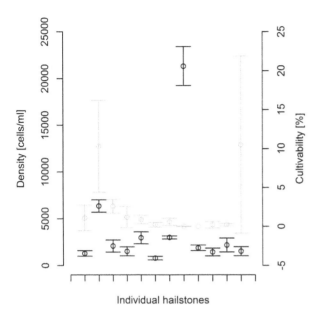

Figure 3. Bacterial density and proportion of cultivable cells. The mean density of bacterial cells as determined with flow cytometry in individual hailstones is presented (gray lines). The proportion of cultivable cells is shown for the same hailstones (dark lines). Error bars denote the standard deviation.

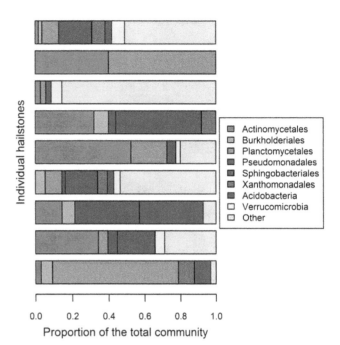

Figure 4. Total community composition in the storm cloud. Proportion of characteristic orders and phyla in 9 out of 12 hailstones, from which clone libraries were made. Characteristic orders and phyla are defined as the ones detected in ≥3 hailstones.

unusually high cultivability has recently been described for epiphytic bacterial community [35], might point to an epiphytic origin of a part of storm cloud community [36], although high cultivability is not a strong proof by itself.

The total bacterial community was highly variable in individual hailstones (Figure 4), which is likely a consequence of storm clouds being a highly dynamic and temporary environment. However, there were taxa that were found in ≥3 hailstones and may represent typical cloud inhabitants. Bacterial orders and phyla, of which representatives were found in at least three individual hailstones, are termed characteristic and are presented in Figure 4. Sequences from *Actinobacteria* (23% of all sequences) were found in all investigated hailstones and representatives from *Plantomycetes* (11%), the *Bacteroidetes* (14%) as well as the γ-*Proteobacteria* (12%) were found in ≥5 hailstones. Also, sequences belonging to *Acidobacteria* (3%), *Verrucomicrobia* (3%), α- (5%) and β-*Proteobacteria* (8%) were present in ≥3 individual hailstones. Terrestrial habitats, and plants in particular, have been identified as major sources of bacteria in the atmosphere [2]. When investigating the spatial variability of airborne bacterial communities, Bowers et al [3] found that the community composition depended on the type of their source environment. Considering the high concentrations of *Actinobacteria* and *Bacteroides* and the low concentrations of *Rhizobiales* (Figure 4), the storm cloud community of our study resembles most closely the airborne communities found by Bowers et al [3] at suburban locations. This agreement fits well with the location of the storm cloud formation, which was over a city. The indicative bacterial taxa of our study pointed to a predominant terrestrial source of the atmospheric bacterial community [3]. The low abundance of *Acidobacteria* and *Rhizobiales* (Figure 4) indicated that soil bacteria were not dominant in the community, whereas the high relative abundance of γ-*Proteobacteria* and *Sphingobacteriales* (Figure 4) suggested that the microbiota in cloud droplets is to a large extent influenced by bacteria of epiphytic origin. This fits well with the high fraction of cultivable cells found for the storm

cloud community and is consistent with the results obtained by others (e.g. [3]).

Common cultivable genera, found in individual hailstones, are presented in Figure 5. Bacterial genera of the cultivable community, which were isolated from ≥3 individual hailstones, were considered characteristic for the cloud. In contrast to the community represented by the clone library, the cultivable bacterial community had a higher proportion (43.5%) of characteristic genera (Figure 5). Some of the characteristic genera (*Bacillus*, *Paenibacillus*, *Bradyrhizobium*) that were represented by isolates are consistent with soil origin of the cultivable community, but there was a remarkable (22%) contribution of typical plant-surface bacteria belonging to the genus *Methylobacterium*. They are adapted to a number of stress factors common for plant surfaces and the atmosphere [6], and therefore predestinated to remain active in the airborne state. We found, for example, that about 90% of the *Methylobacterium* isolates produced reddish, most likely carotenoid-type pigments, which can protect the cells against UV-induced cell damage [37]. In addition to being adapted to atmospheric stress, several *Methylobacterium* isolates have a wide substrate range, which is consistent with an opportunistic ecological strategy [6] and would predispose these cells to growth in the atmosphere. On the contrary, members of typical soil inhabiting genera, *Bacillus* and *Paenibacillus*, most likely get airborne as endospores, which hinders their growth in the atmosphere.

Despite the fact that the total and cultivable bacterial community composition as well as high cultivability all indicated the dominance of plant-associated bacterial groups, the molecular characteristics of DOM pointed to a soil origin of most aerosol particles in the cloud droplets. In fact, very few molecules suggested direct plant-surface origin, as the plant-derived compounds showed the chemical signature of decomposition in soil prior to aerosolization. A likely explanation for the discrepancy between chemical and microbial data is that bacteria originating

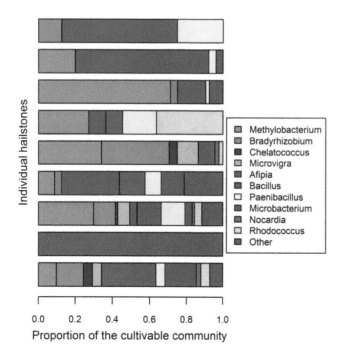

Figure 5. Cultivable genera in the storm cloud. Proportion of characteristic cultivable genera in 9 out of 12 hailstones, which contained cultivable bacteria. Characteristic genera are defined as the ones isolated from ≥3 hailstones.

from plant surfaces are better adapted to survival and growth in the atmosphere, whereas the stressors encountered in the atmosphere such as desiccation and UV radiation act as strong selective barriers against the soil inhabiting bacteria. Consequently, epiphytic bacteria may get enriched in the atmosphere, which could not only affect the chemical composition of the atmosphere,

but also impact precipitation patterns more strongly than previously thought, as most INA bacteria stem from plant surfaces [38].

Conclusions

The unique data sets that we obtained by analyzing individual hailstones provide us with unprecedented insight into the microbial and chemical inventory of storm clouds. They allow us to conclude that while the majority of aerosols were mainly soil-derived, the storm cloud contained microbial communities with a strong plant-surface signature, which links the troposphere to the phyllosphere. Many plant-associated bacteria are efficient in utilizing variable substrates on short timescales as well as in coping with atmospheric stress. Growth of these bacteria can be supported by the trace amounts of carbohydrates, lipids and some nitrogen-containing compounds that we detected among the high molecular mass DOM. The accumulating evidence strongly points to a selection process of bacterial cells in the course of cloud formation, which likely impacts the long-distance transport and the global distribution of bacteria. Our study on hailstones indicates that storm clouds are among the most extreme habitats on Earth, where microbial life can exist.

Acknowledgments

We thank Marijan Govedič for sample collection and Tina Thane, Kathrin Klaproth and Matthias Friebe for excellent technical assistance. We appreciate the helpful advice of Mark A. Lever and Kasper U. Kjeldsen regarding the molecular work with low density bacterial environments.

Author Contributions

Conceived and designed the experiments: TŠT KF BMH UGK. Performed the experiments: TŠT TD RT. Analyzed the data: TŠT TD RT NWN. Contributed reagents/materials/analysis tools: TŠT TD RT KF BMH UGK. Wrote the paper: TŠT KF UGK TD.

References

1. Delort AM, Vaïtilingom M, Amato P, Sancelme M, Parazols M, et al. (2010) A short overview of the microbial population in clouds: potential roles in atmospheric chemistry and nucleation processes. Atmos Res 98(2–4): 249–260.
2. Burrows SM, Elbert W, Lawrence MG, Pöschl U (2009) Bacteria in the global atmosphere – Part 1: Review and synthesis of literature data for different ecosystems. Atmos Chem Phys Discuss 9: 10777–10827.
3. Bowers RM, McLetchie S, Knight R, Fierer N (2011) Spatial variability in airborne bacterial communities across land-use types and their relationship to the bacterial communities of potential source environments. ISME J 5:601–612.
4. Bowers RM, Lauber CL, Wiedinmyer C, Hamady M, Hallar AG, et al. (2009) Characterization of Airborne Microbial Communities at a High-Elevation Site and Their Potential To Act as Atmospheric Ice Nuclei. Appl Environ Microbiol 75(15): 5121–5130.
5. Kourtev PS, Hill KA, Shepson PB, Konopka A (2011) Atmospheric cloud water contains a diverse bacterial community. Atmos Environ 45: 5399–5405.
6. Temkiv TŠ, Finster K, Hansen BM, Nielsen NW, Karlson UG (2012) The microbial diversity of a storm cloud as assessed by hailstones. FEMS Microbiol Ecol 81(3): 684–695. DOI: 10.1111/j.1574-6941.2012.01402.x
7. Sun J, Ariya P (2006) Atmospheric organic and bio-aerosols as cloud condensation nuclei (CCN): A review. Atmos Environ 40: 795–820.
8. Möhler O, DeMott PJ, Vali G, Levin Z (2007) Microbiology and atmospheric processes: the role of biological particles in cloud physics. Biogeosciences 4: 1059–1071.
9. Govindarajan AG, Lindow SE (1988) Size of bacterial ice-nucleation sites measured in situ by radiation inactivation analysis. Proc Nat Acad Sci USA 85(5): 1334–1338.
10. Bauer H, Kasper-Giebl A, Löflund M, Giebl H, Hitzenberger R, et al. (2002) The contribution of bacteria and fungal spores to the organic carbon content of cloud water, precipitation and aerosols. Atmos Res 64: 109–119.
11. Hill KA, Shepson PB, Galbavy ES, Anastasio C, Kourtev PS, et al. (2007) Processing of atmospheric nitrogen by clouds above a forest environment. J Geophys Res 112: 1–16.

12. Herlihy LJ, Galloway JN, Mills AL (1987) Bacterial utilization of formic and acetic acid in rainwater. Atmos Environ 21(11): 2397–2402.
13. Sattler B, Puxbaum H, Psenner R (2001) Bacterial growth in supercooled cloud droplets. Geophys Res Lett 28(2): 239–242.
14. Amato P, Demeer F, Melaouhi A, Fontanella S, Martin-Biesse AS, et al. (2007) A fate for organic acids, formaldehyde and methanol in cloud water: their biotransformation by micro-organisms. Atmos Chem Phys 7(15): 4159–4169.
15. Vaïtilingom M, Amato P, Sancelme M, Laj P, Leriche M, et al. (2010) Contribution of Microbial Activity to Carbon Chemistry in Clouds. Appl Environ Microbiol 76(1): 23–29.
16. Reasoner DJ, Geldreich EE (1985) A new medium for the enumeration and subculture of bacteria from potable water. Appl Environ Microbiol 49: 1–7.
17. Stubbins A, Dittmar T (2012) Low volume quantification of dissolved organic carbon and dissolved nitrogen. Limnol Oceanogr. Methods 10, 347–352.
18. Dittmar T, Koch BP, Hertkon N, Kattner G (2008) A simple and efficient method for the solid-phase extraction of dissolved organic matter (SPE-DOM) from seawater. Limnol Oceanogr-Meth 6: 230–235.
19. Marie D, Brussaard CPD, Thyrhaug R, Bratbak G, Vaulot D (1999) Enumeration of marine viruses in culture and natural samples by flow cytometry. App Environ Microbiol 65(1): 45–52.
20. CD-HIT Suite: Biological Sequence Clustering and Comparison website. Available: http://weizhong-lab.ucsd.edu/cdhit_suite/cgi-bin/index.cgi. Accessed 2012 July 10.
21. Wang Q, Garrity GM, Tiedje JM, Cole JR (2007) Naive Bayesian classifier for rapid assignment of rRNA sequences into the new bacterial taxonomy. Appl Environ Microbiol 73(16): 5261–5267.
22. Marinoni A, Laj P, Sellegeri K, Mailhot G (2004) Cloud chemistry at the Puy de Dôme: variability and relationships with environmental factors. Atmos Chem Phys 4: 715–728.
23. Willey JD, Kieber RJ, Eyman MS, Avery GB Jr (2000) Rainwater dissolved organic carbon: Concentrations and global flux. Global biochem cy 14(1): 139–148.

24. Cornell SE, Jickells TD, Cape JN, Rowland AP, Duce RA (2003) Organic nitrogen deposition on land and costal environments: a review of methods and data. Atmos Environ 37: 2173–2191.
25. Maita Y, Yanada M (1990) Vertical distribution of total dissolved nitrogen and dissolved organic nitrogen in seawater. Geochem J 24: 245–254.
26. Dittmar T, Paeng J (2009) A heat-induced molecular signature in marine dissolved organic matter. Nature Geosci 2: 175–179.
27. Wozniak AS, Bauer JE, Sleighter RL, Dickhut RM, Hatcher PG (2008) Technical Note: Molecular characterization of aerosol-derived water soluble organic carbon using ultrahigh resolution electrospray ionization Fourier transform ion cyclotron resonance mass spectrometry. Atmos Chem Phys 8: 5099–5111.
28. Stubbins A, Spencer RGM, Chen H, Hatcher PG, Mopper K, et al.(2010) Illuminated darkness: molecular signatures of Congo River dissolved organic matter and its photochemical alteration as revealed by ultrahigh precision mass spectrometry. Limnol Oceanogr 55: 1467–1477.
29. Koch BP, Dittmar T (2006) From mass to structure: An aromaticity index for high-resolution mass data of natural organic matter. Rapid Commun Mass Spectrom 20: 926–932.
30. Stenson AC, Marshall AG, Cooper WT (2003) Exact Masses and Chemical Formulas of Individual Suwannee River Fulvic Acids from Ultrahigh Resolution Electrospray Ionization Fourier Transform Ion Cyclotron Resonance Mass Spectra. Anal Chem 75(6): 1275–1284.
31. Tremblay LB, Dittmar T, Marshall AG, Cooper WJ, Cooper WT (2007) Molecular characterization of dissolved organic matter in a north Brazilian mangrove porewater and mangrove-fringed estuary by ultrahigh resolution Fourier transform-ion cyclotron resonance mass spectrometry and excitation/emission spectroscopy. Mar Chem 105: 15–29.
32. Ahrens CD (2009) Meteorology Today. An Introduction to Weather, Climate, and the Environment. Ninth Edition. Belmont: Brooks/Cole. 549 p.
33. Amato P, Menager M, Sancelme M, Laj P, Mailhot G, et al. (2005) Microbial population in cloud water at the Puy de Dome: Implications for the chemistry of clouds. Atmos Environ 39(22): 4143–4153.
34. Oliver JD (2005) The Viable but Nonculturable State in Bacteria. J Microbiol 43: 93–100.
35. Niwa R, Yoshida S, Furuya N, Tsuchiya K, Tsushima S (2011) Method for simple and rapid enumeration of total epiphytic bacteria in the washing solution of rice plants. Can J Microbiol 57: 62–67.
36. Garland JL, Cook KL, Adams JL, Kerkhof L (2001) Culturability as an Indicator of Succession in Microbial communities. Microbiol Ecol 42(2): 150–158.
37. Jacob JL, Carroll TL, Sundin GW (2004) The Role of Pigmentation, Ultraviolet Radiation Tolerance, and Leaf Colonization Strategies in the Epiphytic Survival of Phyllosphere Bacteria. Microbial Ecol 49: 104–113.
38. Morris CE, Sands DC, Vinatzer BA, Glaux C, Guilbaud C, et al. (2008) The life history of the plant pathogen *Pseudomonas syringae* is linked to the water cycle. ISME J: 1–14.
39. Perdue EM (1984) Analytical constraints on the structural features of humic substances. Geochim Cosmochim Ac 48: 1435–1442.

16

Estimating Global "Blue Carbon" Emissions from Conversion and Degradation of Vegetated Coastal Ecosystems

Linwood Pendleton[1,⑤], **Daniel C. Donato**[2*,⑤], **Brian C. Murray**[1], **Stephen Crooks**[3], **W. Aaron Jenkins**[1], **Samantha Sifleet**[4], **Christopher Craft**[5], **James W. Fourqurean**[6], **J. Boone Kauffman**[7], **Núria Marbà**[8], **Patrick Megonigal**[9], **Emily Pidgeon**[10], **Dorothee Herr**[11], **David Gordon**[1], **Alexis Baldera**[12]

1 Nicholas Institute for Environmental Policy Solutions, Duke University, Durham, North Carolina, United States of America, 2 Ecosystem & Landscape Ecology Lab, University of Wisconsin, Madison, Wisconsin, United States of America, 3 ESA Phillip Williams & Associates, San Francisco, California, United States of America, 4 United States Environmental Protection Agency, Research Triangle Park, North Carolina, United States of America, 5 School of Public and Environmental Affairs, Indiana University, Bloomington, Indiana, United States of America, 6 Department of Biological Sciences and Southeast Environmental Research Center, Florida International University, North Miami, Florida, United States of America, 7 Department of Fisheries and Wildlife, Oregon State University, Corvallis, Oregon, United States of America and Center for International Forest Research, Bogor, Indonesia, 8 Department of Global Change Research, Mediterranean Institute for Advanced Studies, Esporles, Illes Balears, Spain, 9 Smithsonian Environmental Research Center, Edgewater, Maryland, United States of America, 10 Conservation International, Arlington, Virginia, United States of America, 11 International Union for the Conservation of Nature, Washington, District of Columbia, United States of America, 12 The Ocean Conservancy, Baton Rouge, Louisiana, United States of America

Abstract

Recent attention has focused on the high rates of annual carbon sequestration in vegetated coastal ecosystems—marshes, mangroves, and seagrasses—that may be lost with habitat destruction ('conversion'). Relatively unappreciated, however, is that conversion of these coastal ecosystems also impacts very large pools of previously-sequestered carbon. Residing mostly in sediments, this 'blue carbon' can be released to the atmosphere when these ecosystems are converted or degraded. Here we provide the first global estimates of this impact and evaluate its economic implications. Combining the best available data on global area, land-use conversion rates, and near-surface carbon stocks in each of the three ecosystems, using an uncertainty-propagation approach, we estimate that 0.15–1.02 Pg (billion tons) of carbon dioxide are being released annually, several times higher than previous estimates that account only for lost sequestration. These emissions are equivalent to 3–19% of those from deforestation globally, and result in economic damages of $US 6–42 billion annually. The largest sources of uncertainty in these estimates stems from limited certitude in global area and rates of land-use conversion, but research is also needed on the fates of ecosystem carbon upon conversion. Currently, carbon emissions from the conversion of vegetated coastal ecosystems are not included in emissions accounting or carbon market protocols, but this analysis suggests they may be disproportionally important to both. Although the relevant science supporting these initial estimates will need to be refined in coming years, it is clear that policies encouraging the sustainable management of coastal ecosystems could significantly reduce carbon emissions from the land-use sector, in addition to sustaining the well-recognized ecosystem services of coastal habitats.

Editor: Simon Thrush, National Institute of Water & Atmospheric Research, New Zealand

Funding: Funding for this effort was provided by the Linden Trust for Conservation (lindentrust.org) and Roger and Victoria Sant. The funders had no role in study design, data collection and analysis, decision to publish, or preparation of the manuscript.

Competing Interests: Appointment funding for one author (SC) comes from ESA Phillip Williams & Associates, a commercial source.

* E-mail: ddonato@wisc.edu

⑤ These authors contributed equally to this work.

Introduction

Anthropogenic contributions to atmospheric greenhouse gases (GHG) are due largely to the combustion of fossil fuels. Land-use activities, especially deforestation, are also a major source of GHG, accounting for ~8–20% of all global emissions [1]. While the role of terrestrial forests as a source and sink of greenhouse gases is well known, new evidence indicates that another source of GHG is the release, via land-use conversion, of carbon (C) stored in the biomass and deep sediments of vegetated ecosystems such as tidal marshes, mangroves, and seagrass beds. These coastal carbon stocks are increasingly referred to as "blue carbon" [2,3]. The exact amount of carbon stored by these ecosystems is still an active area of research, but the potential contribution to GHG from their loss is becoming clear. Yet these emissions are so far relatively unappreciated or even neglected in most policies relating to climate change mitigation [4]. Here, we estimate the potential magnitude and economic impact of these previously unaccounted emissions.

Carbon is stored in vegetated coastal ecosystems throughout the world (Figure 1). Seagrass beds are found from cold polar waters to the tropics. Mangroves are confined to tropical and sub-tropical areas, while tidal marshes are found in all regions, but most commonly in temperate areas. Combined, these ecosystems cover approximately 49 million hectares (Figure 1, Table 1) and provide a diverse array of ecosystem services such as fishery production, coastline protection, pollution buffering, and high rates of carbon sequestration [5].

Rapid loss of vegetated coastal ecosystems through land-use change has occurred for centuries, and has accelerated in recent decades. Causes of habitat conversion vary globally and include conversion to aquaculture, agriculture, forest over-exploitation, industrial use, upstream dams, dredging, eutrophication of overlying waters, urban development, and conversion to open water due to accelerated sea-level rise and subsidence [6–12]. Estimates of cumulative loss over the last 50–100 years range from 25–50% of total global area of each type [12]. This decline continues today, with estimated losses of ~0.5–3% annually depending on ecosystem type, amounting to ~8000 km^2 lost each year [7,11,13–19]. At current conversion rates, 30–40% of tidal marshes and seagrasses [20] and nearly 100% of mangroves [8] could be lost in the next 100 years.

An emerging body of literature recognizes the importance of coastal habitat loss to climate change [2,15,21,22]. However this research has focused almost exclusively on the lost carbon sequestration potential (annual uptake), while the conversion of large standing carbon pools (previously sequestered and stored C) associated with vegetated coastal ecosystems has been relatively overlooked. Only in the most recent studies and reviews has the release of standing carbon pools begun to gain more attention [12,23,24].

Quantitative estimates of these emissions are scarce. Indications are that such 'pulse' releases may have the largest and most immediate impact on green house gas (GHG) emissions, possibly amounting to 50 times the annual net carbon sequestration rate [12,25]. Similar greenhouse gas emissions from the conversion or degradation of freshwater wetlands (e.g., peatlands) are recognized by scientists and international policy-making bodies [1,26], while blue carbon remains largely unaccounted.

Vegetated coastal ecosystems typically reside over organic-rich sediments that may be several meters deep and effectively 'lock up' carbon due to low-oxygen conditions and other factors that inhibit decomposition at depth [27]. These C stocks can exceed those of terrestrial ecosystems, including forests, by several times [24,28]. When coastal habitats are degraded or converted to other land uses, the sediment carbon is destabilized or exposed to oxygen, and subsequent increased microbial activity releases large amounts of GHG to the atmosphere or water column [25,27,29–32]. For example, sediment C was reduced by 50% within 8 years after land clearing in a Panamanian mangrove [29]. Lovelock et al. [33] reported large short-term CO_2 efflux from the sediment surface of cleared mangroves of approximately 29 Mg CO_2 ha^{-1} yr^{-1}. Eventually the majority of carbon in disturbed coastal ecosystems can be released to the atmosphere (in the form of CO_2, CH_4, or other carbon species) with the timeframe highly variable and dependent on the specific land use and nature of the sediment [23].

The potential economic impacts that come from releasing stored coastal blue carbon to the atmosphere are felt worldwide. Economic impacts of GHG emissions in general stem from associated increases in droughts, sea level, and frequency of extreme weather events [34]. Costs are believed to be borne most acutely in low-income countries. However, the potentially large carbon emissions from degraded vegetated coastal ecosystems may also offer a new carbon mitigation opportunity that is currently unrealized—similar to, or even part of, Reduced Emissions from Deforestation and Degradation (REDD+), in which economic

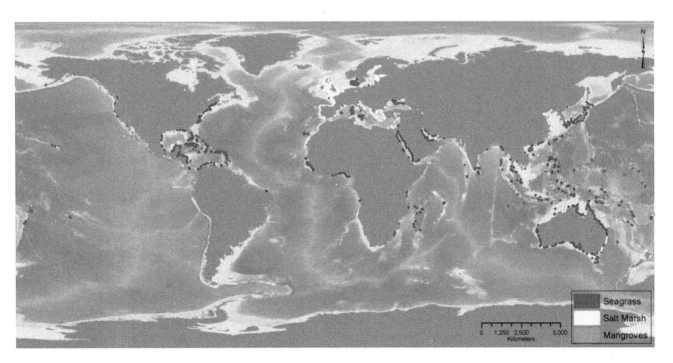

Figure 1. Global distribution of seagrasses, tidal marshes, and mangroves. Data sources: Seagrass and saltmarsh coverage data are from the United Nations Environment Programme World Conservation Monitoring Centre (UNEP-WCMC); mangrove coverage data are from UNEP-WCMC in collaboration with the International Society for Mangrove Ecosystems (ISME).

Table 1. Estimates of carbon released by land-use change in coastal ecosystems globally and associated economic impact.

| Ecosystem | Inputs | | | Results | |
	Global extent (Mha)	Current conversion rate (% yr^{-1})	Near-surface carbon susceptible (top meter sediment+biomass, Mg CO$_2$ ha^{-1})	Carbon emissions (Pg CO$_2$ yr^{-1})	Economic cost (Billion US$ yr^{-1})
Tidal Marsh	2.2–40 (5.1)	1.0–2.0 (1.5)	237–949 (593)	**0.02–0.24 (0.06)**	**0.64–9.7 (2.6)**
Mangroves	13.8–15.2 (14.5)	0.7–3.0 (1.9)	373–1492 (933)	**0.09–0.45 (0.24)**	**3.6–18.5 (9.8)**
Seagrass	17.7–60 (30)	0.4–2.6 (1.5)	131–522 (326)	**0.05–0.33 (0.15)**	**1.9–13.7 (6.1)**
Total	33.7–115.2 (48.9)			**0.15–1.02 (0.45)**	**6.1–41.9 (18.5)**

Notes: 1 Pg = 1 billion metric tons. To obtain values per km^2, multiply by 100. See Methods section for detailed description of inputs and their sources. In brief, data for global extent and conversion rate are recently published ranges (minimum - maximum, and central estimate in parentheses). For near-surface carbon susceptible to land-use conversion (expressed in potential CO$_2$ emissions [48–50]), uncertainty range is based on assumption of 25–100% loss C upon land-use impact; thus, the high-end estimate is the literature-derived global mean carbon storage in vegetation and the top meter of sediment only (central estimate is thus 63% loss). Results for carbon loss are non-parametric 90% confidence intervals (median in parentheses) from Monte Carlo uncertainty propagation of the three input variables (see Methods). Economic estimates apply a multiplier of US$ 41 per ton of CO$_2$ to lower, upper, and central emission estimates (see Methods).

incentives encourage the maintenance of forest ecosystem C storage [4].

Our objective here is to provide the most comprehensive estimate to date of the global carbon emissions and economic impacts of the ongoing conversion of standing carbon stocks in coastal ecosystems, including carbon emissions from sediments – the first analysis to do so. Policy makers need at least an order of magnitude estimate of the potential importance of coastal habitat change as a contributor to global GHG emissions. Although uncertainties exist in the available data underlying such estimates, there is strong need at the international level for the most up-to-date assessment; sufficient information is available to evaluate the importance of coastal blue carbon in both absolute and relative terms. Given the scientific uncertainties present, we used a parsimonious uncertainty/sensitivity framework to a) establish bookends that very likely contain the true value of global emissions from coastal ecosystem conversion, and b) identify the key data gaps relevant to moving forward with their inclusion in carbon policies.

Methods

Analytical framework

To gauge potential carbon emissions from the conversion of coastal ecosystems, we combined estimates of global area, current conversion rate (% of area lost per year), and near-surface carbon stocks susceptible to loss in each of the three habitat types (Table 1). Each of the input multipliers has varying degrees of uncertainty owing to ranges reported in the literature or limited available data. Therefore we used a Monte Carlo approach [35] to propagate uncertainties in each factor using the best available ranges from the literature (see below). Simulations comprised 50,000 iterations for each habitat type and assumed a normal distribution of input variables within reported ranges, except when ranges were heavily right skewed (i.e., a minority of extremely high estimates in the literature for a given input). In the latter case, we applied a simple gamma distribution with parameters corresponding to the minimum and maximum reported values in order to account for, but avoid undue influence of, possible high-end extremes. This distribution applied to global area estimates of tidal marsh (gamma shape 1.6, scale 6) and seagrass (shape 4, scale 4). Fifth and 95th percentiles were extracted from the 50,000 Monte Carlo iteration outputs to obtain non-parametric 90% confidence intervals for emissions in each type.

Data inputs

Global area. Global area inputs were derived from international monitoring databases and recently published literature. For tidal marshes, we applied a central estimate of 5.1 Mha (obtained from the United Nations Environment Programme-World Conservation Monitoring Centre [UNEP-WCMC] spatial data, 2005) and a range of 2.2 to 40 Mha [12,15,21]. The low-end estimate may be too low, but until improved estimates are available, we took the conservative approach of including the full range of data sources. It should also be noted that, in some cases, these published estimates were derived from the same primary sources, so not all values are truly independent of each other. Of the three habitat types considered, mangroves have perhaps the best global extent data and a fairly narrow range of reported values; we applied the recently reported range of 13.8 to 15.2 Mha [12,19,36] and a corresponding central estimate of 14.5 Mha. For seagrasses, we applied a central estimate of 30 Mha (obtained from UNEP-WCMC spatial data, 2005) and a range of 17.7 to 60 Mha [12,37–39].

Annual area loss. Current rates of global annual loss (land-use conversion) were derived from recently published literature. We assigned a global annual loss rate of 1–2% for tidal marshes [12,16,18]; 0.7–3% for mangroves [7,12,14,17,19]; and 0.4–2.6% for seagrasses [11–13,15,38].

Carbon loss upon conversion. Carbon loss per hectare converted has not been well quantified in coastal ecosystems, but likely bounds can be derived. The loss of vegetation biomass is the most common and readily apparent result of conversion, but there are also losses from the surface sediment carbon pool (<1 m deep; [29,32]) as well as potentially large, but not well understood, C losses from deep sediments [23], [40]. We therefore took a conservative approach by focusing only on carbon in vegetation and the top meter of sediment. These pools are most susceptible to land-use change and are termed here 'near-surface' carbon. For the uncertainty range used in the simulations, we used the best available estimate of global mean near-surface C in each ecosystem type, with a range of possible fates of this pool upon conversion, from 25% to 100% emission to the atmosphere depending on disturbance type, possible re-burial of disturbed material, and degree of C recalcitrance. The high end of 100% would apply if most land uses tend toward extreme impacts that convert the system to a qualitatively different state that removes and prevents recovery of near-surface carbon. The low end of 25% would apply if most land uses are relatively light-handed and retain, bury, or merely redistribute most near-surface carbon. Blue carbon that is oxidized via disturbance and exposure (converted to species such as CO_2, HCO_3^-, or CO_3^{2-}) increases the effective CO_2 concentration of the ocean-atmosphere system. Because of the partial pressure equilibrium of CO_2 between air and water, atmospheric CO_2 levels are affected by either direct ocean-to-atmosphere gas exchange, or by reductions in the ability of the ocean to absorb atmospheric CO_2 [41].

Mechanisms of disturbance to sediment carbon vary by ecosystem type, but often affect near-surface carbon to at least one meter depth. In tidal marshes, a primary land-use activity is the creation of arable land via diking and draining, an effect that may persist for decades and lead to the loss of several meters of sediment, along with its carbon, due to oxidation [23]. For mangroves, conversion to aquaculture is widespread, with the excavation of mangrove sediments to depths of about one meter exposing a large portion of the sediment carbon to oxygen; system degradation through over-harvest can also lead to sediment erosion and exposure [25]. In seagrass systems, water quality impairment, generally from excess nutrients or sediments from terrestrial sources, is a leading cause of ecosystem decline and loss, and ultimately exposure of sediment carbon to the water column or atmosphere [42]. Direct impacts such as dredging, trawling, and anchoring also affect seagrass beds [42].

For near-surface carbon stocks (including just the top meter of sediment), studies suggest conservative carbon storage estimates of approximately 250 Mg of carbon per hectare for tidal marshes [16,21]; 280 Mg C ha^{-1} for mangroves [24,28,43]; and 140 Mg C ha^{-1} for seagrasses [44–47]. Following IPCC protocol for tracking changes in carbon stocks [48–50], and to facilitate comparison among most other assessments, we express ecosystem carbon in terms of potential CO_2 emissions – obtained by multiplying C stocks by 3.67, the molecular weight ratio of CO_2 to C. The values for tidal marshes, mangroves, and seagrasses therefore become 917, 1028, and 512 Mg of potential CO_2 emissions per hectare, respectively. These estimates are conservative since larger amounts of carbon are often held in as much as 6 meters of sediment and biomass beneath the emergent vegetation [21,24,51]. The carbon in emergent living biomass of these ecosystems ranges widely, from estimated mean values of 1 to 129 Mg C ha^{-1} (2 to 474 Mg of potential CO_2 emission ha^{-1}) depending on habitat type [16,24,28,43,45,52–54]. This vegetation biomass increases the near-surface carbon estimates to global means of 259, 407, and 142 Mg C ha^{-1} (949, 1492, and 522 Mg of potential CO_2 emissions ha^{-1}) for tidal marshes, mangroves, and seagrasses, respectively (Table 1).

For each ecosystem, we focus on the total amount of CO_2 that could be released from annual rates of conversion, but we do not attempt to estimate over what course of time these releases would be made. (At the scale of the individual site, the rate of release likely follows a negative exponential curve with time—initially high and tapering in later years. The temporal dynamic of near-surface carbon pools after conversion is a significant research need, but some studies suggest it may have a half-life on the order of 5–10 years [29].) It is important to note that any assumption of the temporal period of release within a degraded site, whether 5–10 years or much longer, is inconsequential to the results of this analysis. When summed over the globe and integrated over time, as long as ecosystem conversion rates are stable or increasing over time, the total amount of carbon released annually would be greater than or equal to our estimates.

Conservative approach

We emphasize that the analysis above should be considered conservative in its estimate of emissions. First, we reduced the emphasis on high-end estimates of global area by using gamma distributions to minimize the impact of especially high estimates. Second, we did not include any potential impacts on deep sediment C (>1 m depth), in part because of limited available science. These layers often contain more C per hectare than all the near-surface carbon combined [24] and have been found to be impacted by land-use change in the few cases studied [23]. This means that even our high-end scenario of 100% C loss upon conversion is actually much less than all of the ecosystem carbon. Third, the low-end scenario of 25% C loss upon conversion effectively assumes that all land-use changes in coastal systems across the entire globe could retain 75% of all near-surface carbon (if most C in disturbed systems is merely buried or redistributed) – an extremely conservative assumption. Fourth, we did not include the loss of annual sequestration of sediment carbon that occurs due to vegetation removal or hydrological isolation that reduces new sediment inputs.

Regarding other greenhouse gases such as methane (CH_4) or nitrous oxide (N_2O), excluding changes in these components is likely either a neutral or conservative approach. In highly saline wetlands (>18 ppt), sediment C sequestration rates exceed CH_4 emission rates in CO_2-equivalent units [55], suggesting that the net effect of losing both sequestration and CH_4 emissions with disturbance should be an increase in greenhouse gas emissions. In lower salinity wetlands (salinity 5–18 ppt), CH_4 emissions and sequestration are approximately in balance [56], except perhaps for oligohaline systems (<5 ppt) that are a small portion of the global area we evaluated. Finally, we conservatively did not consider evidence that common disturbances, such as conversion to shrimp ponds, that cause eutrophication have been shown to stimulate CH_4 emissions [27]. Eutrophication is likely to also increase N_2O emissions if the system receives high nitrate loading; otherwise it is not necessary to account for changes in N_2O fluxes because emissions from anaerobic sediments are negligible in the absence of nitrate loading.

Economic impact

Finally, we calculated the estimated cost to the global economy of the estimated emissions resulting from coastal ecosystem conversion. We multiplied the global emissions estimates for each type by a recent estimate of the global economic cost of new atmospheric carbon of $41 per ton of CO_2 (2007 U.S. dollars) [57]. This cost is a central estimate of the "social cost of carbon" (SCC), which is defined as the marginal value of economic damages of the climate change attributable to an additional ton of CO_2 in the atmosphere in 2020 (2007 dollars) [57]. The SCC estimate is an estimate of the environmental damages that can be avoided by reducing emissions, but does not necessarily equal the price that the market will pay for reducing emissions, since that market price is determined by the avoided cost of regulatory controls on carbon and not avoided damages per se [57].

Results and Discussion

CO_2 emissions

We estimate that the conversion and degradation of coastal ecosystems each year may ultimately release between 0.15 and 1.02 Pg (billion tons) of CO_2 to the atmosphere, with a central estimate of 0.45 Pg CO_2 (Table 1). Mangroves contain the largest per-hectare carbon stocks and contribute approximately half the estimated total blue carbon emissions. Seagrasses, despite containing the lowest per-hectare carbon stocks, contribute the second most to global blue carbon emissions, due to their larger global area. Tidal marshes contain moderate to high carbon stocks, but their relatively small total area results in the lowest— although still substantial—global emissions.

To put these emissions in perspective, the central estimate of 0.45 Pg CO_2 yr^{-1} approaches the annual fossil fuel CO_2 emissions of the United Kingdom (the world's 9th ranked country by emissions), while the low estimate of 0.15 Pg is roughly equivalent to those of Venezuela (ranked 30th) and the high estimate of 1.02 Pg approaches those of Japan (ranked 5th) [20]. Comparing to other ecosystem C fluxes, the loss of vegetated coastal ecosystems may contribute an additional 3–19% above the most recent estimates of global emissions from deforestation (5.5 Pg CO_2 yr^{-1} including freshwater peatlands) [1], or offset 12–80% of the carbon sink in the ocean's continental shelves globally (1.26 Pg CO_2 yr^{-1}) [58]. The lost annual sequestration potential of coastal ecosystems, which is considerable, would push these estimates higher [22].

Worth noting is that these estimates account only for changes in ecosystem C *in situ*, and do not account for possible exchanges among different ecosystems – e.g., the transfer of C from one system in another, which would effectively reduce the atmospheric emissions result. The degree to which some disturbed blue carbon is merely redistributed (e.g., exported from a disturbed mangrove to adjacent seagrass) just means that the true value of global emissions may be more toward the lower end of our uncertainty range, which assumes as much as 75% retention of near-surface carbon. While the amount of C transferred to other habitats is likely to be small compared to the C gas emissions described here, we recommend care be taken when aggregating carbon budgets across multiple habitats should they include assumptions on the transfer and deposition of carbon from one habitat to another.

Although tidal marshes, mangroves, and seagrasses occupy only a thin coastal fringe, they play a disproportionally large role in land-use carbon gas emissions. For example, compared to the highly publicized loss of tropical forests, the combined area of the three coastal ecosystems equates to only 2–6% of tropical forest area but contributes up to an additional 19% over current estimates of deforestation emissions. Disturbance of the carbon stored in the biomass and top meter of sediment in a typical hectare of mangrove could contribute as much emissions as three to five hectares of tropical forest [24,28,43,48,59]. Even a hectare of seagrass meadow, with its small living biomass, may hold as much near-surface carbon as a hectare of tropical forest [47,48,59].

The emissions estimates derived here are considerably higher than previous estimates of the potential greenhouse impact of coastal ecosystem loss that have only considered lost sequestration potential. Bridgham et al. [16] estimated that the destruction of mangroves and tidal marshes has resulted in reduced sequestration of 0.076 Pg CO_2 per year. Pidgeon [60] estimated that 0.003 Pg CO_2 per year of sequestration potential are lost due to current rates of mangrove and seagrass loss. Irving et al. [22] provided an analysis of the large sequestration potential of restoring degraded coastal ecosystems. Those studies focused on the annual new sequestration that is lost (gained) when the ecosystem is converted (restored). Our estimates focus on the loss of carbon stocks in coastal ecosystem sediments that have accumulated over hundreds to thousands of years and are lost, upon disturbance, within a period of decades [23]. These emissions (summed over all converted area and assuming a relatively constant or increasing conversion rate globally) are additional to the lost sequestration potential just referenced.

Economic impacts

Combining the uncertainty range in emissions with a central estimate for the social cost of carbon gas emissions of $41 per Mg of CO_2, we estimate the current global cost of coastal ecosystem conversion to be between $6.1 and $42 billion incurred annually (Table 1). The range would be even wider if we considered the full range of SCC values from $7–81 [57]. However, even at the low end of the range there is relatively high economic value in maintaining sediment carbon beneath coastal ecosystems and out of the atmosphere. The high ongoing cost of coastal ecosystem loss also supports the conclusion of Irving et al. [22], that management efforts focused on reducing coastal habitat loss may be more beneficial than the extensive restoration efforts being conducted in many regions which have smaller carbon benefits.

Around the globe, coastal ecosystems are lost because market forces give landowners incentive to profitably convert habitat. Elsewhere, ecosystems are lost because governments have been unwilling or unable to enforce clean water regulations and other measures that would help guarantee the continued ecological sustainability of these systems. There are, however, only a few mechanisms currently in place that would pay landowners, managers, or governments to protect the carbon stored in coastal ecosystems,

The cost of coastal ecosystem protection includes the expense of creating and managing protected areas, improving water quality, and particularly the opportunity costs of foregone alternative uses (e.g., aquaculture, real estate development). These costs can be quite high in some cases; therefore strong economic incentive would be required to counteract conversion. Absent payment mechanisms for the protection of coastal carbon, the degradation and loss of coastal ecosystems will likely continue. The global economic consequences will exceed the social cost of increased greenhouse gases as the loss of the array of ecosystem services they provide, such as fishery nurseries, biodiversity support, and coastal protection have tremendous economic value in their own right [5,8].

A global market for greenhouse gas emission reductions could help remedy this situation. Such "carbon markets" have been operating throughout the world since the adoption of the United Nation's Framework Convention on Climate Change's Kyoto Protocol, but there has been a very limited role for terrestrial carbon reductions (e.g., forests), and no role for carbon in coastal ecosystems. Recent efforts may create a global market opportunity for reduced emissions from deforestation and degradation (REDD+). Guidance on modalities relating to deforestation emissions [61] highlight the need to include significant carbon pools in forest reference emission levels and/or forest reference levels, or to otherwise provide reasons for omitting these pools. These guidelines may be applied to mangrove forests and their belowground carbon [62], providing one step toward inclusion of a major source of coastal blue carbon in such programs.

Other opportunities have been outlined to include coastal carbon management such as ecosystem conservation, restoration, and sustainable use into the UNFCCC [4,63]. Nationally Appropriate Mitigation Actions (NAMAs) could be an opening for developing countries to reduce carbon gas emissions while increasing national capacity-building and data collection activities. The newly adopted definition of wetland drainage and rewetting under the Kyoto Protocol provides an incentive to account for anthropogenic greenhouse gas emissions and removals by Annex-I Parties [64]. These represent further potential mechanisms for reducing emissions of coastal blue carbon to the atmosphere.

Remaining Uncertainties

Scientific understanding differs among the various coastal ecosystems. Based on a sensitivity analysis within the Monte Carlo simulations, the largest contributions to uncertainty in emissions stemmed from wide published ranges for global area and conversion rates. Uncertainty is relatively high for emissions estimates for tidal marsh systems largely due to limited information on spatial extent, which had the widest influence on total emissions estimates of any input variable (accounting for 30% of total uncertainty). For mangroves, global area is better quantified, but uncertainty in conversion rates is substantial and had a large influence on total emissions estimates (18%). For seagrasses, the range in conversion rate was the most important influence on total uncertainty (14%). The proportion of C lost when converted had variable influence: the range for tidal marshes contributed only 2% total uncertainty, that for seagrasses contributed 9%, and that for mangroves contributed 18%. The value is largest for mangroves because they contain the largest near-surface C stocks. However, because of the limited number of studies of whole-ecosystem blue carbon stocks in these systems, we did not apply ranges in carbon stock estimates, focusing instead on the proportion released as applied to the best available central estimates. Further studies across a broad geographic range will allow development of likely ranges of C stocks and a more complete accounting of the uncertainty in blue carbon (gas) emissions. Overall, the most important information needs relevant to moving forward with blue carbon conservation (e.g., REDD+) include better quantification of the global area of tidal marshes and seagrasses, the actual areal conversion rates of mangrove and seagrass ecosystems, and the fate of blue carbon when disturbed in all systems.

We focused on potential CO_2 emissions from conversion of standing stocks (in this case, determined by areal rates of conversion) and did not address the separate effects of changes in background flux rates which have been covered in other analyses [16,22,57]. Lost annual C sequestration would effectively increase the emissions consequences of conversion. In the most saline systems (salinity >18), this is true even if the disturbance were to decrease emissions of CH_4 [55]. In addition, common disturbances such as conversion to shrimp ponds may increase CH_4 due to euthrophication [27]. Thus, we are potentially underestimating greenhouse consequences of conversion. In oligohaline tidal marshes, however, natural methane efflux is often present in undisturbed conditions and may decrease when altered [56], which diminishes the emissions consequences of conversion [65]. Although methane is a strong greenhouse gas, changes in its contribution within the context of blue carbon may be less than ~10–15% of our estimates of increased CO_2 emissions due to conversion [55,65]. Nevertheless, further refinement of methane dynamics in response to ecosystem conversion remains a research need.

Conclusion

We currently know that coastal ecosystems contain substantial quantities of blue carbon. To our knowledge this analysis is the first to a) combine the best available estimates of global area, conversion rates, and ecosystem C *stocks* (not simply lost sequestration potential) to estimate blue carbon emissions on a global scale; b) use an uncertainty analysis to identify key data uncertainties relevant to moving forward with conservation of blue carbon; and c) estimate the global economic impacts of blue carbon emissions. Our analysis suggests that the greenhouse consequences of conversion of these ecosystems are larger than previously appreciated, by as much as an order of magnitude. These emissions add considerably to existing estimates of land-use carbon gas emissions such as tropical deforestation. Although these ecosystems occur as relatively thin coastal fringes, the economic impacts of $US 6–42 billion per year are borne globally.

This analysis establishes bookends and highlights the likely importance of blue carbon conversion. Information available to support these estimates, however, has high uncertainty. New research is needed to improve our estimates of how much carbon is trapped in these ecosystems, how much carbon is released into the atmosphere by their conversion, and where on the planet carbon loss is occurring most rapidly. Our analysis incorporated widely varying inputs and therefore shows that, regardless of how the science is ultimately refined, the unaccounted carbon gas emissions from coastal conversions are quite likely very high.

While more natural science research is underway, the development of policies and protocols that allow existing and emerging carbon markets to compensate stewards for conserving these ecosystems and reducing the amount of carbon gas emissions to the atmosphere could move forward. If markets and policies are in place, emerging science can translate into action for coastal blue carbon. Such policies could have a significant impact on greenhouse gas emissions, and a transformational impact on the ecosystems themselves.

Acknowledgments

All authors contributed directly to this manuscript as part of a workshop held at Duke University in October 2010 and through the contribution of data, interpretation, and writing. We thank S. Copeland for valuable insights on this topic and 2 anonymous reviewers for constructive comments.

Author Contributions

Conceived and designed the experiments: LP DCD BCM SC WAJ SS CC JWF JBK NM PM EP DH DG AB. Analyzed the data: LP DCD BCM SC WAJ CC JWF JBK NM PM. Wrote the paper: LP DCD BCM SC WAJ SS CC JWF JBK NM PM EP DH DG AB.

References

1. van der Werf GR, Morton DC, DeFries RS, Olivier JGJ, Kasibhatla PS, et al. (2009) CO_2 emissions from forest loss. Nature Geoscience 2: 737–738.

2. Nellemann C, Corcoran E, Duarte CM, Valdes L, DeYoung C, et al. (Eds) (2009). Blue Carbon. A Rapid Response Assessment. United Nations Environment Programme, GRID-Arendal website. www.grida.no. Accessed 2011 Nov 11.

3. Gordon D, Murray BC, Pendleton L, Victor B (2011) Financing Options for Blue Carbon: Opportunities and Lessons from the REDD+ Experience. Report NI R 11-11, Nicholas Institute for Environmental Policy Solutions, Duke University, Durham. Duke University website. Available at: http:// nicholasinstitute.duke.edu/economics/naturalresources/financing-options-for-blue-carbon/at_download/paper. Accessed 2012 January 5.

4. Climate Focus (2011) Blue Carbon Policy Options Assessment. Washington DC.

5. Barbier EB, Hacker SD, Kennedy C, Koch EW, Stier AC, et al. (2011) The value of estuarine and coastal ecosystem services. Ecological Monographs 81: 169–193.

6. Short FT, Wyllie-Echeverria S (1996) Natural and human-induced disturbance of seagrasses. Environmental Conservation 23: 17–27.

7. Valiela I, Bowen JL, York JK (2001) Mangrove forests: One of the world's threatened major tropical environments. BioScience 51: 807.

8. Duke NC, Meynecke JO, Dittmann S, Ellison AM, Anger K, et al. (2007) A world without mangroves? Science 317: 41–42.

9. Giri C, Zhu Z, Tieszen LL, Singh A, Gillette S, et al. (2008) Mangrove forest distributions and dynamics (1975–2005) of the tsunami-affected region of Asia. Journal of Biogeography 35: 519–528.

10. Giri C, Muhlhausen J (2008) Mangrove forest distributions and dynamics in Madagascar (1975–2005). Sensors 8, 2104–2117.
11. Waycott M, Duarte CM, Carruthers TJB, Orth RJ, Dennison WC, et al. (2009) Accelerating loss of seagrasses across the globe threatens coastal ecosystems. Proceedings of the National Academy of Science USA 106: 12377–12381.
12. McLeod E, Chmura GL, Bouillon S, Salm R, Bjork M, et al. (2011) A blueprint for blue carbon: toward an improved understanding of the role of vegetated coastal habitats in sequestering CO_2. Frontiers in Ecology and the Environment 9: 552–560.
13. Costanza R, d'Arge R, de Groot R, Farber S, Grasso M, et al. (1997) The value of the world's ecosystem services and natural capital. Nature 387: 253–60.
14. Alongi DM (2002) Present state and future of the world's mangrove forests. Environmental Conservation 29: 331–49.
15. Duarte CM, Middelburg J, Caraco N (2005) Major role of marine vegetation on the oceanic carbon cycle. Biogeosciences 2: 1–8.
16. Bridgham SD, Megonigal JP, Keller JK, Bliss NB, Trettin C (2006) The carbon balance of North American wetlands. Wetlands 26: 889–916.
17. Food and Agriculture Organization (FAO) of the United Nations (2007) The World's Mangroves 1980–2005. FAO, Rome.
18. Duarte CM, Dennison WC, Orth RJW, Carruthers TJB (2008) The charisma of coastal ecosystems: Addressing the imbalance. Estuaries and Coasts 31: 233–238.
19. Spalding MD, Kainuma M, Collins L (2010) World atlas of mangroves. Earthscan, London.
20. Intergovernmental Panel on Climate Change (IPCC) (2007) Climate change 2007: Assessment Report. IPCC, Valencia.
21. Chmura GL, Anisfeld SC, Cahoon DR, Lynch JC (2003) Global carbon sequestration in tidal saline wetland sediments. Global Biogeochemical Cycles 17: 1111.
22. Irving AD, Connell SD, Russell BD (2011) Restoring coastal plants to improve global carbon storage: Reaping what we sow. PLoS ONE 6: e18311.
23. Crooks S, Herr D, Laffoley D, Tamelander J, Vandever J (2011) Regulating Climate Change Through Restoration and Management of Coastal Wetlands and Near-shore Marine Ecosystems: Mitigation Potential and Policy Opportunities. World Bank, IUCN, ESA PWA, Washington, Gland, San Francisco.
24. Donato DC, Kauffman JB, Murdiyarso D, Kurnianto S, Stidham M, et al. (2011) Mangroves among the most carbon-rich forests in the tropics. Nature Geoscience 4: 293–297.
25. Eong OJ (1993) Mangroves – a carbon source and sink. Chemosphere 27: 1097–1107.
26. Murdiyarso D, Hergoualc'h K, Verchot LV (2010) Opportunities for reducing greenhouse gas emissions in tropical peatlands. Proceeding of the National Academy of Science USA 107: 19655–19660.
27. Kristensen E, Bouillon S, Dittmar T, Marchand C (2008) Organic carbon dynamics in mangrove ecosystems. Aquatic Botany 89: 201–219.
28. Donato DC, Kauffman JB, Mackenzie RA, Ainsworth A, Pfleeger AZ (2012) Whole-island carbon stocks in the tropical Pacific: Implications for mangrove conservation and upland restoration. Journal of Environmental Management 97: 89–96.
29. Granek E, Ruttenberg BI (2008) Changes in biotic and abiotic processes following mangrove clearing. Estuarine and Coastal Shelf Science 80: 555–562.
30. Sjöling S, Mohammed SM, Lyimo TJ, Kyaruzi JJ (2005) Benthic bacterial diversity and nutrient processes in mangroves: impact of deforestation. Estuarine and Coastal Shelf Science 63: 397–406.
31. Strangmann A, Bashan Y, Giani L (2008) Methane in pristine and impaired mangrove sediments and its possible effects on establishment of mangrove seedlings. Biology and Fertility of Soils 44: 511–519.
32. Sweetman AK, Middleburg JJ, Berle AM, Bernardino AF, Schander C, et al. (2010). Impacts of exotic mangrove forests and mangrove deforestation on carbon remineralization and ecosystem functioning in marine sediments. Biogeosciences 7: 2129–2145.
33. Lovelock CE, Reuss RW, Feller IC (2011) CO2 efflux from cleared mangrove peat. PLoS ONE 6(6): e21279.
34. Tol RSJ (2009) The Economic Effects of Climate Change, Journal of Economic Perspectives 23: 29–51.
35. Doucet A, de Freitas N, Gordon N (2001) Sequential Monte Carlo methods in practice. Springer, New York.
36. Giri C, Ochieng E, Tieszen LL, Zhu Z, Singh A, et al. (2010) Status and distribution of mangrove forests of the world using earth observation satellite data. Global Ecology and Biogeography 20: 154–159.
37. Duarte CM, Borum J, Short FT, Walker DI (2005) Seagrass ecosystems: their global status and prospects. In: (ed. Polunin, N.V.C.) Aquatic ecosystems: trends and global prospects. Cambridge University Press, Cambridge.
38. Green EP, Short FT (2003) World atlas of seagrasses. California University Press, Berkeley.
39. Charpy-Roubaud C, Sournia A (1990) The comparative estimation of phytoplanktonic and microphytobenthic production in the oceans. Marine Microbial Food Webs 4: 31–57.

40. Deverel SJ, Leighton DA (2010) Historic, Recent, and Future Subsidence, Sacramento-San Joaquin Delta, California, USA. San Francisco Estuary and Watershed Science 8(2): 1–23.
41. Drever JI (1997) The Geochemistry of Natural Waters: Surface and Groundwater Environments. Prentice Hall, New Jersey.
42. Orth RJ, Carruthers TJB, Dennison WC, Duarte CM, Fourqurean JW, et al. (2006) A global crisis for seagrass ecosystems. Bioscience 56: 987–996.
43. Kauffman JB, Heider C, Cole TG, Dwyer K, Donato DC (2011) Ecosystem carbon stocks of Micronesian mangrove forests. Wetlands 31: 343–352.
44. Mateo MA, Romero J, Perez MM, Littler DS (1997) Dynamics of millenary organic deposits resulting from the growth of the Mediterranean seagrass Posidonia oceanica. Estuarine and Coastal Shelf Science 44: 103–110.
45. Duarte CM, Chiscano CL (1999) Seagrass biomass and production: a reassessment. Aquatic Botany 65: 159–174.
46. Vichkovitten T, Holmer M (2005) Dissolved and particulate organic matter in contrasting Zostera marina (eelgrass) sediments. Journal of Experimental Marine Biology and Ecology 316: 183–201.
47. Fourqurean JW, Duarte CM, Kennedy H, Marba N, Holmer M, et al. (2012) Seagrass ecosystems as a globally significant carbon stock. Nature Geoscience 5: 505–509.
48. Intergovernmental Panel on Climate Change (IPCC) (2003) Good Practice Guidance for Land use, Land-use Change, and Forestry (Eds Penman, J. et al.). Institute for Global Environmental Strategies.
49. Pearson T, Walker S, Brown S (2005) Sourcebook for land use, land-use change and forestry projects. BioCF and Winrock International website. Available at: http://www.winrock.org/ecosystems/tools.asp?BU=9086. Accessed 2010 October 10.
50. Pearson TRH, Brown SL, Birdsey RA (2007) Measurement guidelines for the sequestration of forest carbon. General Technical Report-NRS-18, USDA Forest Service, Northern Research Station.
51. Golley F, Odum HT, Wilson RF (1962) The structure and metabolism of a Puerto Rican red mangrove forest in May. Ecology 43: 9–19.
52. Morgan PA, Short FT (2002) Using functional trajectories to track constructed salt marsh development in the Great Bay Estuary, Maine/New Hampshire USA. Restoration Ecology 10: 461–473.
53. Komiyama A, Ong JE, Poungparn S (2008) Allometry, biomass, and productivity of mangrove forests: A review. Aquatic Botany 89: 128–137.
54. Yu OT, Chmura GL (2009) Sediment carbon may be maintained under grazing in a St. Lawrence Estuary Tidal Marsh. Environmental Conservation 36: 312–320.
55. Poffenbarger HJ, Needelman BA, Megonigal JP (2011) Methane emissions from tidal marshes. Wetlands 31: 831–842.
56. Krithika K, Purvaja R, Ramesh R (2008) Fluxes of methane and nitrous oxide from an Indian mangrove. Current Science 94(2): 218–224.
57. United States Government (USG) (2010) Technical Support Document: Social Cost of Carbon for Regulatory Impact Analysis Under Executive Order 12866. United States Environmental Protection Agency website. Available at: http://www.epa.gov/otaq/climate/regulations/scc-tsd.pdf. Accessed 2011 May 8.
58. Chen CTA, Borges AV (2009) Reconciling opposing views on carbon cycling in the coastal ocean: continental shelves as sinks and near-shore ecosystems as sources of atmospheric CO_2. Deep-Sea Research II 56: 578–590.
59. Pan Y, Birdsey RA, Fang J, Houghton R, Kauppi PE, et al. (2011) A large and persistent carbon sink in the world's forests. Science 333: 988–993.
60. Pidgeon E (2009) Carbon Sequestration by Coastal Marine Habitats: Important Missing Sinks. In: (eds. Laffoley, D., Grimsditch, G.) The Management of Natural Coastal Carbon Sinks. IUCN, Gland.
61. UNFCCC (United Nations Forum Convention on Climate Change (2011) Draft Decision -/CP.17 Guidance on systems for providing information on how safeguards are addressed and respected and modalities relating to forest reference emission levels and forest reference levels as referred to in decision 1/CP.16. United Forum Convention on Climate Change website. Available at: http://unfccc.int/files/meetings/durban_nov_2011/decisions/application/pdf/cop17_safeguards.pdf. Accessed 2012 January 15.
62. Alongi DM (2011) Carbon payments for mangrove conservation: ecosystem constraints and uncertainties of sequestration potential. Environmental Science and Policy 14: 462–470.
63. Herr D, Pidgeon E, Laffoley D (Eds.) (2011). Blue Carbon Policy Framework: International Blue Carbon Policy Working Group. Gland, Arlington.
64. UNFCCC (2011) Decision -/CMP.7 Land use, land-use change and forestry. United Nations Forum Convention on Climate Change website. Available at: http://unfccc.int/files/meetings/durban_nov_2011/decisions/application/pdf/awgkp_lulucf.pdf. Accessed 2012 January 15.
65. Murray BC, Pendleton L, Jenkins WA, Sifleet S (2011) Green payments for blue carbon: Economic incentives for protecting threatened coastal habitats. Report NI R 11-04, Nicholas Institute for Environmental Policy Solutions, Duke University, Durham.

17

Alpine Grassland Soil Organic Carbon Stock and Its Uncertainty in the Three Rivers Source Region of the Tibetan Plateau

Xiaofeng Chang[1,2], Shiping Wang[3]*, Shujuan Cui[2,4], Xiaoxue Zhu[2,4], Caiyun Luo[2], Zhenhua Zhang[2], Andreas Wilkes[5]

1 State Key Laboratory of Soil Erosion and Dryland Farming on Loess Plateau, Institute of Soil and Water Conservation, Northwest A&F University, Yangling, China, 2 Key Laboratory of Adaptation and Evolution of Plateau Biota, Northwest Institute of Plateau Biology, Chinese Academy of Science, Xining, China, 3 Laboratory of Alpine Ecology and Biodiversity, Institute of Tibetan Plateau Research, Chinese Academy of Sciences, Beijing, China, 4 University of Chinese Academy of Science, Beijing, China, 5 World Agroforestry Centre East Asia Programme, Beijing, China

Abstract

Alpine grassland of the Tibetan Plateau is an important component of global soil organic carbon (SOC) stocks, but insufficient field observations and large spatial heterogeneity leads to great uncertainty in their estimation. In the Three Rivers Source Region (TRSR), alpine grasslands account for more than 75% of the total area. However, the regional carbon (C) stock estimate and their uncertainty have seldom been tested. Here we quantified the regional SOC stock and its uncertainty using 298 soil profiles surveyed from 35 sites across the TRSR during 2006–2008. We showed that the upper soil (0–30 cm depth) in alpine grasslands of the TRSR stores 2.03 Pg C, with a 95% confidence interval ranging from 1.25 to 2.81 Pg C. Alpine meadow soils comprised 73% (i.e. 1.48 Pg C) of the regional SOC estimate, but had the greatest uncertainty at 51%. The statistical power to detect a deviation of 10% uncertainty in grassland C stock was less than 0.50. The required sample size to detect this deviation at a power of 90% was about 6–7 times more than the number of sample sites surveyed. Comparison of our observed SOC density with the corresponding values from the dataset of Yang et al. indicates that these two datasets are comparable. The combined dataset did not reduce the uncertainty in the estimate of the regional grassland soil C stock. This result could be mainly explained by the underrepresentation of sampling sites in large areas with poor accessibility. Further research to improve the regional SOC stock estimate should optimize sampling strategy by considering the number of samples and their spatial distribution.

Editor: Kurt O. Reinhart, USDA-ARS United States of America

Funding: This work was supported by the National Basic Research Program (2013CB956000), Strategic Priority Research Program (B) of the Chinese Academy of Sciences (XDB03030403), Special Program of Carbon Sequestration of the Chinese Academy of Sciences (XDA05070205), the National Science Foundation for Young Scientists (41303062), the West Light Foundation of the Chinese Academy of Science. The funders had no role in study design, data collection and analysis, decision to publish, or preparation of the manuscript.

Competing Interests: The authors have declared that no competing interests exist.

* E-mail: wangsp@itpcas.ac.cn

Introduction

Soil stores more carbon (C) than the vegetation and atmosphere pools combined, and minor changes in soil organic carbon (SOC) stock could have momentous effects on atmospheric CO_2 concentrations [1]. As the Earth's third pole, the Tibetan Plateau is mostly covered by typical alpine grasslands, which contain large soil C stocks [2,3]. Alpine grasslands in the Tibetan Plateau could feedback to accelerate the current warming trend by releasing large amounts of this stored C to the atmosphere [4,5]. Therefore, estimates of organic C stocks in alpine grasslands are crucial for understanding the regional and global greenhouse gas balance [2]. Despite considerable research over the past 20 years, much uncertainty exists regarding the SOC stock in the alpine grasslands. For example, Yang et al. [2] used a satellite-based approach and estimated that the SOC stock in the top 1 m in alpine grasslands was 7.4 Pg C, with an average density of 6.5 kg C m^{-2}. Wang et al. [3], using the First National Soil Survey dataset and field measurements surveyed in the eastern part of the

Tibetan Plateau, estimated the SOC stock at 33.52 Pg for alpine grasslands, with an average SOC density of 20.9 kg C m^{-2}. Therefore, precise quantification of soil C stocks in alpine grasslands of the region is needed to make credible conclusions about the potential scale of feedback between the terrestrial C cycle and climate.

Regional scale assessments of SOC have typically been supported by data from soil inventories [3]. An important issue with soil C stock inventories is spatial heterogeneity [6]. Increased SOC variability causes decreased sampling representativeness and increased sample size is needed to estimate the true SOC distribution [7,8]. Previous studies also found that a large number of sampling plots is useful to assess the spatial variation of C stocks in a heterogeneous landscape and to reduce the uncertainty in the final SOC estimates [9]. Yu et al. [10] examined spatial variability of SOC in a red soil region of South China varying in land use and soil type, using six sampling densities (14, 34, 68, 130, 255 and 525 points in 927 km^2). They found that high sampling densities gradually decreased the variation in SOC. Similarly, Muukkonen et al. [11] showed that the spatial variation in C stock in boreal forest soil decreased with increasing number of samples, without further increase in the precision of the estimate after 20–30 samples in a 6.25 m² area. Such results suggest incentives for soil studies to increase the number of samples to reduce variability and improve soil C stock estimates. Despite these small–scale efforts, there is a lack of information on the effects of sampling effort on SOC estimates at regional scale [10,12].

As the variability within a relatively homogeneous stratum is lower than the variability within a broad heterogeneous landscape, stratification of soil sampling can improve the SOC stock estimate [13,14]. These results were the basis for conducting a stratified, random sampling design in the present study. Because grassland type is the most important variable driving the spatial pattern of SOC [15], the Three Rivers Source Region (TRSR) was stratified by grassland type, and this strategy is expected to capture a large part of SOC variation. In this study, we assessed alpine grassland soil C stock and its uncertainty. Specifically, our study objectives were (i) to quantify SOC density from alpine grasslands at three depths (0–10, 0–20 and 0–30 cm), and (ii) to investigate the statistical power and sample size requirement to detect a deviation of 10% uncertainty.

Materials and Methods

Ethics Statement

All necessary permits were obtained for the described regional soil inventory from the Qinghai Province Environmental Protection Bureau, which is responsible for the Three Rivers' Headwaters National Nature Reserve. No specific permissions are required by individuals since land in China belongs to the state. No endangered or protected species have been disturbed in our field sampling. The geographic information of sampling sites is provided in Table S1.

Study Region

The TRSR is composed of the water source region of the Yangtze River, Yellow River and Lancang (Mekong) River. The TRSR covers 30.23×10^4 km², which ranges in longitude from 89.75 to 102.38°E and in latitude from 31.65 to 36.20°N. Climate variation in the region is represented by a mean annual temperature range of −5.38 to 4.14°C, with mean annual precipitation ranging 262.2 to 772.8 mm, and annual evaporation from 730 to 1700 mm [16]. The elevation ranges from 3500 to 4800 m. Most of the region is dominated by alpine steppe and alpine meadow, with some areas of sparse alpine shrub and alpine marsh (Fig. 1). Alpine steppe is dominated by hardy perennial xeric herbs such as *Stipa purpure*, *Carex moorcrofii* and *Dalea racemosa*. Alpine meadow is dominated by *Kobresia pygmaea*, *K. humilis* and *K. tibetica*. Alpine shrub is dominated by *Salix oritrepha var. amnematch-inensis* (L.). Alpine marsh has formed in permanently waterlogged areas or where the soil has been over-saturated, and supports hardy perennial hydro-philous or hydro-mesophytic herbs such as *K. tibetica* [17–18]. Soils in this region are shallow, with a depth of about 30–50 cm [19].

Field Sampling and Laboratory Measurements

A total of 298 soil profiles from 35 sites were sampled from August to September 2006 and 2008 (Fig. 1). Of the 35 sites, 9 were alpine steppe, 21 were alpine meadow, 2 alpine shrubland and 3 alpine marsh (Table 1). At each site of alpine steppe and meadow, eight soil profiles along a 200 m transect were collected. For alpine shrub and marsh, 11 and 12 soil profiles were assigned to each transect, respectively. Soil samples were collected from

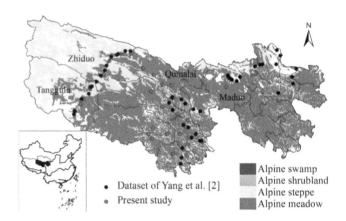

Figure 1. Soil sampling locations from our survey and the extracted dataset of Yang et al. [2].

each profile at every 10 cm to a depth of 30 cm. These soil samples were then mixed to yield one composite sample for each depth interval. All 105 composite samples (35 sites × 3 soil depths) were later air-dried, sieved (2 mm), and analyzed for SOC (measured with a Shimadzu 5000 SOC analyzer, Kyoto, Japan). Bulk density was taken using intact cores of 100 cm³ for each depth in the soil profile. The fine earth (< 2 mm) and stone content were weighed by oven-drying at 105°C for 48 h in the laboratory. The bulk density of the fine soil was calculated after correction for the mass of stone fragments. Then, bulk densities within each site were averaged for each depth. SOC densities (kg C m⁻²) were calculated in each depth interval as the product of soil C concentration (%), bulk density (g cm⁻³) and fixed soil depth (10 cm). SOC density for the 30 cm profile was calculated by summing stocks for the individual 10 cm layers.

Comparison Dataset Collection

In order to identify the effect of sample size on estimates of soil C stocks and their uncertainties on a regional scale, we began with a search of the dataset previously collected by Yang et al. [2]. In this dataset, soil samples were collected from 135 sites in July and August 2001–2004. SOC densities at different depths (30, 50, and 100 cm) were determined at each site. Climate data were also examined using spatial interpolation from the records of 43 meteorology stations across the Tibetan Plateau. Site-specific information on location, soil texture and grassland type are available in the dataset. From this dataset, we extracted records of 49 sites (23 in alpine meadow and 26 in alpine steppe) located in the TRSR for subsequent analysis.

Data Analysis

One-way ANOVA was performed to compare the SOC density values of alpine grasslands in the current study with the corresponding dataset extracted from Yang et al. [2]. Geostatistical methods were used to examine the spatial variation in SOC across the study region. Experimental variograms were computed and the appropriate mathematical function was fitted to the semi-variogram. Maps of predicted SOC density were obtained for the study region using ordinary kriging integrated with the parameters of the appropriate variogram model. We then overlaid the 1:1000 000 vegetation map of the Tibetan Plateau (Chinese Academy of Science 2001) on the SOC density prediction maps to determine the SOC densities for each grassland type. Based on the statistics

Table 1. The study area, distribution of sampling sites and soil organic carbon density by grassland type in the Three River Source Region.

Grassland type	Area (10⁴km²)	Proportion (%)	No. of sample sites	Soil organic carbon density (kg C m⁻²)		
				0–10cm	10–20cm	20–30cm
Alpine meadow	20.18	67.9	21	2.93±0.83	2.44±0.61	2.02±0.53
Alpine steppe	7.52	25.3	9	1.71±0.51	1.57±0.38	1.48±0.35
Alpine shrubland	2.03	6.8	2	3.68±0.39	3.01±0.22	2.52±0.34
Alpine marsh	-	-	3	3.84±0.50	3.07±0.36	2.69±0.98
Total	29.73	100	35			

on SOC density, the regional SOC stock in a given depth interval can be calculated by the following equations:

$$SOCS_j = SOCD_j \times AREA_j \times 10^{-12}, \qquad (1)$$

$$SOCS_i = \sum SOCS_j, \qquad (2)$$

where $SOCS_j$ is the SOC stock (Pg), $SOCD_j$ is the SOC density (kg m⁻²), and $AREA_j$ is the area (m²) for each grassland type j. $SOCS_i$ is the regional SOC stock in layer i (0–10 cm, 0–20 cm and 0–30 cm). Based on the statistics on SOC density estimated using the formulas above, the average value and corresponding standard deviation were estimated using a Monte Carlo approach with 10 000 iterations. From these 10 000 runs, we obtained the average value and the 2.5 and 97.5 percentiles as the final estimate of the mean and of uncertainty (i.e., 95% confidence interval). A percentage uncertainty was estimated based on half of the 95% confidence interval divided by the average SOC stock estimate. Furthermore, we tested whether the number of sampling sites in the current study was sufficient to detect a 10% uncertainty of the SOC stock estimate with power analysis. All power calculations were based on a 0.05 probability of Type I error.

The data were analyzed using different software packages. The descriptive statistical parameters and statistical analyses were performed with SPSS 16.0 (SPSS Inc., USA). Spatial analyses were performed using ArcGIS 9.3 (ESRI Inc., USA). Monte Carlo simulation was calculated using RiskAMP software (http://www.thumbstacks.com/) added in Microsoft Excel 2003 (Microsoft Corporation, USA). Power analysis was conducted using the PROC MIXED procedure of SAS (SAS Institute, 2001).

Results

Overall Patterns of SOC Density

The experimental variogram for SOC revealed an evident spatial structure at the regional scale, with a nugget effect of about 15% (Table S2). This was reflected in the maps of the kriged estimates. The predicted value of SOC density decreased from the southeastern to the northwestern areas (Fig. S1). Spatial variation in SOC values was also observed for soil profiles. Table 1 illustrates the results, showing clearly that more C was stored in the top of the profile than at depth. Generally, alpine marsh soils had

the largest SOC density at the three soil depths, followed by alpine shrubland and alpine meadow, and then alpine steppe.

Regional SOC Stock and its Uncertainty

There were clear differences in total SOC stocks between grassland types (Fig. 2). Alpine meadow soils had the largest SOC stock, accounting for about 73% of the regional SOC stock for the three soil depths because of its extensive area (Table 1). By contrast, alpine marsh soils had the smallest areal extent, and lowest C stocks. The SOC stocks for alpine steppe and alpine shrubland soils were 0.36 and 0.19 Pg C, respectively. When the grassland area is examined overall, the estimate of the total stock for the TRSR is 2.03 Pg C. The 95% confidence interval around this estimate is 1.25 to 2.81 Pg C. Alpine meadow soils contributed most to the uncertainty in the regional C stock estimate (Fig. 2).

Power Analysis of SOC

Following Allen et al. [6], we used tolerable uncertainty of 10% above and below the mean of SOC density. Fig. 3 shows the statistical power to detect a deviation within 10% uncertainty for our current study. The probabilities of detecting the tolerable uncertainty were extremely low for all grassland types at three soil depths. Alpine meadow had the largest sample size but had less than 40% chance of detecting the tolerable uncertainty. The

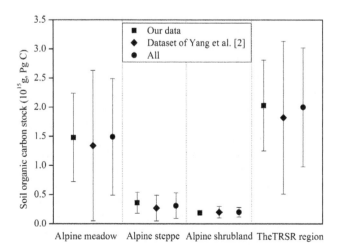

Figure 2. Mean SOC stocks with uncertainty estimates (95% confidence intervals) for alpine grasslands in the Three Rivers Source Region from our data, the dataset of Yang et al. [2], and the pooled dataset.

statistical power was much lower for the other three grassland types. There were clear differences in the power values between soil depths. The larger power values were observed in the subsurface layers (10–20 and 20–30 cm) for alpine meadow and alpine steppe, while the other grassland types had the lowest values at 20–30 cm soil depth. Generally, the required number of sites to detect a deviation of 10% uncertainty with a statistical power of 0.90 was much higher for alpine meadow and alpine steppe than for the other grassland types. The power analyses showed that 90 sample sites are required to meet the tolerated uncertainty for estimating mean soil C stocks for alpine meadow, while 30 sites would be adequate for alpine shrubland at three soil depths. In comparison, about 100 sites would be necessary for alpine steppe at the surface 0–10 cm depth and 150 sites for alpine marsh at a deeper depth (20–30 cm).

Discussion

Comparisons of SOC Density with Earlier Observations

Our current estimates (Table 1) were comparable to that reported by Guo et al. [20] (alpine steppe = 3.88 kg C m^{-2}; alpine meadow = 5.18 kg C m^{-2} for 0–30 cm). To compare regional datasets, data points geographically distributed in the TRSR were extracted from the dataset of Yang et al. [2]. SOC densities derived from our dataset and the extracted dataset were compared for alpine steppe and alpine meadow. We found that there were no significant differences (alpine steppe $t = 0.132$, $P = 0.895$; alpine meadow $t = -1.175$, $P = 0.246$). We also compared the differences in SOC density estimates after kriging of our data vs. Yang et al. [2]. By contrast, local estimates were relatively divergent (Fig. 4). Likewise, SOC values (both concentration and recalculated density) were extracted from our prediction maps for site-specific comparison with previous studies [21]. There were large differences in SOC concentration and density at the small scale level (Fig. 5). These results indicated that regional estimates of SOC are very similar but local predictions based on kriging are not. Part of this variation of SOC density could be explained by varying methods used to determine soil bulk density. Because bulk

density was ignored in previous studies, we estimated the bulk density using two pedotransfer functions for alpine grassland developed by Yang et al. [22] and Zhong et al. [23]. As shown in Fig. 5, there were relatively large differences in the SOC densities calculated using different bulk estimates. Some studies have shown that indirect bulk density estimates based on pedotransfer functions can lead to errors from 9% up to 36% of the SOC density [24,25]. The most potential explanation for this deviation in SOC values may be attributed to the current sampling regime, which is not intensive enough to reveal the spatially explicit patterns of SOC in the study area. In spite of strongly spatial autocorrelation at the regional scale, the experimental variogram showed a relatively large nugget effect (15%), indicating that there was considerable short range variability.

Sample Size Effect on Estimate Uncertainty

Despite a relatively large number of detailed surveys of grassland soil C stocks undertaken to date on the Tibetan Plateau, calculated errors remain relatively high, largely due to insufficient sample size and great sample variance [2]. Our analysis showed that the largest power value observed in any of the grassland types or soil depths exceeded 50%, and more than six times as many sites as those used in our survey were required to detect a 10% uncertainty in regional SOC stocks (Fig. 3). Hence we expected estimated uncertainty in SOC stock would decrease with increasing sample size. However, the estimated uncertainty of regional SOC stocks was not reduced when the two datasets were pooled together (Fig. 2). This result is perhaps unsurprising given that the spatial distribution of sampling sites across the study region was different between the two datasets. Because of bad weather and inaccessibility to vehicles, the previous survey by Yang et al. [2] was mainly conducted along the major roads, while more observations were distributed in the Lancang (Mekong) river basin in our field survey (Fig. 1). Areas that were intensively sampled captured local-scale spatial heterogeneity [26]. However, a sparse sampling distribution increases the standard deviation of the SOC density due to their large spatial coverage. Therefore, increasing sample size only without considering spatial represen-

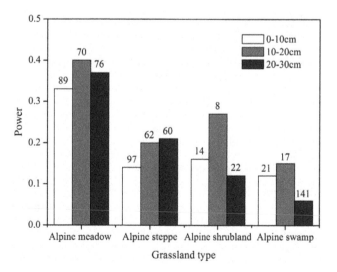

Figure 3. Probability (power) of detecting a 10% uncertainty in soil organic carbon density with a 0.05 level of significance for grassland types at three soil depths. The numbers above each bar represent the number of sampling sites needed to be taken at three depths for each alpine grassland to detect the tolerated uncertainty with 90% probability.

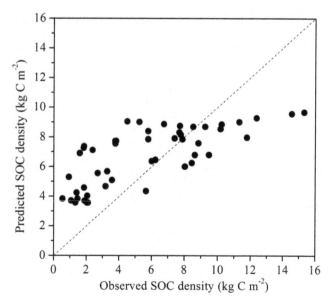

Figure 4. Predicted SOC densities by kriging of our data against observed values of Yang et al. [2].

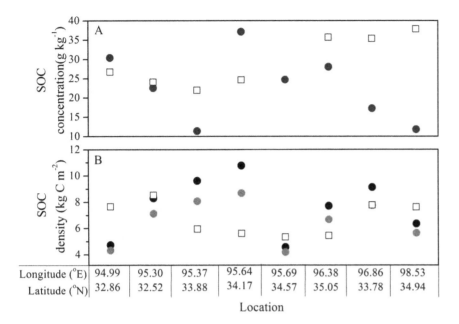

Figure 5. Comparisons of soil organic carbon concentration (A) and density (B) between present and previous studies. The present SOC values (open squares) were extracted from our kriged prediction map. The dark grey circles (A) indicated the measured SOC concentration values from previous study, while the previous corresponding SOC densities (B) were re-calculated using different bulk density values derived from pedotransfer functions of Yang et al. [22] (dark circles) and Zhong et al. [23] (gray circles), since bulk density was not measured in the previous study [21].

tativeness would not necessarily reduce the level of uncertainty and improve statistical power.

Implication for Sampling Design

Given the large uncertainty in estimation of regional SOC stocks, there is considerable interest in quantifying the magnitude and designing an efficient sample scheme to reduce uncertainty [26,27]. One way of improving the efficiency and precision of a future survey is to tailor the sample size to the expected variability in soil C stock [28]. An additional ca. 200 sample sites across the TRSR would be required for accurate C stock estimate within 10% uncertainty, which would greatly increase costs and survey effort. Previous soil C research has produced a wealth of information that can be synthesized into a comprehensive and quantitative dataset, which offers an alternative that is likely to decrease sample size requirements for a subsequent inventory [7]. However, we found that almost all current studies concentrated sampling in easily accessed areas along the main roads on the Tibetan Plateau, while large areas that are poorly accessible due to limited road networks or poor security were substantially underrepresented [2,3,20,23]. Future sampling effort on the Tibetan Plateau would be better directed to exploring these underrepresented areas, such as the northern Tanggula Mountains and the Hoh Xil region. However, a sound sampling design may be impractical and prohibitively expensive. Therefore, future inventories should incorporate the use of geoinformatics (GIS/ Remote Sensing) tools or extrapolative spatial models of regional C stock estimates, as these have the potential to reduce costs [2,29,30]. As the greatest uncertainty in the regional SOC estimate originated from the variance in SOC values assigned to alpine meadow (Fig. 2), a number of additional sample sites to increase representation of alpine meadow soils would significantly reduce the uncertainties in regional soil C stock estimates in the TRSR.

Conclusions

This work presents, to the best of our knowledge, the first regional estimate of SOC stock in the TRSR region. The SOC stored in alpine grassland soils of the TRSR region at 0–30 cm depth ranged from 1.25 to 2.81 Pg C at 95% confidence, with a mean of 2.03 Pg C. We observed that the largest source of uncertainty affecting regional SOC estimates derives from alpine meadow soils. SOC stocks varied with grassland type. Mean SOC stocks were 0.36 and 0.19 Pg C in alpine steppe and alpine shrubland, respectively. Approximately 73% (about 1.48 Pg C) of the regional SOC storage occurred in alpine meadow, which covers about 68% of the grassland area. Our result also indicated that uncertainty in the SOC stock estimates did not reduce when our dataset was pooled with the extracted dataset of Yang et al. [2], even though this provided more than twice as many sample sites as those in our survey. This most likely resulted from the underrepresentation of soils sampled in large areas that are relatively inaccessible in the northwest and southeastern part of the TRSR. Therefore, improvement in regional SOC stock estimates requires the addition of a number of sampling sites targeted to these known gaps and to address the high variability of estimated SOC stocks in alpine meadow soils.

Acknowledgments

We thank Chao Zengguo and Xu Guangping for supporting the field campaign and Hu Yigang for his assistance during laboratory analysis. We would also like to thank two anonymous reviewers for careful revision and critical comments on a earlier version of the manuscript.

Author Contributions

Conceived and designed the experiments: SW. Performed the experiments: XC. Analyzed the data: SC XZ CL ZZ. Wrote the paper: XC SW AW.

References

1. Johnston CA, Groffman P, Breshears DD, Cardon ZG, Currie W, et al. (2004) Carbon cycling in soil. Frontiers in Ecology and the Environment 2: 522–528.
2. Yang YH, Fang JY, Tang YH, Ji CJ, Zheng CY, et al. (2008) Storage, patterns and controls of soil organic carbon in the Tibetan grasslands. Global Change Biology 14: 1592–1599.
3. Wang GX, Qian J, Cheng GD, Lai YM (2002) Soil organic carbon pool of grassland soils on the Qinghai-Tibetan Plateau and its global implication. The Science of the Total Envionment 291: 207–217.
4. Wang GX, Wang YB, Li YS, Cheng HY (2007) Influences of alpine ecosystem responses to climatic change on soil properties on the Qinghai-Tibet Plateau, China. Catena 70: 506–514.
5. Wang GX, Li YS, Wang YB, Wu QB (2008) Effects of permafrost thawing on vegetation and soil carbon pool losses on the Qinghai-Tibet Plateau, China. Geoderma 143: 143–152.
6. Allen DE, Pringle MJ, Page KL, Dalal RC (2010). A review of sampling designs for the measurement of soil organic carbon in Australian grazing lands. Rangeland Journal 32: 227–246.
7. Conant RT, Paustian K (2002) Spatial variability of soil organic carbon in grasslands: implications for detecting change at different scales. Environmental Pollution 116: S127–S135.
8. Kravchenko A, Robertson G (2011) Whole-profile soil carbon stocks: The danger of assuming too much from analyses of too little. Soil Science Society of America Journal 75: 235–240.
9. Sierra CA, Del Valle JI, Orrego SA, Moreno FH, Harrmon ME, et al. (2007) Total carbon stocks in a tropical forest landscape of the Porce region, Colombia. Forest Ecology and Management 243: 299–309.
10. Yu DS, Zhang ZQ, Yang H, Shi XZ, Tan MZ, et al. (2011). Effect of soil sampling density on detected spatial variability of soil organic carbon in a red soil region of China. Pedosphere 21: 207–213.
11. Muukkonen P, Häkkinen M, Mäkipää R (2009) Spatial variation in soil carbon in the organic layer of managed boreal forest soil-implications for sampling design. Environmental Monitoring and Assessment 158: 67–76.
12. Zhang W, Weindorf DC, Zhu Y (2011) Soil Organic Carbon Variability in Croplands: Implications for Sampling Design. Soil Science 176: 367–371.
13. Pringle M, Allen D, Dalal R, Payne J, Mayer D, et al. (2011) Soil carbon stock in the tropical rangelands of Australia: Effects of soil type and grazing pressure, and determination of sampling requirement. Geoderma 167: 261–273.
14. Zhang ZQ, Yu DS, Shi XZ, Weindorf DC, Wang XX, et al. (2010) Effect of sampling classification patterns on SOC variability in the red soil region, China. Soil and Tillage Research 110: 2–7.
15. Shi Y, Baumann F, Ma Y, Song C, Kühn P, et al. (2012) Organic and inorganic carbon in the topsoil of the Mongolian and Tibetan grasslands: pattern, control and implications. Biogeosciences 9: 2287–2299.
16. Yi X, Li G, Yin Y (2012) Temperature variation and abrupt change analysis in the Three-River Headwaters Region during 1961–2010. Journal of Geographical Sciences 22: 451–469.
17. Zhou X, Wang Z, Du Q (eds) (1987) *The vegetaion of Qinghai Province*. Qinghai People Press, Qinghai, China.
18. Zhou X (2001) *Chinese Kobresia pygmaea meadow*. Science Press, Beijing, China.
19. Zhao X, Cao G, Li Y (2009) *Alpine meadow ecosystem and global change*. Science Press, Beijing, China.
20. Guo X, Han D, Zhang F, Li Y, Lin L, et al. (2011) The response of potential carbon sequestration capacity to different land use patterns in alpine rangeland. Acta Agrestia Sinica 19 (5): 740–745.
21. Liu Y, Li X, Li C, Sun H, Lu G, et al. (2009) Vegetation decline and reduction of soil organic carbon stock in high-altitude meadow grasslands in the source area of three major reviers of China. Journal of Agro-Enviroment Science 28 (12): 2559–2567.
22. Yang YH, Fang JY, Smith P, Tang YH, Chen AP, et al. (2009) Changes in topsoil carbon stock in the Tibetan grasslands between the 1980s and 2004. Global Change Biology 15: 2723–2729.
23. Zhong C, Yang ZF, Xia XQ, Hou QY, Jiang W (2012) Estimation of soil organic carbon storage ans analysis of soil carbon source/sink factors in Qinghai province. Geoscience 26: 896–909.
24. Boucneau G, Van Meirvenne M, Hofman G (1998) Comparing pedotransfer functions to estimate soil bulk density in northern Belgium. Pedologie-Themata 5: 67–70.
25. De Vos B, Van Meirvenne M, Quataert P, Deckers J, Muys B (2005) Predictive quality of pedotransfer functions for estimating bulk density of forest soils. Soil Science Society of America Journal 69: 500–510.
26. Yuan Z, Gazol A, Lin F, Ye J, Shi S, et al. (2013) Soil organic carbon in an old-growth temperate forest: Spatial pattern, determinants and bias in its quantification. Geoderma 195: 48–55.
27. Goidts E, van Wesemael B, Crucifix M (2009) Magnitude and sources of uncertainties in soil organic carbon (SOC) stock assessments at various scales. European Journal of Soil Science 60: 723–739.
28. Vanguelova E, Nisbet T, Moffat A, Broadmeadow S, Sanders T, et al. (2013) A new evaluation of carbon stocks in British forest soils. Soil Use and Management 29: 169–181.
29. Akumu CE, McLaughlin JW (2013) Regional variation in peatland carbon stock assessments, northern Ontario, Canada. Geoderma 209: 161–167.
30. Cambule A, Rossiter D, Stoorvogel J (2013) A methodology for digital soil mapping in poorly-accessible areas. Geoderma 192: 341–353.

18

Atmospheric Reaction Systems as Null-Models to Identify Structural Traces of Evolution in Metabolism

Petter Holme[1,2]*, Mikael Huss[3,4], Sang Hoon Lee[1]

1 IceLab, Department of Physics, Umeå University, Umeå, Sweden, **2** Department of Energy Science, Sungkyunkwan University, Suwon, Korea, **3** Science for Life Laboratory Stockholm, Solna, Sweden, **4** Department of Biochemistry and Biophysics, Stockholm University, Stockholm, Sweden

Abstract

The metabolism is the motor behind the biological complexity of an organism. One problem of characterizing its large-scale structure is that it is hard to know what to compare it to. All chemical reaction systems are shaped by the same physics that gives molecules their stability and affinity to react. These fundamental factors cannot be captured by standard null-models based on randomization. The unique property of organismal metabolism is that it is controlled, to some extent, by an enzymatic machinery that is subject to evolution. In this paper, we explore the possibility that reaction systems of planetary atmospheres can serve as a null-model against which we can define metabolic structure and trace the influence of evolution. We find that the two types of data can be distinguished by their respective degree distributions. This is especially clear when looking at the degree distribution of the reaction network (of reaction connected to each other if they involve the same molecular species). For the Earth's atmospheric network and the human metabolic network, we look into more detail for an underlying explanation of this deviation. However, we cannot pinpoint a single cause of the difference, rather there are several concurrent factors. By examining quantities relating to the modular-functional organization of the metabolism, we confirm that metabolic networks have a more complex modular organization than the atmospheric networks, but not much more. We interpret the more variegated modular arrangement of metabolism as a trace of evolved functionality. On the other hand, it is quite remarkable how similar the structures of these two types of networks are, which emphasizes that the constraints from the chemical properties of the molecules has a larger influence in shaping the reaction system than does natural selection.

Editor: Matjaz Perc, University of Maribor, Slovenia

Funding: PH was supported by the Swedish Research Council and the World Class University program through National Research Foundation Korea funded by Ministry of Education, Science and Technology R31-2008-000-10029-0. The funders had no role in study design, data collection and analysis, decision to publish, or preparation of the manuscript.

Competing Interests: The authors have declared that no competing interests exist.

* E-mail: petter.holme@physics.umu.se

Introduction

Reaction systems are, at many levels of the universe, motors driving the creation of higher structure. From the metabolism in our bodies, via reactions in planetary interiors and atmospheres, to the nuclear reaction systems in stars; these are all systems shaped by the physical properties of constituents—the atoms and molecules. Among these systems, metabolism is special in the sense that its control has evolved by natural selection. But the physical properties of molecules and the relative abundance of elements constrain the evolution of this genetic control. Perhaps these constraints explain that very different reaction systems—reactions in planetary atmospheres and the organismal metabolism—share large-scale features (like the right-skewed probability distributions of degree, which roughly speaking reflects the number of molecules a molecule can react with) [1,2]. Still, as we will see, there are differences between these two types of systems and in this paper we will focus on what these differences are and what they can tell us of the evolution of metabolism. To put it short, we explore the idea that the reaction systems of planetary atmospheres can be null-models for studying metabolic networks in an evolutionary perspective.

The study of reaction-system topology (the set of all participating reactions) has long been restricted, by lack of data, to small subsystems. These systems, like e.g. the citric acid cycle of metabolism [3] or the carbon-nitrogen-oxygen cycle of stellar nuclear reactions [4] (two systems that were, coincidentally, both discovered in the mid-1930's), have been modeled in great detail with e.g. differential equations. It has, however, not until recently been possible to investigate the system-wide organization of any type of reaction system. Since about a decade, we do have methods to infer the entire set of reactions (again coincidentally) both in metabolism and planetary atmospheres. Still these datasets are so crude that our conclusions in this paper will be rather hypothetical in nature. On the encouraging side, however, the early conclusions mentioned above—that reaction network are right-skewed and fat-tailed [1,2]—still hold for contemporary datasets. If we go beyond the topology, even less is known. A full picture of reaction rates and concentrations for a traditional kinetic modeling is far into the future. One complication comes from the fact that metabolites (and also molecular species in atmospheres) are distributed heterogeneously in space [5] and sometimes so few in number that concentration based models do not apply. This means that when investigating the global organization of reaction systems, we will have to rely on graph-based analysis techniques for still some time. Even though graph-based methods need to discard much of the knowledge we have about reaction kinetics, one can still encode much information into the graph. The molecular species present determine the vertices of the network; the catalysts present define the reactions. But what should the edges represent? Should one also include separate vertex-types for reactions and catalysts? The fundamental trade-off is between a graph representation including more information and a simpler representation that suits a larger variety of analysis methods. Much of the recent development in the graph structure of reaction systems has focused on either adapting analysis techniques to complex and informative graph representations [6–9], or to find simple graph representations encoding as much relevant information as possible [10–12]. In this paper, we will focus more on the latter developments and study the topology of two simple graph representations: one *substance graph* where the vertices are molecular species and an edge represents that two vertices participate in the same reaction, and a *reaction graph* where vertices symbolize reactions and two vertices are linked if they share some molecular species. In addition to these representations we also study the reaction systems as a bipartite graph with two classes of vertices, one for reactions and one for molecular species with edges connecting substances to the reactions they participate in. (Note that this representation, although more informative, still means a reduction of the information from the entire reaction system since one no longer can see which reactants that need to be present for a reaction to occur, or which products that are produced.) We investigate several topological properties of such graphs from reaction systems of planetary atmospheres and organismal data sets. Apart from degree distributions, we study network modularity (reflecting how well a graph can be decomposed into dense sub-graphs that are relatively weakly interconnected), currency substances (abundant molecular species that can react with a broad spectrum of other substances) and degree correlations (if edges primarily go between vertices of similar degree, or if the degrees are unbalanced with many edges between high- and low-degree vertices).

Results

The different degree distributions of the human metabolic and Earth atmospheric networks

Since the degree of a vertex count the number of other vertices it interacts with, it is a fundamental network quantity. The high-degree vertices can, and in most situations will, interact with many other vertices. The early findings that reaction systems have fat-tailed degree distributions—i.e. most vertices interacts only with a few others while some interact with a number far larger than the average—points at a diversity of functions among the vertices. For the metabolism, the common interpretation is that the high-degree metabolites are supplying building blocks to metabolites with more specialized functions, and lower degree. For atmospheric reaction networks, the low-degree vertices typically correspond to more complex molecules. We start our comparison of planetary and metabolic reaction system by looking at the substance and reaction graphs of Earth's atmosphere and the human metabolism. In Fig. 1A, we show the degree distributions of the substance graphs of the human metabolism and Earth's atmospheric reaction system. These distributions are rather similar—peaked and right skewed with tails of about the same slope. The degree distributions of the reaction graph, seen in Fig. 1B, are strikingly different. The human reaction graph is skewed and fat-tailed like its substance graph (but with a smaller exponent), whereas the Earth reaction graph has a degree distribution of an entirely different functional form, suggesting a different organization. The graphs are too big, however, for layout programs to give a hint of a deeper explanation of this difference (Fig. 2). Indeed, it is difficult to single out a more fundamental quantity causing the differences in degree distributions, as we will see in the rest of this section.

In our quest for a more detailed explanation of the difference of degree distributions in Fig. 1, we look closer at the bipartite representations mentioned above. In Fig. 3 (panels A, B, E and F), we plot the probability distribution of bipartite degree K_i for the human metabolic (Figs. 3A and B) and Earth atmospheric (Figs. 3E and F) networks in the substance (Figs. 3A and E) and reaction (Figs. 3B and F) projections. (For the other data sets this information can be found in Figure S1 and S2.) For substances, the degree distributions are right skewed in a fashion similar to the substance graph of Fig. 1A. For reactions, the two types of reaction systems both show unimodal degree distributions. A slight difference is that the Earth data gives a left-skewed distribution while the human network is right-skewed. This also means that the bipartite reaction-degree distribution, for the human data, is radically different than the projected distribution of Fig. 1B. To understand this better, we can decompose the degrees of the projected networks into three quantities as follows (where the left-

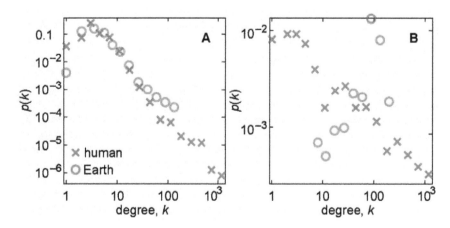

Figure 1. Degree distributions of substance and reaction graphs of the human metabolism and Earth's atmospheric reaction system. Panel A shows the probability mass-function of the degree of the substance graph of the reaction system of the Earth's atmosphere and the human metabolic networks. B shows the same as A, but for the reaction network. The similar behavior in A is drastically different in B. The plots are log-binned and plotted on double logarithmic scales.

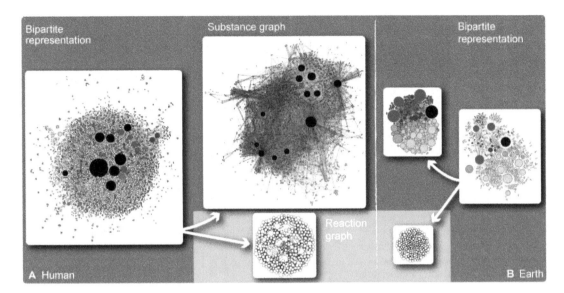

Figure 2. Ridiculograms of the human metabolism and Earth's atmospheric reaction system in bipartite, substance and reaction graph representations. The areas of the vertices are proportional to their degree. White vertices are reaction vertices; black vertices are currency vertices. For the other vertices the color represent different network modules. The colors of the edges are the same as their vertex of largest degree.

hand side is the degree of the projected network and the right-hand side quantities refer to the bipartite representations):

$$k_i = S_i - K_i - X_i = K_i(\kappa_i - 1) - X_i, \qquad (1)$$

where S_i is the sum of degrees of i's neighbors, K_i is i's degree, X_i is

the number of four-cycles that i is a part of, and κ_i is the average degree of i's neighbors. If there are few four-cycles in the bipartite network and there are no strong degree correlations (so κ_i can be assumed constant with respect to k_i), then i's degree in the bipartite network is a linear function of k_i (according to Eq. (1)). This is thus not the case for, at least, the metabolic reaction network where the k- and K-degree distribution, as mentioned, differs much. Indeed,

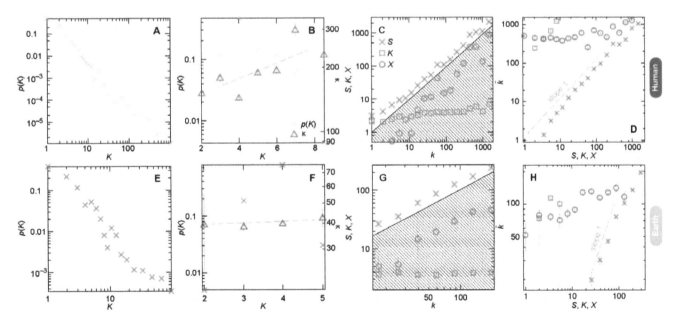

Figure 3. Deeper investigations of the degree distributions. Panel A displays the degree distribution of substances in a bipartite representation of the reaction system, i.e. the probability distribution of the number of reactions a substance participates in. Panel B shows the corresponding plot for reactions and also the average degree of neighbors. The dashed line is a linear-regression line to highlight the trend in κ. C and G displays the values of the three bipartite-network terms of k—S (the sum of the degrees of neighbors), K (the degree) and X (the number of four-cycles the vertex participates in). The diagonal line shows the k-value (so if you subtract the values of circles and squares from the values of crosses you would get this line). Panel D and H shows the average degrees $\bar{k}$ of nodes with certain values of the three terms that contribute to the degree in the projected reaction networks. $\bar{k}$ is averaged over logarithmic bins of S, K, and X values. The dashed line is a reference corresponding to a linear $\bar{k}$-dependence. Panels A–D are for the human metabolic reaction networks, E–H show the corresponding plots for the Earth atmospheric reaction networks.

in Fig. 3B we see a positive correlation between K and κ, stronger than the corresponding correlation for the Earth network in Fig. 3F (which is almost absent). This means that $S = K\kappa$ grows super-linearly with K so the tail of $p(K)$ gets stretched into the distribution of Fig. 1B. Here, we still assume that the number of four-cycles does not contribute to k significantly, which we justify below. This is justified to some extent in Fig. 3C (and 3G for the Earth network)—the k-scaling of S and X is similar, so $S - X$ scales like S (and thus the arguments above still hold). That S (and thus $S - X$) scales like X is also true for the atmospheric network (Fig. 3G), which explains that the shape of Fig. 1B is to a large degree determined by K (so the hump shape of $p(K)$ gives a hump-shaped $p(k)$). Another view of S, K and X is given in panels D and H where, we plot the average degrees of nodes given their S-, K- and X-values. We can see that, as expected, S is the best predictor of $\bar{k}$ (showing close to a linear relationships for the metabolic data, and a clear correlation for the atmospheric network). Another observation is that X shows more structure (apart from the scaling itself) in the metabolic network compared to the atmospheric network. This can perhaps be explained by the more pronounced modular structure of the metabolic network (that we will discuss further below). From Fig. 3D and H we also learn that $\bar{k}$ shows a strong positive K-dependence for the metabolic network, but not for Earth's atmospheric network. This is reflected in Figs. 3B and F too—since κ grows with K for the metabolic network, S and K will be positively correlated, and since $\bar{k}$ grows with S then it will also grow with K.

In summary, the difference between the degree distributions of the reaction graphs of the metabolic and atmospheric networks cannot be explained by one single feature of the original reaction system's topology. Instead it can be traced to a combination of the slightly different skewness of the distribution of a reaction's number of participating substances and the different correlation properties between the degree of a vertex and the average degree of its neighbors. In Figures S3 and S4), we plot the bipartite degree distributions of all the planets and organisms. Essentially, the conclusions for the Earth's atmospheric network extends to other planets, except that the data sets are smaller and the degree distributions does not have the same negative trend similar to power-laws.

Comparing degree distributions of planetary atmospheric and organismal metabolic networks

So far, we focused on finding lower-level causes for the degree distributions of projected networks of the human metabolic and Earth atmospheric networks. We now turn to the question how much these observations can be generalized to the other networks. To this end, we will use more rigorous methods for analyzing probability distributions than we used so far. We will analyze the data using methods from Ref. [11]. First, we test the hypothesis that degrees are power-law distributed by (roughly speaking, details in the Methods section) finding parameter values for the power-law distribution that fits the data best, then draw as many series of numbers from this distribution with the same size as the raw data and check the likelihood that the synthetic and real data come from the same distribution. We also check which is the most likely distribution generating that degree distribution—power-law or log-normal (a right-skewed distribution with a more narrow tail than a power-law that is visually similar to the Earth reaction graph of Figure 1B). The results of these measurements are shown in Table 1. As hypothesized above, the reaction graphs are unanimously inconsistent with power-laws. Of the substance graphs, only planetary atmospheric networks are consistent with

Table 1. Statistical tests of various types of degree distributions.

| | | Atmospheres of planets and moons | | | | | | | | Metabolism of organisms | | | | | | | | |
		Earth	Venus	Titan	Titan 2	Mars	Jupiter	Io	Solar system	Human (KEGG)	Human (BiGG)	M. genitalium	S. cerevisiae (KEGG)	S. cerevisiae (BiGG)	E. coli	M. musculus	D. melanogaster	C. elegans
Substance graph	Power-law?	Y	N	Y	Y	Y	Y	Y	Y	N	N	N	N	N	N	N	N	N
	PL or LN	PL	LN	LN	LN	LN	LN	LN	LN	PL	PL	PL	PL	PL	PL	PL	PL	PL
Reaction Graph	Power-law?	N	N	N	N	N	N	N	N	N	N	N	N	N	N	N	N	N
	PL or LN	LN	LN	LN	LN	LN	LN	LN	LN	PL	LN	PL	PL	PL	PL	PL	PL	PL
Bipartite substances	Power-law?	Y	Y	Y	N	N	Y	Y	Y	N	Y	Y	N	Y	N	N	N	Y
	PL or LN	PL	LN	LN	LN	LN	LN	LN	LN	PL	PL	PL	PL	PL	PL	PL	PL	PL

Statistics of the reactions in the bipartite representation are omitted since they are not fat-tailed. "Y" ("N") indicates that the data set is consistent (inconsistent) with the tested hypothesis. "PL" stands for "power-law" (i.e., testing for a power-law hypothesis); "LN" means "log-normal".

power-laws. This does not mean that it is fair to describe them as power-laws; especially since most of them fit better to a log-normal form. Since the planetary data sets are relatively small, the relative errors are larger and it is harder to refute the possibility of another functional form. The substance graphs are, on the other hand, closer to log-normals than power-laws. The reason is seen for the human metabolic network in Fig. 1 (and for the other datasets sets in Fig S1), that they are even more fat-tailed than a power-law— they have more vertices of highest degrees than the best-fitting power-law does. Thus they are even further from log-normals than power-laws. The reaction graphs are more similar to power-laws than log-normals for the metabolic networks, but the other way around for the planetary atmospheres, which is also in line with our observations. This study cannot, however, strengthen the observation that the substance graphs are similar to the metabolic networks except for Earth's network that falls into the same category as the metabolic networks. There are two possibilities— either the difference can be explained by a difference in sizes and that the other planetary atmospheres have to be measured by indirect methods, or the Earth network is radically different (more than just the sizes). Ref. [2] makes the latter hypothesis, and argues a difference from the influence on the biosphere on the Earth's atmosphere creates a visible difference. On the other hand, many reactions typical for Earth (e.g. involving molecular oxygen) are also present in the other datasets.

The substances' degrees in the bipartite representation do not separate the planetary and metabolic data so well (both types of datasets contain degree distributions consistent with power laws, and not). Similar to the observations in the detailed studies above, the projections to substance or reaction graphs create the difference. However, the planet-network distributions are more similar to log-normal than power-laws, whereas it is the other way around for the metabolic networks.

Modularity and currency metabolites

Biological systems are commonly described as modular—being composed of different subunits, or modules, which perform some specific task relatively independent of the rest of the system. Some modules are quite conspicuous—a cell is a prime example—but also more nebulous systems, like metabolism, are thought to consist of modules. If we treat all reactions equal (the essence of the

graph theoretic approach), then independence means that the connections within the network module should be denser than the connections out of the module. A module on a graph-level resolution of metabolism is thus equal to what is commonly known as a network cluster or community [13]. This is not quite the whole story however. The most abundant metabolites (like water, carbon dioxide and so on) do not put any restriction on the reactions, and would not contribute to the specialized function of a module. It is thus common to preprocess the graph by identifying such *currency metabolites* and removing them from the network, considering only a network of other less frequent molecular species that are more of bottlenecks in the metabolic machinery. There are methods to identify both network clusters and currency metabolites (described in the Methods section) from the topology of substance graphs. Although these definitions have been developed for metabolic networks, there is nothing that stops us from applying them to networks of planetary atmospheres. *A priori*, since atmospheric reaction system has not evolved through natural selection, we expect them to have less distinct modules and currency metabolites. This is indeed the case as can be seen in Fig. 4—there is a size-difference between the metabolic and atmospheric networks, but it is less pronounced than both the relative modularity and the number of currency vertices. Thus there seems to be a stronger tendency for the metabolic networks to be organized into modules supplied by currency vertices than the networks of planetary atmospheres.

Discussion

In this article, we have directly compared functionally informative network characteristics of metabolic reaction systems of a wide variety of organisms and the reaction systems of planets and moons of the solar system. One such quantity is degree—the number of other nodes a node interacts with. (Where "interact" is defined via the network in question.) In most types of networks, degree indicates the importance of a node, but in biochemical networks, where both low- and high-degree vertices can be essential for the cell's functionality, then degree rather separates chemical substances of different functionality—at least in metabolic substance networks, the high-degree vertices are typically light molecules that supply atoms and molecular groups to the functionally more specialized low-degree vertices [14]. For

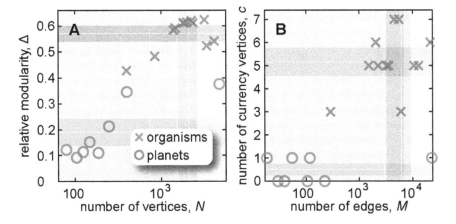

Figure 4. Relative modularity and the number of currency vertices separate networks of metabolism from networks of planetary atmospheres more than their sizes do. To show that the maximal relative modularity separates metabolism from reaction systems of planetary atmospheres, we display (panel A) the relative modularity Δ as a function of the number of vertices N. The shaded areas indicate the standard deviation and means of the respective quantities. Similarly, in B, we show another quantity related to the functional organization, the number of currency vertices c, as a function of the number of edges M in the network. Note that axes are linear and logarithmic respectively.

reaction networks one can assume a similar interpretation—high-degree vertices are reactions supporting many subsystems of the reaction system. All substance projections, for both atmospheric and metabolic networks, do indeed have relatively broad degree distributions. This supports the above-mentioned picture of functional differentiation by degree. Using statistical tests, we can separate organisms from planet fairly well. The networks of planetary atmospheres are typically consistent with power-laws, but the metabolic networks are not. The planetary networks are, however, statistically more similar to log-normal distributions, which suggests that the fact they are deemed consistent with power-laws is an effect that they are, on average, smaller than the metabolic systems (and thus does not provide enough data to give statistical significance).

We note that in the substance-network projection, the Earth atmospheric and human metabolic networks have rather similar degree distributions, but for the reaction-network projection the distributions are strikingly different. We investigate lower-level explanations for this observation in terms of degree distributions of a bipartite representation of the reaction system and degree correlations. It is however not easy to single out a low-level cause for this difference, rather it seems to be a combined effect of a slightly difference in the distribution of reaction-degrees and degree correlations in the bipartite representation.

When we look closer at quantities designed to characterize the modular functionality, we see higher network modularity and more currency metabolites in metabolic networks than atmospheric networks. On the other hand, the differences are not larger than that they can almost be explained by the sizes of the networks alone. Furthermore, fundamental structures such as the shape of some of the degree distributions are skewed in a qualitatively similar way. Our conclusion is thus that the main structure of metabolic networks is probably shaped by the same fundamental stoichiometric constraints as all chemical reaction systems, but there are also traces of evolution in the network structure of metabolism. At the same time the network-modular structure, the traces of evolution, is not so clear as the picture the analogy to engineering paints—there are more than a couple of in- and output terminals. Maybe the largest open question is not why metabolic networks are modular but why they are not more modular? How can we reconcile the logical picture of evolution operating by adding and deleting of modules with the modular-but-not-very-much-so picture of metabolic networks? We believe the approach we take in this paper, to use a natural system as a null-model for the metabolism can be fruitful.

Methods

Datasets for metabolic and chemical networks

Reaction sets for planetary atmospheres are described in Ref. [5], except the "solar system" data that was obtained from the UMIST database [15]. The metabolic networks come from the KEGG [16] and BiGG [17] database and are described in Ref. [7]. We select nine datasets from the KEGG and BiGG databases to match the number of planetary atmosphere datasets. To get a rough error estimate of sampling effects, we also analyze the human data both from BiGG and KEGG, and two independent datasets from Jupiter's atmosphere. Our selection criterion is that the datasets should be a diverse selection among the most well-studied model organisms.

Network representations

To choose the graph representation of a reaction system involves a trade-off between information content and usefulness.

One can use a complex representation with substances, catalysts and reactions as separate classes of vertices and directed edges representing the general direction of the matter flow. The advantage with such a representation is that all topological aspects of the reaction system are encoded into the graph. But the price for this is that there few general analysis methods can be applied to it; they would need to be modified, something that is not always possible. Alternatively, one chooses a simple-graph representation with one type of vertices and one type of (undirected) edges, without multiple edges or self-edges. Such a representation can be analyzed by a multitude of off-the-shelf methods. A disadvantage with simple graphs, except that they encode less information, is that there is no obvious way of reducing the reaction system to a simple graph. We choose a substance graphs as our main graph simple-graph representation. In such a graph one put an edge between all substances that can participate in the same reaction, so if the reaction $2H_2O \rightarrow 2H_2 + O_2$, would contribute with three edges—(H_2O, H_2), (H_2O, O_2) and (O_2, H_2O)—to a substance graph. There is some evidence that substance graphs are good simple-graph representations of metabolic networks [11,18], but to the best of our knowledge, no corresponding studies for other categories of reaction systems. In addition, we use a reaction graph representation that is in some sense dual to the substance graphs—every reaction is a vertex in this network and two reactions that have a substance in common is connected.

Testing degree distributions

We use the approach in Clauset et al. [11] to test the degree distributions for the hypothesis that they follow power-laws. This method starts from the real data and obtains the exponent of a best-fitting power-law, α, by maximum likelihood estimation. Then one draws sets of random numbers, of the same cardinality as the original data, from the probability distribution

$$p_k = \begin{cases} \Lambda k^{-\alpha} & \text{if } 0 < k \leq k_{max} \\ 0 & \text{otherwise} \end{cases} \quad (2)$$

where Λ is a normalization constant. Finally, one use the Kolmogorov–Smirnov test statistics (the maximal difference, for all k-values, between the cumulative density functions of the real and synthetic data) to estimate the p-value of the hypothesis that the real data was drawn from p_k.

Ref. [19] also adapts a method by Vuong [20] to compare different heavy-tailed distributions. We use it to test which distribution of power-law and log-normal distribution functions that best fits our data. The log-normal distribution is defined by the probability density function

$$p_k = \frac{A'}{k} \exp\left[-a'(\ln k - \mu)^2\right] \quad (3)$$

where A', a' and μ are positive constants (A' is a normalization factor, a' and μ are parameters giving the shape of the curve). Vuong's method takes the likelihoods, L_1 and L_2, of the two functional forms generating the observed data as its starting point. The method uses the result that $V = \ln(L_1/L_2)$ is normally distributed for large data sets to compute a p-value for the hypothesis that the data was generated by distribution 1 rather than distribution 2.

Network Modularity

The concept of network modularity, cluster, or community structure strives to capture the large-scale organization of networks into dense subnetworks that are relatively weakly interconnected [21]. There is no unique way of deriving a measure for network modularity or dividing a graph into such dense subgraphs; rather, there is a number of different methods each capturing some certain aspect of network modularity. The method in this work is based on the popular method of maximizing Newman and Girvan's Q-modularity. For this measure, one assume the graph is divided into a number of subgraphs and let e_{ij} be the fraction of all edges going between subgraph i and j, and defines

$$Q = \sum_i \left[e_{ii} - \left(\sum_j e_{ij} \right)^2 \right] \qquad (4)$$

A class of module-detection methods starts by assuming that the division maximizing Q is a sensible decomposition into subgraphs. Already from the Equation (4) one can see that edges within a subgraph give a positive contribution to Q, and edges between communities decrease Q. The advantages with this clustering algorithm are that Q is easy to interpret and closely matching the verbal definition of a network module above; and furthermore the maximal Q, $\hat{Q}$, is a crude measure of the network modularity of an entire graph. The two disadvantages with Q-maximization methods are the following. First, it fails to divide some subgraphs into what looks like obvious clusters. This is roughly speaking because the second sum compares a division i with all other divisions j, even if it does not matter (for a visually good clustering) if i and j are far apart [22]. Second, it is technically hard to find the maximizing division— Q is a very flat function (in sub-division space) near its maximum [23]. For our purpose these latter two objections are not so serious— there is no general biological argument that the modules that look like they can be further subdivided are not sensible clusters, and there is no need to find the actual subdivision into modules, we just want a good estimate of $\hat{Q}$, which we do have if we only get close to the mentioned plateau in subdivision space.

As a measure of the modularity of a graph, $\hat{Q}$, is not ideal. On one hand $\hat{Q}$ close to zero would mean a low modularity and $\hat{Q}$ close to one would imply modularity. On the other hand, the intermediate values depend on many factors regarded as more fundamental (like the number of vertices and edges and the degree distribution) than modularity. To compensate for such effects as much as possible we rather measure $\hat{Q}$ relative to the average value of $\hat{Q}$ in an ensemble, or null-model, of graphs (obtained by standard edge rewiring [24]) with the same sizes N and M and the same degrees as the substance graph G, but everything else random. So we define

$$\Delta = \hat{Q} - \bar{Q} \qquad (5)$$

where $\bar{Q}$ is the average of the maximal modularity over 1000 rewired graphs.

Currency vertices

The hubs in metabolic networks—e.g. H_2O, NADH, ATP and CO_2—are typically also the most abundant metabolites through-out the cell. These are the workhorses of metabolism, supplying functional groups to proteins and other molecules with more specialized functions. Since these currency metabolites are present throughout the cell and do not put much of constraints on the reactions they participate in, one can learn more about the functionality of the network if one exclude them from the graph representation. The circumstance that they are common through-out the cell and participate in many reactions also means that they connect network modules and effectively lower the modularity. This observation, along with the fact they have a high degree, has been used as a definition of currency metabolites [10]. If one deletes vertices in order of their degree (starting from large degrees) and monitor Δ, then for metabolic networks, Δ typically first increase to a maximum and later decrease. Ref. [10] defines currency metabolites as those that give the largest Δ before Δ reached a value larger than in the original graph. This definition is general enough to apply to other reaction-system networks, and one can speak of *currency vertices* also for atmospheric or nuclear reaction systems [14].

Supporting Information

Figure S1 Degree distributions for the substance networks. The data is log-binned and plotted in log–log scale.

Figure S2 Degree distributions for the reaction networks. The data is log-binned and plotted in log–log scale.

Figure S3 Degree distributions for the substances in the bipartite representations. The data is log-binned and plotted in log–log scale.

Figure S4 Degree distributions for the reactions in the bipartite representations. The data is log-binned and plotted in log–log scale.

Figure S5 A plot corresponding to Fig. 3C, D, G and H for substance networks. Panels A and C display the values of the three terms of k—S, K and X. The diagonal line shows the k-value. Panels B and D show the average degrees $\bar{k}$ of nodes with certain values of the three terms that contribute to the degree in the projected networks. $\bar{k}$ is averaged over logarithmic bins of S, K, and X values. Panels A and B is data for the human network; C and D are the corresponding plots for the Earth atmospheric network.

Acknowledgments

The authors thank Andreea Munteanu for help with the data acquisition.

Author Contributions

Conceived and designed the experiments: PH MH. Analyzed the data: PH SHL. Wrote the paper: PH MH SHL.

References

1. Jeong H, Tombor B, Oltvai ZN, Barabási AL (2000) The large-scale organization of metabolic networks. Nature 407: 651–654.
2. Solé RV, Munteanu A (2004) The large-scale organization of chemical reaction networks in astrophysics. Europhys Lett 68: 170–176.
3. Krebs HA, Johnson (1937) The role of citric acid in intermediate metabolism in animal tissues. Enzymologia 4: 148 156.
4. Bethe HA (1939) Energy production in stars. Phys Rev 55: 434 456.
5. Yung YL, Demore WB (1999) Photochemistry of planetary atmospheres. New York: Oxford University Press.
6. Veeramani B, Bader JS (2010) Predicting functional associations from metabolism using bi-partite network algorithms. BMC Systems Biology 4: 95.
7. Miyake S, Takenaka Y, Matsuda H (2004) A graph analysis method to detect metabolic sub-networks based on phylogenetic profile. IEEE Computational Systems Bioinformatics Conference. pp 634–635.

8. Klamt S, Haus U-U, Theis F (2009) Hypergraphs and cellular networks. PLoS Comput Biol 5: e1000385.

9. Holme P, Huss M, Jeong H (2003) Subnetwork hierarchies of biochemical pathways. Bioinformatics 19: 532–538.

10. Huss M, Holme P (2007) Currency and commodity metabolites: Their identification and relation to the modularity of metabolic networks. IET Systems Biology 1: 280–288.

11. Holme P, Huss M (2010) Substance networks are optimal simple-graph representations of metabolism. Chinese Sci Bull 55: 3161–3168.

12. Serrano MA, Sagues F (2010) Network-based confidence scoring system for genome-scale metabolic reconstruction. e-print arXiv: 1008.3166.

13. Newman MEJ (2010) Networks: An Introduction. Princeton, NJ: Princeton University Press.

14. Holme P (2009) Signatures of currency vertices. J Phys Soc Jpn 78: 034801.

15. Woodall J, Agúndez M, Markwick-Kemper AJ, Millar TJ (2007) The UMIST database for astrochemistry. Astronomy and Astrophysics 466: 1197–1204.

16. Kanehisa M, Goto S (2000) KEGG: Kyoto Encyclopedia of Genes and Genomes. Nucleic Acids Res 28: 27–30.

17. Duarte NC, Becker SA, Jamshidi N, Thiele I, Mo ML, et al. (2007) Global reconstruction of the human metabolic network based on genomic and bibliomic data. Proc Natl Acad Sci USA 104: 1777–1782.

18. Holme P (2011) Metabolic robustness and network modularity: A model study. PLoS ONE 6, e16605.

19. Clauset A, Shalizi CR, Newman MEJ (2009) Power-law distributions in empirical data. SIAM Rev 51: 661–703.

20. Vuong QH (1989) Likelihood ratio tests for model selection and non-nested hypotheses. Econometrica 57: 307–333.

21. Fortunato S (2009) Community detection in graphs. Phys Rep 486: 75–174.

22. Fortunato S, Barthélemy M (2006) Resolution limit in community detection. Proc Natl Acad Sci USA 104: 36–41.

23. Good BH, de Montjoye Y-A, Clauset A (2010) The performance of modularity maximization in practical contexts. Phys Rev E 81: 046106.

24. Sneppen K, Maslov S (2002) Specificity and stability in topology of protein networks. Science 296: 910–913.

PERMISSIONS

All chapters in this book were first published in PLOS ONE, by The Public Library of Science; hereby published with permission under the Creative Commons Attribution License or equivalent. Every chapter published in this book has been scrutinized by our experts. Their significance has been extensively debated. The topics covered herein carry significant findings which will fuel the growth of the discipline. They may even be implemented as practical applications or may be referred to as a beginning point for another development.

The contributors of this book come from diverse backgrounds, making this book a truly international effort. This book will bring forth new frontiers with its revolutionizing research information and detailed analysis of the nascent developments around the world.

We would like to thank all the contributing authors for lending their expertise to make the book truly unique. They have played a crucial role in the development of this book. Without their invaluable contributions this book wouldn't have been possible. They have made vital efforts to compile up to date information on the varied aspects of this subject to make this book a valuable addition to the collection of many professionals and students.

This book was conceptualized with the vision of imparting up-to-date information and advanced data in this field. To ensure the same, a matchless editorial board was set up. Every individual on the board went through rigorous rounds of assessment to prove their worth. After which they invested a large part of their time researching and compiling the most relevant data for our readers.

The editorial board has been involved in producing this book since its inception. They have spent rigorous hours researching and exploring the diverse topics which have resulted in the successful publishing of this book. They have passed on their knowledge of decades through this book. To expedite this challenging task, the publisher supported the team at every step. A small team of assistant editors was also appointed to further simplify the editing procedure and attain best results for the readers.

Apart from the editorial board, the designing team has also invested a significant amount of their time in understanding the subject and creating the most relevant covers. They scrutinized every image to scout for the most suitable representation of the subject and create an appropriate cover for the book.

The publishing team has been an ardent support to the editorial, designing and production team. Their endless efforts to recruit the best for this project, has resulted in the accomplishment of this book. They are a veteran in the field of academics and their pool of knowledge is as vast as their experience in printing. Their expertise and guidance has proved useful at every step. Their uncompromising quality standards have made this book an exceptional effort. Their encouragement from time to time has been an inspiration for everyone.

The publisher and the editorial board hope that this book will prove to be a valuable piece of knowledge for researchers, students, practitioners and scholars across the globe.

LIST OF CONTRIBUTORS

Andrew D. Irving and Sean D. Connell and Bayden D. Russell
Southern Seas Ecology Laboratories, School of Earth and Environmental Sciences, The University of Adelaide, Adelaide, South Australia, Australia

Hongguang Cheng
School of Environment, Beijing Normal University, Beijing, China

Tan Zhou
School of Environment, Beijing Normal University, Beijing, China
Spatial Science Laboratory, Texas A&M University, College Station, Texas, United States of America

Qian Li
School of Environment, Beijing Normal University, Beijing, China

Lu Lu
School of Environment, Beijing Normal University, Beijing, China

Chunye Lin
School of Environment, Beijing Normal University, Beijing, China

Eric S. Melby, Douglas J. Soldat and Phillip Barak
Department of Soil Science, University of Wisconsin-Madison, Madison, Wisconsin, United States of America

Simon J. Thuss, Jason J. Venkiteswaran and Sherry L. Schiff
Department of Earth and Environmental Sciences, University of Waterloo, Waterloo, Ontario, Canada

Barbara Gworek
Institute of Environmental Protection – National Research Institute, Warsaw, Poland

Katarzyna Klimczak
Warsaw University of Life Sciences – SGGW, Department of Soil Environment Sciences, Warsaw, Poland

Marta Kijeń ska
Institute of Environmental Protection – National Research Institute, Warsaw, Poland

Benjamin D. Duval
Energy Biosciences Institute, University of Illinois at Urbana-Champaign, Urbana, Illinois, United States of America

Global Change Solutions, Urbana, Illinois, United States of America

Kristina J. Anderson-Teixeira
Energy Biosciences Institute, University of Illinois at Urbana-Champaign, Urbana, Illinois, United States of America

Sarah C. Davis
Energy Biosciences Institute, University of Illinois at Urbana-Champaign, Urbana, Illinois, United States of America

Cindy Keogh
Natural Resource Ecology Laboratory, Fort Collins, Colorado, United States of America

Stephen P. Long
Energy Biosciences Institute, University of Illinois at Urbana-Champaign, Urbana, Illinois, United States of America
Global Change Solutions, Urbana, Illinois, United States of America
Department of Plant Biology, University of Illinois at Urbana- Champaign, Urbana, Illinois, United States of America

William J. Parton
Natural Resource Ecology Laboratory, Fort Collins, Colorado, United States of America

Evan H. DeLucia
Energy Biosciences Institute, University of Illinois at Urbana-Champaign, Urbana, Illinois, United States of America
Global Change Solutions, Urbana, Illinois, United States of America
Department of Plant Biology, University of Illinois at Urbana- Champaign, Urbana, Illinois, United States of America

Tianxing Wang, Jiancheng Shi, Yingying Jing, Tianjie Zhao, Dabin Ji and Chuan Xiong
State Key Laboratory of Remote Sensing Science, Institute of Remote Sensing and Digital Earth, Chinese Academy of Sciences. Beijing, China

Yi Liu
Laboratory of Aquatic Botany and Watershed Ecology, Wuhan Botanical Garden, Chinese Academy of Sciences China, Wuhan, China

Kai-yuan Wan
Laboratory of Aquatic Botany and Watershed Ecology, Wuhan Botanical Garden, Chinese Academy of Sciences China, Wuhan, China

Yong Tao
Laboratory of Aquatic Botany and Watershed Ecology, Wuhan Botanical Garden, Chinese Academy of Sciences China, Wuhan, China

Zhi-guo Li
Laboratory of Aquatic Botany and Watershed Ecology, Wuhan Botanical Garden, Chinese Academy of Sciences China, Wuhan, China

Guo-shi Zhang
Laboratory of Aquatic Botany and Watershed Ecology, Wuhan Botanical Garden, Chinese Academy of Sciences China, Wuhan, China

Shuang-lai Li
Institute of Plant Protection and Soil Fertilizer, Hubei Academy of Agricultural Sciences, Wuhan, China

Fang Chen
Laboratory of Aquatic Botany and Watershed Ecology, Wuhan Botanical Garden, Chinese Academy of Sciences China, Wuhan, China

Frida Sidik and Catherine E. Lovelock
The School of Biological Sciences, The University of Queensland, St Lucia, Queensland, Australia

Sergey Zimov and Nikita Zimov
Northeast Science Station, Pacific Institute for Geography, Russian Academy of Sciences, Cherskii, Russia

Nathalie Gypens
Ecologie des Systèmes Aquatiques, Université Libre de Bruxelles, Brussels, Belgium

Alberto V. Borges
Unité d'Océanographie Chimique, Université de Liège, Liège, Belgium

Gaelle Speeckaert
Ecologie des Systèmes Aquatiques, Université Libre de Bruxelles, Brussels, Belgium

Christiane Lancelot
Ecologie des Systèmes Aquatiques, Université Libre de Bruxelles, Brussels, Belgium

Alberto V. Borges
Chemical Oceanography Unit, Université de Liège, Liège, Belgium

Cédric Morana
Department of Earth and Environmental Sciences, KU Leuven, Leuven, Belgium

Steven Bouillon
Department of Earth and Environmental Sciences, KU Leuven, Leuven, Belgium

Pierre Servais
Ecologie des Systèmes Aquatiques, Université Libre de Bruxelles, Bruxelles, Belgium

Jean-Pierre Descy
Research Unit in Environmental and Evolutionary Biology, University of Namur, Namur, Belgium

François Darchambeau
Chemical Oceanography Unit, Université de Liège, Liège, Belgium

J. Alexander Maxwell
Institute for a Sustainable Environment, Clarkson University, Potsdam, New York, United States of America

M. Holsen
Department of Civil and Environmental Engineering, Clarkson University, Potsdam, New York, United States of America

Sumona Mondal
Department of Mathematics, Clarkson University, Potsdam, New York, United States of America

Jason J. Venkiteswaran
Department of Geography and Environmental Studies, Wilfrid Laurier University, Waterloo, Ontario, Canada

Sherry L. Schiff
Department of Earth and Environmental Sciences, University of Waterloo, Waterloo, Ontario, Canada

Marcus B. Wallin
Department of Ecology and Genetics/Limnology, Uppsala University, Uppsala, Sweden

Tina S˘antl-Temkiv
Department of Environmental Science, Aarhus University, Roskilde, Denmark
Microbiology Section, Department of Bioscience, Aarhus University, Aarhus, Denmark
Stellar Astrophysics Centre, Department of Physics and Astronomy, Aarhus University, Aarhus, Denmark

Kai Finster
Microbiology Section, Department of Bioscience, Aarhus University, Aarhus, Denmark

Stellar Astrophysics Centre, Department of Physics and Astronomy, Aarhus University, Aarhus, Denmark

Thorsten Dittmar
Max Planck Research Group for Marine Geochemistry, Institute for Chemistry and Biology of the Marine Environment, University of Oldenburg, Oldenburg, Germany

Bjarne Munk Hansen
Department of Environmental Science, Aarhus University, Roskilde, Denmark

Runar Thyrhaug
Department of Biology, University of Bergen, Bergen, Norway

Niels Woetmann Nielsen
Danish Meteorological Institute, Copenhagen, Denmark

Ulrich Gosewinkel Karlson
Department of Environmental Science, Aarhus University, Roskilde, Denmark,

Linwood Pendleton
Nicholas Institute for Environmental Policy Solutions, Duke University, Durham, North Carolina, United States of America,

Daniel C. Donato
Ecosystem & Landscape Ecology Lab, University of Wisconsin, Madison, Wisconsin, United States of America

Brian C. Murray
Nicholas Institute for Environmental Policy Solutions, Duke University, Durham, North Carolina, United States of America

Stephen Crooks
ESA Phillip Williams & Associates, San Francisco, California, United States of America

W. Aaron Jenkins
Nicholas Institute for Environmental Policy Solutions, Duke University, Durham, North Carolina, United States of America

Samantha Sifleet
United States Environmental Protection Agency, Research Triangle Park, North Carolina, United States of America

Christopher Craft
School of Public and Environmental Affairs, Indiana University, Bloomington, Indiana, United States of America

James W. Fourqurean
Department of Biological Sciences and Southeast Environmental Research Center, Florida International University, North Miami, Florida, United States of America

J. Boone Kauffman
Department of Fisheries and Wildlife, Oregon State University, Corvallis, Oregon, United States of America and Center for International Forest Research, Bogor, Indonesia

Nú ria Marbà
Department of Global Change Research, Mediterranean Institute for Advanced Studies, Esporles, Illes Balears, Spain

Patrick Megonigal
Smithsonian Environmental Research Center, Edgewater, Maryland, United States of America

Emily Pidgeon
Conservation International, Arlington, Virginia, United States of America

Dorothee Herr
International Union for the Conservation of Nature, Washington, District of Columbia, United States of America

David Gordon
Nicholas Institute for Environmental Policy Solutions, Duke University, Durham, North Carolina, United States of America

Alexis Baldera
The Ocean Conservancy, Baton Rouge, Louisiana, United States of America

Index

Printed in the USA
CPSIA information can be obtained
at www.ICGtesting.com
JSHW051446221024
72173JS00006B/1591